46.00 100k

The State of
High Energy Physics
(BNL/SUNY Summer School, 1983)

1983 Summer School on High Energy Particle Accelerators

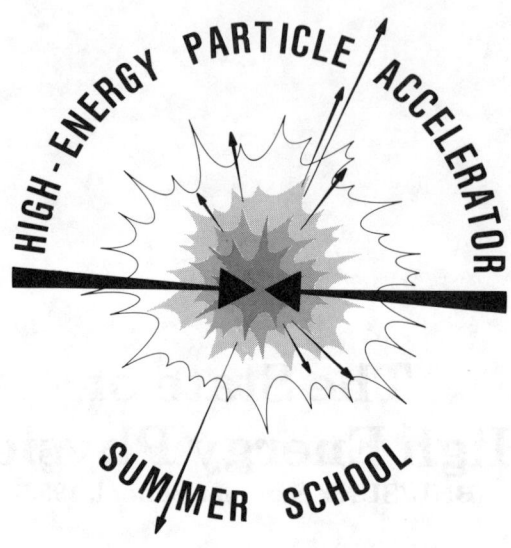

The US Summer School on High Energy Particle Accelerators is dedicated to the future of High Energy Physics.

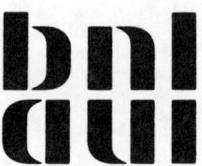

Brookhaven National Laboratory and
State University of New York at Stony Brook
July 6-16, 1983

AIP Conference Proceedings
Series Editor: Rita G. Lerner
Number 134

The State of High Energy Physics
(BNL/SUNY Summer School, 1983)

Edited by
Melvin Month, Per F. Dahl,
and Margaret Dienes

American Institute of Physics
New York   1985

Copy fees: The code at the bottom of the first page of each article in this volume gives the fee for each copy of the article made beyond the free copying permitted under the 1978 US Copyright Law. (See also the statement following "Copyright" below.) This fee can be paid to the American Institute of Physics through the Copyright Clearance Center, Inc., 21 Congress Street, Salem, MA 01970.

Copyright © 1985 American Institute of Physics

Individual readers of this volume and non-profit libraries, acting for them, are permitted to make fair use of the material in it, such as copying an article for use in teaching or research. Permission is granted to quote from this volume in scientific work with the customary acknowledgment of the source. To reprint a figure, table or other excerpt requires the consent of one of the original authors and notification to AIP. Republication or systematic or multiple reproduction of any material in this volume is permitted only under license from AIP. Address inquiries to Series Editor, AIP Conference Proceedings, AIP, 335 E. 45th St., New York, N.Y. 10017.

L.C. Catalog Card No. 85-73170
ISBN 0-88318-333-1
DOE CONF-8307126- -Pt2

Printed in the United States of America

TABLE OF CONTENTS

Preface vii

Acknowledgments xi

Summer School Organization xii

High Energy Physics in the United States 1
    M. Month

CERN and High-Energy Physics 74
    H. Schopper

DESY and High-Energy Physics 98
    G.-A. Voss

Physics of e^+e^- Colliders: Present, Future, and Far Future 122
    M.E. Peskin

Hadron-Hadron Colliders 168
    A.V. Tollestrup

Physics Results of the UA1 Collaboration at the CERN Proton-Antiproton Collider 205
    C. Rubbia

How Does the Standard Model Stand Up to the Real World? 298
    C.H. Llewellyn Smith

The Future of High-Energy Physics   320
    J.D. Bjorken

An Ultrahigh Energy Hadron Collider: Round Table Discussion 331
    Introduction, M. Month
    Particle Physics of Multi-TeV Collisions, J.D. Bjorken
    Accelerator Technology for a Multi-TeV Collider, M. Tigner
    Experiments at Multi-TeV Energies, C. Rubbia
    Prospects of an Ultrahigh Energy Collider: A View from
        Washington, N.D. Pewitt
    Planning for a Super Collider, P.J. Reardon
    Discussion of $\bar{p}p$ vs. pp and the Question of Aperture and
        Luminosity Appendix: The SSC Two Years Later, P. Dahl

Where is the SSC Today? 366
    M. Tigner

PREFACE

The material contained in this volume covers lectures presented at the Symposium on the State of High Energy Physics, which was part of the third annual U.S. Summer School on High Energy Particle Accelerators, held at Brookhaven National Laboratory (BNL) and the State University of New York at Stony Brook (SUNY), July 6-16, 1983. The school, sponsored by the Department of Energy (DOE) and the National Science Foundation (NSF), is one of a continuing series of such schools organized at different high energy physics laboratories across the country. Past and planned schools are:

Fermilab	July 13-24, 1981
SLAC	August 2-13, 1982
BNL/SUNY	July 6-16, 1983
Fermilab	August 13-24, 1984
SLAC	July 15-26, 1985

Although the school symposium was held in July, 1983, much of the material has been updated. Thus, this review substantially represents a picture of high energy physics as it currently exists with a flavor of the 1983 viewpoint.

This third summer school was planned by an organizing committee consisting of M. Month (BNL, Chairman), J.D. Bjorken (Fermilab), H. Grunder (DOE/LBL), V.W. Hughes (Yale), F.R. Huson (Fermilab), B. McDaniel (Cornell), C. Pellegrini (BNL), B. Richter (SLAC), R. Schwitters (Harvard/Fermilab), and R.R. Wilson (Columbia). P.J. Reardon (BNL) served as the local school director, and, with P. Dahl as his deputy, had the responsibility for administering the school. N.P. Samios, BNL Director, hosted the school and significantly contributed to its success.

The purpose of these schools derives from a recommendation made by a subpanel of the High Energy Physics Advisory Panel (HEPAP) that convened in 1979-1980 in order to assess the current state of accelerator R&D. The subpanel issued a strong appeal to the high energy physics community to attempt to encourage a greater number of scientists and students to work in the field of high energy particle accelerators. These national summer schools constitute one response to that appeal. Indeed, it is the main purpose of the school to attract scientists and students and to enhance their education in accelerator physics.

To carry out its mission, the school is guided in its operation by the following objectives: (i) to present in a thorough and up-to-date manner the entire spectrum of knowledge pertaining to particle accelerators; (ii) to help in the training of scientists who plan to work in accelerator physics, thereby building a base of particle accelerator specialists in this country; (iii) to encourage development of accelerator physics programs in American universities by providing text materials and training for the potential faculty of such programs; and (iv) to foster a more extensive dialogue between accelerator physicists and scientists and engineers working in particle physics and other accelerator-based sciences. Success in achieving these goals could be an important factor in continuing the advances in accelerator development necessary for a vigorous program in high energy physics and other sciences.

Each year the school produces a volume of its proceedings structured so that it can be read as a comprehensive textbook on accelerator physics and technology. The text for the 1983 BNL/SUNY school appears as Volume 127 of the American Institute of Physics Conference Series.

The field of accelerator physics and technology finds a place within the larger scientific enterprise of high energy physics. This high energy physics endeavor is an exciting adventure, probing the ultimate mysteries of physical nature; and to attempt to put this into perspective, the school offers an annual Symposium on the State of High Energy Physics, of which this volume represents the proceedings. The Symposium in general consists of a series of seminars on a broad range of subjects such as developments in particle theory and experiments, detector development, the nature and operation of high energy physics laboratories, and the status of ongoing and planned future projects. This provides a picture of the broad cultural framework of high energy physics within which the field of particle accelerators coexists. The general theme for the 1983 BNL/SUNY Symposium is related to the very large colliders envisioned by the scientific community and prompted by the current predictions of exciting new physics in the few-TeV energy region. The Symposium included a Round Table on an Ultrahigh Energy Collider, chaired by G.-A. Voss, with the following agenda:

Particle Physics of Multi-TeV Collisions	J.D. Bjorken, Fermilab
Accelerator Technology for a Multi-TeV Collider	M. Tigner, Cornell
Experiments at Multi-TeV Energies	C. Rubbia, CERN/Harvard
Prospects of an Ultrahigh Energy Collider: A View from Washington	N.D. Pewitt, Office of Sci. & Tech. Policy
Planning for a Super Collider	P.J. Reardon, BNL

These proceedings of the Symposium are the second in a series entitled The State of High Energy Physics, the first being AIP Conference Proceedings 92 (Fermilab, 1981).

Participation in the school, in terms of both lecturers and students, continues to be excellent. As anticipated, the major U.S. high energy physics laboratories (SLAC, Fermilab, and BNL) provide about 50% of the students. There have been efforts to improve university and foreign participation, primarily by greater interaction between the school, on the one hand, and U.S. universities and foreign institutions, on the other. Among other features, the following participation table shows the success that has been achieved in raising the university and foreign participation, and indeed it shows a striking increase in attendance in 1984.

Participation in Summer Schools

Source of Students	1981 Fermilab	1982 SLAC	1983 BNL/SUNY	1984 Fermilab
Major HEP labs, U.S.	65	92	78	94
Universities, U.S.	32	14	37	55
Other, U.S.	13	24	17	15
Foreign	10	19	16	31
Total students	120	149	148	195
Total lecturers	24	19	33	50

The school functions through an organizing committee, a school office, and a local school administration. The organizing committee meets once or twice per year as needed, determines overall school policy, and determines the school program and lecturers. The local school administration is established each year at the institution where the school is to take place. It coordinates and operates the school and implements all school functions. The general administrative functions of the school are carried out by a central school office located at Fermilab. This office coordinates the activities of the organizing committee, maintains a school file, and is responsible for collecting and reviewing the manuscripts and organizing the publication of the school text. The office also initiates and supports various activities which advance education and documentation in the field of particle accelerators, such as the 1985 U.S./CERN Topical Course on Nonlinear Dynamics, held in Sardinia, Italy, January 31 to February 5, 1985.

The past decade has proved to be one of the most fertile in the history of high energy physics, with the many great experimental discoveries of this period having been made possible largely by the new generation of high energy accelerators. These new machines have greatly increased the maximum energy range of particle beams, and have thus opened a window to exciting regimes of higher particle interaction energies. This has culminated in the proton-antiproton ($p\bar{p}$) collider, completed more than two years ago at CERN, which has achieved an interaction energy of 540 GeV, almost an order of magnitude more than in previous experiments. But still higher energies wait to be explored. In western Europe, the CERN Sp$\bar{p}$S Collider (the $p\bar{p}$ collider at the large CERN synchrotron, the SPS) is being upgraded in both energy and luminosity, the CERN LEP Project is in construction (LEP: e^+e^- collisions at more than 200 GeV center-of-mass energy), and the DESY HERA Project has been initiated (HERA: ep collisions at about 30 GeV x 800 GeV beam energies). Meanwhile, in the United States, two new colliders in the process of development stand ready to take up the task. One is the hadron collider with colliding proton and antiproton beams being constructed at Fermilab (Tevatron I: $p\bar{p}$ collisions at about 2 TeV in the center of mass); the other is the SLAC electron-positron collider (SLC: e^+e^- collisions at about 100

GeV in the center of mass.) Beyond this, we can imagine in the U.S. a program of study evolving in the next decade that is backed by a hadron-hadron collider in the multi-TeV energy range, such as the recently proposed SSC (Superconducting Super Collider) capable of collision energies up to 20 TeV x 20 TeV, and an e^+e^- collider in the energy range of many hundreds of GeV. In addition, with an existing SSC, one might imagine a tandem facility which could include very high energy ep collisions or e^+e^- collisions in a circular storage ring reaching perhaps ½ TeV in center-of-mass energy.

In recent years, theoretical understanding of high energy physics has exceeded the capability of experimental physics to verify many of the predictions of these theories. But this situation has been suddenly turned around with unexpected rapidity. Almost unseen, the age of the high energy hadron colliders is suddenly upon us, signalled by the great experimental achievements of the CERN Sp$\bar{\text{p}}$S Collider. In the past two years the CERN p$\bar{\text{p}}$ collider has led to the discovery of the vector bosons ($W^\pm$, Z_0) and the top quark and has produced hints of new phenomena. As experimentation with the new colliders opens up unexplored energy regimes, theory must readjust continuously in order to conform to the rapidly unfolding physical reality emerging. This is true for the operating colliders and will undoubtedly continue to be so for the ones just on the horizon.

The process of accelerator energy increase followed by new elementary particle discoveries has been going on since the 1930s, with accelerator energies increasing at the rate of about an order of magnitude every seven years. In the past, technological innovation leading to new accelerators has been at the heart of this process, with the strong-focusing synchrotrons of today capable of achieving energies six orders of magnitude higher than those achieved by the cyclotrons of forty years ago. If our quest into the nature of matter is to continue into the future, we need abundant ideas to further the advances in technology. But to make this happen, the need above all is for new ideas, which is a challenge that can be met only by new and younger people entering the accelerator field. It is our hope that the summer school will stimulate participants to think about and enter this bold venture to conceive and build new accelerators so as to push back the frontiers of energy. By so doing, they will join a new generation of high energy physicists dedicated to the study of new mass regions heretofore impenetrable.

Melvin Month
Chairman, Organizing Committee
U.S. Summer School on
High Energy Particle Accelerators
October   1984

ACKNOWLEDGMENTS

The 1983 Summer School on High-Energy Particle Accelerators owed much of its success to the dedicated and enthusiastic behind-the-scene contributions of a large number of individuals and supporting groups at Brookhaven National Laboratory and the State University of New York at Stony Brook. Thus, in addition to the members of the school organizing committee, the speakers whose lectures appear in this volume, the symposium speakers whose seminars will appear in a forthcoming companion volume, and members of both the local school administration and the editorial committee, we wish to acknowledge especially the following: D. Barton, R. Blumberg and C. Woody for their efforts as scientific secretaries for the lecture program; A. Forkin and A. Brody for coordinating school activities between BNL and SUNY; M. Heimerle for her continuing expert help in the transaction of school business; A. McClain for coordinating school attendees and handling manuscripts; G. Walczyk for coordinating audiovisual instrumentation; H. Boyd and P. Glenn and the excellent BNL Staff Services Division for arranging physical facilities and numerous supporting services; and N. Gargliardo and her staff in the BNL Travel Office for their patient and expert assistance with travel arrangements to and from the school. We must also acknowledge the invaluable expertise of various groups within the BNL Photography and Graphic Arts Division in manuscript preparation and processing under the general coordination of K.E. Boehm: Word Processing under M. Wigger, Quick Copy Services under J.P. Hanson, Illustration under E.J. Caiazza, Photography under B.S. Style, and Composition under L.J. Casey.

Finally, we thank Professor J.H. Marburger, President of SUNY, for his warm welcoming remarks during the opening session, and N.P. Samios, BNL Director, who closed the school with a most interesting discussion of the history of accelerators both at BNL and in high-energy physics.

P.J. Reardon
M. Month

SUMMER SCHOOL ORGANIZATION

LOCAL SCHOOL ADMINISTRATION

P.J. Reardon, School Director
P.F. Dahl, Deputy School Director
M. Month, Chairman, Organizing Committee
P. Hughes, School Administrator
A. Greene, Administrative Staff
W.T. Weng, Administrative Staff
C. Conrad, Secretary
M. Good, SUNY Liaison
H. Kinney, Public Relations
E. Dahl, Social Activities
S. Edwards, Social Activities
M. Dienes, Proceedings

ORGANIZING COMMITTEE

M. Month, Chairman, Brookhaven National Laboratory
J.D. Bjorken, Fermilab
H. Grunder, Lawrence Berkeley Laboratory
V.W. Hughes, Yale University
F.R. Huson, Fermilab
B. McDaniel, Cornell University
C. Pellegrini, Brookhaven National Laboratory
B. Richter, Stanford Linear Accelerator Center
R. Schwitters, Harvard/Fermilab
R.R. Wilson, Columbia University

LOCAL EDITORIAL COMMITTEE

P.F. Dahl
M. Dienes
A. Flood
J.C. Herrera
M. Month

HIGH ENERGY PHYSICS IN THE UNITED STATES*

Melvin Month
Brookhaven National Laboratory
Upton, NY 11973

TABLE OF CONTENTS

I. INTRODUCTION 3
   A. Overview of the Field 3
   B. Planning for the U.S. High Energy Physics Program. 4
   C. Goals and Status of High Energy Physics 8
   D. Evolution of Accelerator and Detector Facilities... 14
      1. Accelerators... 14
      2. Detectors.. 16
II. U.S. PROGRAM 1985-1995 28
   A. Overview 28
   B. Facilities at National Laboratories 29
      1. Details of the Program........................... 33
      2. Stanford Linear Accelerator Center (SLAC) 34
         a. SLAC Fixed-Target Program 34
         b. PEP Program at SLAC 35
         c. SPEAR Program at SLAC 35
         d. SLAC Linear Collider (SLC) 36
      3. Fermi National Accelerator Laboratory
         (Fermilab) 38
         a. TEVATRON II Program at Fermilab 39
         b. TEVATRON I Program at Fermilab 40
      4. Cornell Laboratory for Nuclear Studies 42
      5. Non-Accelerator Program 43
   C. University Based-Research 44
III. INTERNATIONAL COOPERATION IN HIGH ENERGY PHYSICS 48
   A. General 48
   B. The European Program 52
      1. CERN 52
         a. Fixed-Target Operation of the Super
            Proton Synchrotron (SPS) 52
         b. The SPS Proton-Antiproton Collider 52
         c. Large Electron-Positron Facility (LEP) ... 54
         d. The Next Step at CERN 55

---

*The basis for this paper was a lecture given by W.A. Wallenmeyer (Director of the Division of High High Energy Physics of the Department of Energy) at the 1983 Summer School at BNL. The author of this paper would like to express his appreciation to Dr. Wallenmeyer for his encouragement and support and to the program office staff at DOE for their cooperation and help. The actual material was accumulated from the DOE congressional budget writeup and briefings, from descriptive information prepared by the national laboratories, from various surveys and reports commissioned by the DOE and its scientific advisory body HEPAP, and from other documents.

		2.	DESY	55

```
              2.   DESY ............................................. 55
                   a.   DORIS ....................................... 55
                   b.   PETRA ....................................... 55
                   c.   HERA ........................................ 56
         C.   The Japanese Program ................................. 56
         D.   The U.S.S.R. Program ................................. 57
              1.   Institute for High Energy Physics (IHEP),
                       Serpukhov ................................... 57
              2.   Institute for Nuclear Physics (INP),
                       Novosibirsk ................................. 58
              3.   Joint Institute for Nuclear Research (JINR),
                       Dubna ....................................... 58
         E.   The Program of the People's Republic of China .... 59
  IV.  NEW FACILITY NEEDS FOR THE U.S. HIGH ENERGY PHYSICS
         PROGRAM ................................................. 59
         A.   The Next Step ........................................ 59
         B.   Accelerator R and D in Superconductivity.......... 64
         C.   Longer Range ......................................... 66
         D.   Advanced Technology Research and Development ..... 67
         E.   Funding and the Transition to the SSC ............ 69
              1.   Evolution of the Research Program ............ 69
              2.   Funding Required for the High Energy
                       Physics Program ............................ 70
              3.   Scientific Manpower .......................... 71
```

HIGH ENERGY PHYSICS IN THE UNITED STATES

Melvin Month
Brookhaven National Laboratory. Upton, NY 11973

I. INTRODUCTION

A. Overview of the Field

High energy physics is the field of basic research which addresses the most fundamental questions concerning the nature of the physical universe, i.e., the basic nature of matter, energy, space, and time. Its objective is to find the fundamental constituents of matter (the elementary particles) and the forces that act between them. Recent developments in experimental and theoretical physics point the way to an increasing understanding of the basic structure of matter and to an overall synthesis encompassing all the forces observed in nature.

Exploration of the ultimate constituents of matter requires two essential tools: particle beams of high enough energy and intensity to probe the structure within the nucleons, and detectors sensitive and complex enough to detect and decipher that structure.

The particle beams are generated by complex and large accelerators of various types, including linear accelerators, circular accelerators (synchrotrons), and colliding beam machines. The more fundamental the structure to be probed, the higher are the energies needed; therefore, attempts to probe deeper into the structure of matter require new accelerator capabilities and often new accelerator technologies. In recent years, devlopments in superconducting accelerator magnets and in colliding beam technology have provided the base for the next step in major facility development and construction.

Accelerating particles and bringing them into collision with targets and other beams is only half the task. The other half is to observe and distinguish the particles that emerge from these collisions with particle detectors. Much ingenuity has gone into the conception, development, and fabrication of these devices that can simultaneously register the passage of many subatomic particles traveling at essentially the speed of light, recognize their nature, and measure their energy and other properties.

Particle detectors have come to have a highly sophisticated set of capabilities due to the rapid development of electronics and other technology developments in recent years. Conversely, R and D to meet detector requirements has contributed to developments in a variety of technologies. This process, in response to the increasingly stringent requirements of experimentation at higher energies, has resulted in great improvements in precision and sensitivity and given rise to the modern detector, a large, complex multicomponent instrument.

The increased detector capability coupled with the higher energy and intensity of accelerator beams has resulted in massive amounts of data for analysis, which is done with powerful computers. Some computers are integrated into the detectors and are used to control the apparatus and to analyze data in real time in order to provide rapid feedback of results to guide the conduct of the experiment. Theoretical physicists and accelerator physicists have also come to rely on computers for the complex calculations needed to solve forefront theoretical problems and to simulate the properties of accelerators under design.

B. Planning for the U.S. High Energy Physics Program

High energy physics research is dependent on large complex particle accelerators, colliding beams, and detector facilities, and requires long lead times for planning and implementing intricate experiments and for designing and constructing advanced facilities. Typically, the time from the original concept for an experiment to the publication of results is 3 to 6 years and the time from conceptual design to operation for a major facility is 5 to 10 years or more. In an endeavor with such long lead times effective long-range planning is essential. High energy physics has a long record of efficient long-range planning. Since 1967 planning for the U.S. High Energy Physics Program has benefited substantially from advice from the High Energy Physics Advisory Panel (HEPAP) and its subpanels. Figure 1 indicates the role of the Department of Energy (DOE) as the lead agency responsible for this Program.

Institutionally, the program structure has at its core the large national accelerator laboratories (BNL, Fermilab, and SLAC managed by DOE and Cornell by NSF). Experimental support for the High Energy Physics Program is provided by 120 groups from 64 universities and laboratories with DOE funding and by 90 groups from 52 institutions with NSF funding; theoretical expertise derives form 57 DOE funded

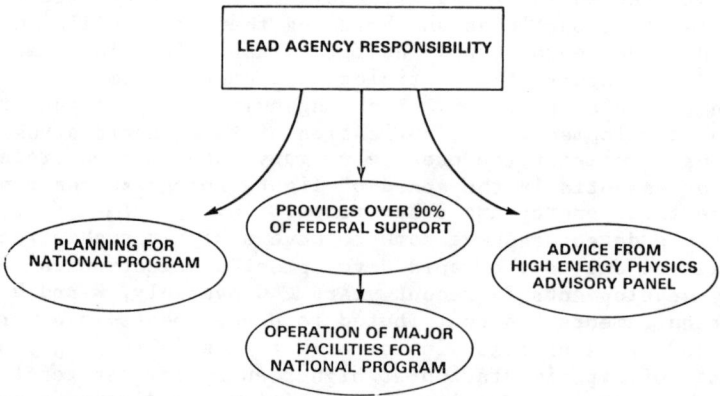

Figure 1. The role of the Department of Energy as the lead agency responsible for the U.S. High Energy Physics Program. The other U.S. government agency responsible for support to High Energy Physics is the National Science Foundation (NSF).

universities and laboratories and 47 NSF funded institutions. The
DOE management of the program is governed by a program philosophy
which can be summarized as follows:
- Ideas and proposals are generated by scientists in the field.
- The agency, with input from the field, establishes policy, plans, and budgets.
- The agency provides funding allocations and general guidance to the field.
- The laboratory management or university principal investigator is entrusted with the responsibility for the day-to-day detailed management of the program.
- The agency reviews and monitors progress and takes corrective action where appropriate.

The annual budget process by which funds get allocated is a rather elaborate one, beginning with proposals from the field and culminating in a funding decision process involving the Congress, the Office of Managment and Budget (OMB), and the Department of Energy (DOE).

Within the DOE, the Division of High Energy Physics (DHEP) first puts together a proposed High Energy Physics budget. This budget then moves to the Office of High Energy and Nuclear Physics (OHENP) where it gets folded in with the Nuclear Physics Budget. As the budget proceeds up the organizational ladder, there is negotiation and budget fitting and reworking. After OHENP, the Office of Energy Research (OER) has a budget which includes Fusion and Basic Energy Sciences. Finally, the budget goes through the Undersecretary of Energy and the Comptroller and becomes a total DOE Budget including other parts of the Department's concerns such as Nuclear Energy and Nuclear Weapons. The OMB acts for the President and pulls together from all the agencies the President's Budget, which is then sent to the Congress. The House and the Senate each act on the President's Budget through three committees: the Budget Committee, the Authorization Committee, and the Appropriation Committee. Hearings are held before subcommittees of each of these in both the House and Senate. Coming out of the hearings, House and Senate bills provide budget figures. If they differ there is a House and Senate conference, and from this conference comes a joint House and Senate bill. The bill returns to the President and with his signature it becomes law. The bill then returns down the line essentially the same way it went up, and apportionment is made at each level.

The government operates on a fiscal year basis. The fiscal year begins on October 1. The budget process begins with the receipt of requests from the field about 18 months before the beginning of the fiscal year. The DOE then sends its request to the OMB about 13 months before the fiscal year begins. Sometime in January, the President's Budget is released and sent to Congress about 9 months before the fiscal year starts. Under ideal circumstances, Congress will return the budget bill to the President about 3 months before October 1; sometimes this doesn't happen, and no budget bill is passed into law by the beginning of the fiscal year. Since government operations must continue, the government then proceeds on the basis of a continuing resolution, a temporary bill that Congress must pass just for this purpose.

The steps in the DOE process are reviewed in Table I, and the annual budget process is outlined in Figure 2. An example of a High Energy Physics Budget allocation, including figures for the FY 1986 President's Budget, is shown in Table II.

Planning for the DOE High Energy Physics Program during FY 1979 -FY 1982 was based on a long-range funding plan agreed to in early 1978 by the DOE and the Office of Management and Budget (OMB). The objectives were to maintain a productive and viable High Energy Physics Program in the U.S. effectively utilizing three accelerator centers (Brookhaven, Fermilab, and SLAC); to maintain U.S. world leadership in the field; to permit construction of new facilities as required; and to accomplish these goals within an approximately constant budget level of $300 million per year in FY 1979 dollars.

Since the time of the DOE/OMB agreement, many profound scientific and technical insights have been gained. Major physics developments include the discovery at CERN of the W and Z particles, which are crucial to theories that unify the weak and electromagnetic forces; the emergence of the theory of quantum chromodynamics (QCD), which describes the strong force in terms of the interactions between quarks and gluons; and progress toward grand unified theories which link the strong force with the electroweak force in one comprehensive framework. Breakthroughs in understanding are expected to occur in the mass region where elementary subnuclear constituents interact with energies of a few TeV (10^{12} electron volts). Many theoretical approaches point to the existence of hitherto unobserved forms of matter in this mass region. The discovery of such new forms of matter would have a profound impact on our understanding of the underlying symmetries of nature and the mechanism for generating the masses of elementary particles.

Major technology and facility developments have also occurred. These include the successful operation at Fermilab of the Energy Saver, the world's first high energy accelerator using superconducting magnets, the TEV II fixed-target project, and the TeV I collider project; the construction of the SLAC SLC and the CERN LEP electron-positron collider projects; the successful achievement of adequate luminosity in the $\bar{p}p$ collider at CERN; and the decision not to proceed with the CBA project at Brookhaven. The success of the Energy Saver, together with R and D progress in the CBA superconducting magnet program, give confidence that very large accelerators using superconducting magnets are feasible. In view of these developments it was prudent that the goals and directions of the U.S. High Energy Physics Program be reassessed.

First steps toward a review and reevaluation of the U.S. High Energy Physics Program plan were taken in mid-1981. A HEPAP Subpanel on Long-Range Planning reviewed and evaluated in depth the high energy physics facilities currently in operation, those under construction, and those proposed for construction. A report with long-range planning recommendations at two assumed budget levels was completed in January 1982. In 1983, a HEPAP Subpanel on New Facilities conducted an in-depth review of the scientific requirements and opportunities

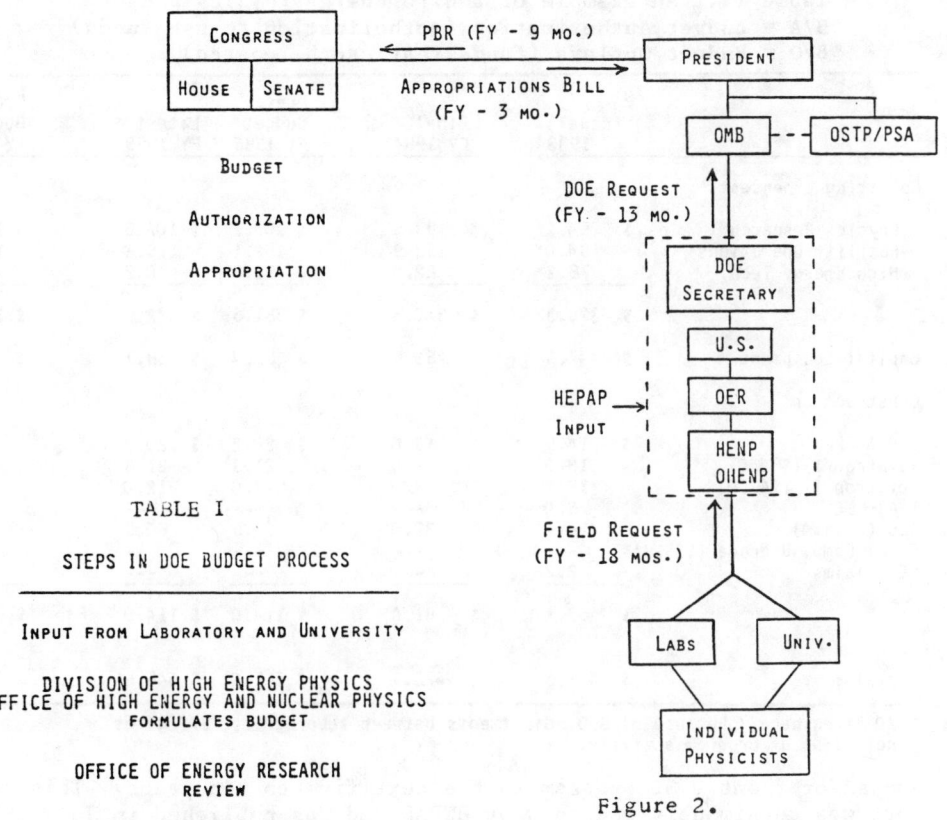

Figure 2.
Annual budget process.

TABLE I

STEPS IN DOE BUDGET PROCESS

INPUT FROM LABORATORY AND UNIVERSITY

DIVISION OF HIGH ENERGY PHYSICS
OFFICE OF HIGH ENERGY AND NUCLEAR PHYSICS
FORMULATES BUDGET

OFFICE OF ENERGY RESEARCH
REVIEW

U.S. DEPARTMENT OF ENERGY
INTERNAL REVIEW

OFFICE OF MANAGEMENT AND BUDGET
REVIEW

PRESIDENT'S BUDGET TO CONGRESS

HOUSE ACTION   SENATE ACTION

CONFERENCE
CONGRESSIONAL BILL

BILL

PRESIDENT
FOR SIGNATURE

OFFICE OF MANAGEMENT AND BUDGET
APPORTIONMENT

U.S. DEPARTMENT OF ENERGY
ALLOCATION

Table II. An example of a High Energy Physics Budget
B/A = Budget Authorization (authorization to use funds)
B/O = Budget Outlays (funds that can be costed)

	Actual FY 1983	Final FY 1984	Pres. Budget FY 1985	Latest FY 1985	Pres. Budget FY 1986
Operating Expenses					
Physics Research	$ 94.1*	$ 99.5	$ 107.7	$ 107.5	$ 111.3
Facility Operations	154.0*	158.9	182.1	175.2	187.8
High Energy Tech.	78.2*	82.0	91.8	90.2	92.4
	$ 326.3	$ 340.4	$ 381.6	$ 372.9	$ 391.5
Capital Equipment	$ 47.5	$ 51.5	$ 65.4	$ 58.7	$ 63.1
Construction					
AIP & GPP	$ 14.5	$ 16.0	$ 20.2	$ 20.2	$ 20.9
Tevatron I ($84.0)	18.0	20.0	21.3	21.3	8.6
Tevatron II ($49.0)	18.0	13.0	12.0	12.0	---
ISABELLE	5.0	---	---	---	---
SLC ($115.4)	---	32.0	60.5	60.5	22.9
Fermi Comp. Upgrade ($23.9)	---	---	---	---	3.1
PEP Claims	2.2	---	---	---	---
	$ 57.7	$ 81.0	$ 114.0	$ 114.0	$ 55.5
Total	$ 431.5	$ 472.9	$ 561.0	$ 545.6	$ 510.1

*B/O given here. Because of GSO adjustments between subprograms, B/O gives indication of program activity.

for a forefront U.S. program in the next five to ten years. This report was unanimously endorsed by HEPAP and was published in July 1983. More recently, a HEPAP in-depth examination of the entire U.S. program was made. This study culminated in a meeting at Berkeley Springs, WV, June 2-8, 1985. Its conclusions are detailed in Report of the 1985 High Energy Physics Advisory Panel Study of the U.S. High Energy Physics Program 1985-1995, published by the U.S. Department of Energy Division of High Energy Physics. It reaffirmed the highest priority for the SSC, identified top priority elements in the present Program, and considered the transition to the SSC.

These reports provided valuable input on high energy physics research and facility status, future facility needs, and program priorities to the HEPAP, which reviewed the reports and forwarded its advice and recommendations to the Department of Energy.

C. Goals and Status of High Energy Physics

The goal of high energy physics is to find the ultimate constituents of matter and the forces acting between them. The purpose of the program is to examine the transformations and interactions among the ultimate constituents of matter, to search for new fundamental laws of nature, and to seek a better understanding of the established laws of nature. Research into the inner structure of matter has proceeded from the level of molecules to atoms, to nuclei, to nucleons, and recently to constituents inside the nucleon called "quarks." Quarks and leptons (electrons, neutrinos, and massive electron-like particles) form simple families and are the most basic constituents of matter so far understood (Figure 3). High energy physics deals with distance scales (Figure 4) of $<10^{-13}$ cm. Such deep penetration into matter can be accomplished only by using particle beams of very high energy (and thus very short wave-length), available from today's giant particle accelerators.

As newer accelerators have probed to smaller distances in matter, a theory has evolved to account for the multitude of observations which has become so widely accepted that it is called the "standard picture." The elemental constituents of matter are currently viewed as six quarks and six leptons, which have distinct physical properties but no internal structure and no size. Quarks combine under the strong force, a residue of which binds protons and neutrons into atomic nuclei and is responsible for fusion processes that power the

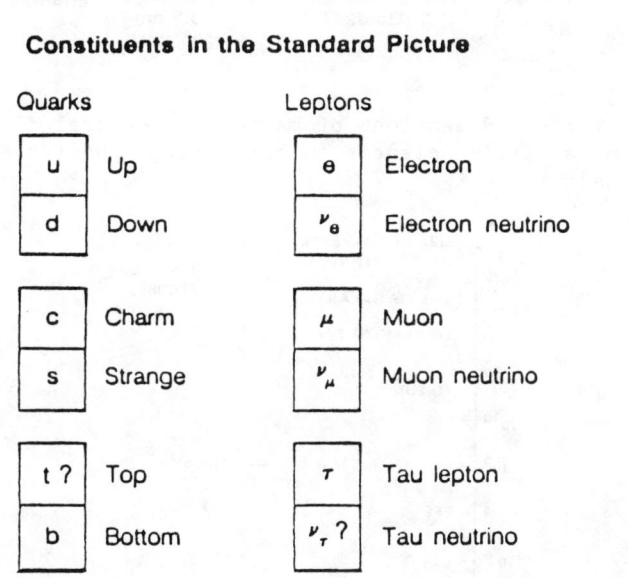

Figure 3. Successive levels in the nested structure of matter discovered through more than a century of research at the "high energy" frontier, and the elementary constituents of matter as indicated by the standard picture, grouped into families of leptons and quarks and into three generations.

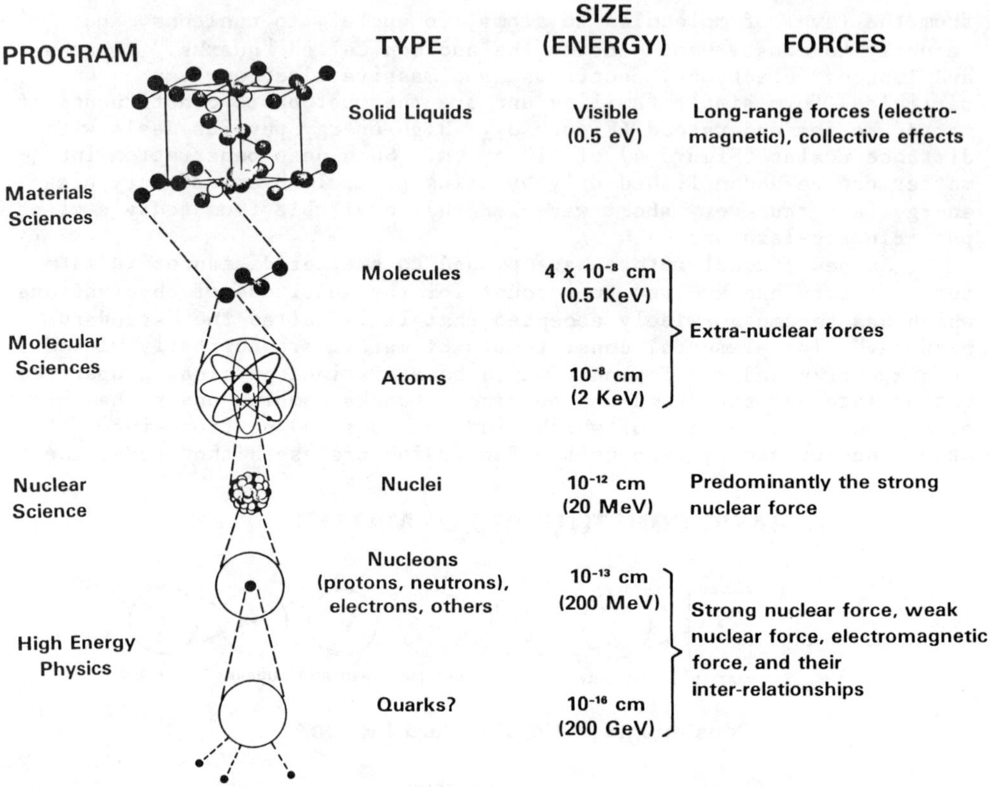

PROGRAM
TYPE
SIZE (ENERGY)
FORCES
Materials Sciences
Solid Liquids
Visible (0.5 eV)
Long-range forces (electro-magnetic), collective effects
Molecular Sciences
Molecules
4 x 10⁻⁸ cm (0.5 KeV)
Atoms
10⁻⁸ cm (2 KeV)
Extra-nuclear forces
Nuclear Science
Nuclei
10⁻¹² cm (20 MeV)
Predominantly the strong nuclear force
High Energy Physics
Nucleons (protons, neutrons), electrons, others
10⁻¹³ cm (200 MeV)
Quarks?
10⁻¹⁶ cm (200 GeV)
Strong nuclear force, weak nuclear force, electromagnetic force, and their inter-relationships

Figure 4a. Typical dimensions of matter for several disciplines ranging from materials sciences to high energy physics at the subatomic scale.

CM
10²⁵
10²⁰
10¹⁵
10¹⁰
10⁵
1
10⁻⁵
10⁻¹⁰
10⁻¹⁵
SIZE OF UNIVERSE (10¹⁰ LIGHT YEARS)
SIZE OF GALAXY
TO NEAREST STAR
TO SUN
ONE KILOMETER
MAN
BEE
CELL
VIRUS
ATOM
NUCLEUS
PROTON
QUARK?
THE REACH OF TELESCOPES
MIDWAY
THE REACH OF ACCELERATORS

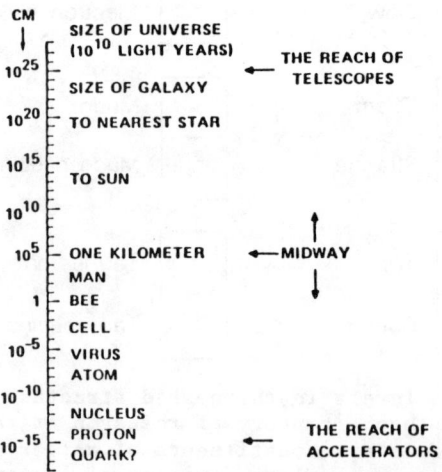

Figure 4b. The scale of Nature.

stars; leptons are unaffected by this strong force but are subject, as are the quarks, to the electromagnetic force that binds leptons (i.e. electrons) to nuclear matter to form atoms and to a weak force that generates transitions among quarks and leptons and manifests itself as radioactivity. Gravity is a fourth force, relatively so weak that it would require the combined effect of an enormous number of particles to make it measurable at the scale of distance so far studied (see Table III). Physicists hope ultimately to unify and so describe all fundamental forces by a single unified theory (Einstein's dream).

This standard picture has brought high energy physics to a turning point. During the last decade, work in the gauge theory of particle interactions has shown how the fundamental forces of nature might be unified. The electromagnetic force has been unified with the weak nuclear interaction even though these two forces differ in strength by a factor of nearly 100,000. This has encouraged the hope that unification of all the fundamental forces may be within reach during this century (Figure 5). Gauge theory now dominates nearly all phases of high energy physics, and the decision to perform new experiments is often based on their relevance for testing its predictions. The potential applications for gauge theory extend far beyond elementary particle physics, to areas as diverse as condensed matter physics, nonlinear wave phenomena, and even pure mathematics.

The primary principle of modern gauge theory is that forces have a property designated "local gauge invariance," a mathematical symmetry such that the various basic forces have a common mathematical structure so that the different forces appear to be manifestations of a greater unified whole. A second principle is that a phenomenon called "spontaneous symmetry breaking" generates the large differences between the strong, weak nuclear, and electromagnetic forces. This mechanism generates the masses of the elementary particles which reflect the differing natures of the forces observed; the photon, carrier of the electromagnetic force, is massless, whereas its analogues, the vector bosons ($W^{\pm}$ and Z^0), which carry the weak nuclear force, are quite massive. The recent discovery of these bosons at CERN confirms their huge mass--almost 100 times that of the proton--in agreement with the prediction.

To understand the generation of the masses of vector bosons and other elementary particles requires that one understand how spontaneous symmetry breaking is manifested in nature. A likely manifestation is an as yet unseen type of subnuclear particle--one with no intrinsic spin (as contrasted with quarks and leptons, which have internal spins of half units, and the carriers of the strong, weak, and electromagnetic forces, which have a spin of one unit). This particle, called the Higgs boson, is predicted by gauge theory. A primary goal of research in the next decade is to discover the Higgs particle or some manifestation of the mechanism for spontaneous symmetry breaking. This could lead to the discovery of a class of matter with fundamentally new properties which would complement the quarks, leptons, photons, and other gauge bosons found at the smallest scale of distance now attainable. This would have far-reaching intellectual and scientific consequences.

Table III The characteristics of the four basic forces

Type of force:	Gravitational	Electromagnetic	Strong nuclear	Weak nuclear
Behavior over distance:	Extends to very large distances	Extends to very large distances	Limited to $\lesssim 10^{-13}$ cm	Limited to $\lesssim 10^{-10}$ cm
Strength relative to strong force at a distance of 10^{-13} cm:	10^{-38}	10^{-2}	1	10^{-13}
Time for a typical hadron to decay via these forces:	10^{17} yr	10^{-20} sec	10^{-23} sec	10^{-10} sec
Particle that carries the force:	Gravitron not discovered	Photon	Gluon (has been identified indirectly but has not been, and perhaps can not be, isolated)	W^+, W^-, and Z^0 intermediate bosons
Mass of particle:	Not known	0	Assumed 0	~ 90 GeV

- **GRAVITATIONAL**
  - CARRIED BY GRAVITONS
  - AFFECTS ALL MASSIVE PARTICLES
  - EXPLAINS FALLING APPLES, TIDES, SOLAR SYSTEMS, GALAXIES, BLACK HOLES
- **ELECTROMAGNETIC**
  - CARRIED BY PHOTONS
  - AFFECTS ALL CHARGED PARTICLES
  - EXPLAINS LIGHTNING, LIFE PROCESSES, RADIO, ATOMS AND MOLECULES
- **WEAK NUCLEAR**
  - CARRIED BY WEAK BOSONS
  - AFFECTS ALL PARTICLES
  - EXPLAINS RADIOACTIVE DECAYS OF PARTICLES, NEUTRINO INTERACTIONS, EARTH'S MOLTEN CORE
- **STRONG NUCLEAR**
  - CARRIED BY GLUONS
  - AFFECTS QUARKS BUT NOT LEPTONS
  - EXPLAINS STARFIRE, EARLY UNIVERSE, NUCLEAR STRUCTURE, ATOMIC ENERGY

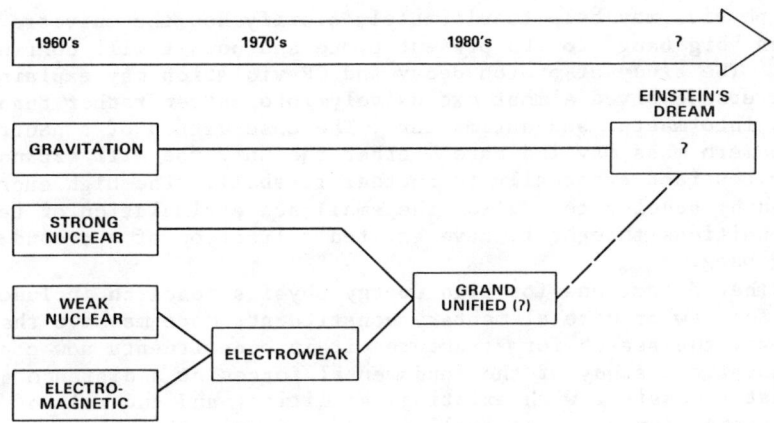

Figure 5. The process of unification of the basic forces of nature over the past quarter century.

All current theories predict fundamental new phenomena at mass values of a few TeV or less. The need to probe the mass region up to a few TeV sets the basic requirements for beam energy and luminosity (beam collision rate) for the next step in the construction of high energy accelerator and colliding beam facilities. The overall priority in high energy physics for exploring the few-TeV mass region arises from the critical question of spontaneous symmetry breaking and the strong expectation that major breakthroughs can be made when the right energy range is probed.

Theory has pushed even further into the unknown and several trial models of "grand unification" unify the description of quark and lepton matter at higher energies and predict that quarks may turn into leptons and vice-versa. This implies the exact equality of lepton and nucleon electric charges and leads to the prediction that protons, long thought to be stable, will be found to decay. These models further indicate that magnetic monopoles, trillions of times heavier than protons, may exist; and some predict that matter may change to antimatter, which implies that neutrinos, previously thought to be massless, may have very small masses that have escaped detection. Finally, these models may provide an approach to understanding why certain microscopic processes do not proceed forward and backward exactly reversibly (CP violation). Many of the predictions are subject to test in experiments already in progress or being designed for facilities now under construction.

A primary goal of the next decade is therefore to continue unifying the four basic forces of nature. Present theories are expected to be developed and extended, and their predictions will be experimentally tested at facilities existing and under construction. Early in the next decade a new colliding beam accelerator giving access to the TeV mass region will provide insight into the breaking of gauge symmetry and hence allow physicists to formulate the scientific foundation of higher-order unification schemes. Research in high energy physics may help to ultimately clarify how the universe evolved from the "big bang" to its present state and how it will continue to evolve. The study of proton decay and CP violation may explain why the universe evolved almost exclusively into matter rather than equally into matter and antimatter. The observation of a neutrino with nonzero mass may indicate whether the universe will expand forever or contract eventually to another fireball. The high energies achieved by accelerators allow the small-scale simulation of temperature conditions thought to have existed a fraction of a second after the big bang.

Other directions for high energy physics research include the search for new or more elementary constituents more massive than any now known; the search for structure within constituents now considered elementary; the study of the fundamental forces at a distance smaller than that accessible with existing facilities; and the role of the gravitational force in the small-distance domain of the elementary particles. The most significant discoveries could be totally unexpected. High energy physics research, as it probes deeper into matter has in the past seen many of the surprises nature has to offer, and in the next step--to higher energies and smaller subnuclear distances--we can only wonder what nature has in store for us.

D. Evolution of Accelerator and Detector Facilities

  1. Accelerators

In this century the push to identify the ultimate constituents of matter and the basic forces that bind them has led to the construction of large and powerful research instruments such as the high energy accelerators at CERN (near Geneva), at DESY (near Hamburg, West Germany), at Stanford University (SLAC; in California), at Brookhaven National Laboratory (near New York City), and at Fermilab (near Chicago). Since the 1940s, significant discoveries about nuclear and subnuclear structure have been closely related to the availability of increasingly more powerful accelerators providing particles of higher and higher energies. The energy growth of accelerators between 1930 and the present is shown in Figure 6.

The smaller the domain to be studied, the more powerful must be the instrument. The tremendous binding forces within and between subatomic particles require the high energy of a particle accelerator to split the particles into new components, which can then be observed individually and as they interact. Part of the immense energy is converted into matter, producing a profusion of particles of varying degrees of stability.

Particle accelerators can be linear accelerators (such as the 30-GeV electron linac at SLAC) in which the particles go through only once, reaching maximum energy at the end, or circular synchrotrons (such as the 800-GeV Tevatron at Fermilab or the 30-GeV Alternating Gradient Synchrotron at Brookhaven), in which the particles travel around a circular path, gaining energy with each revolution. Electrical energy is used to accelerate the particles; whereas in synchro-

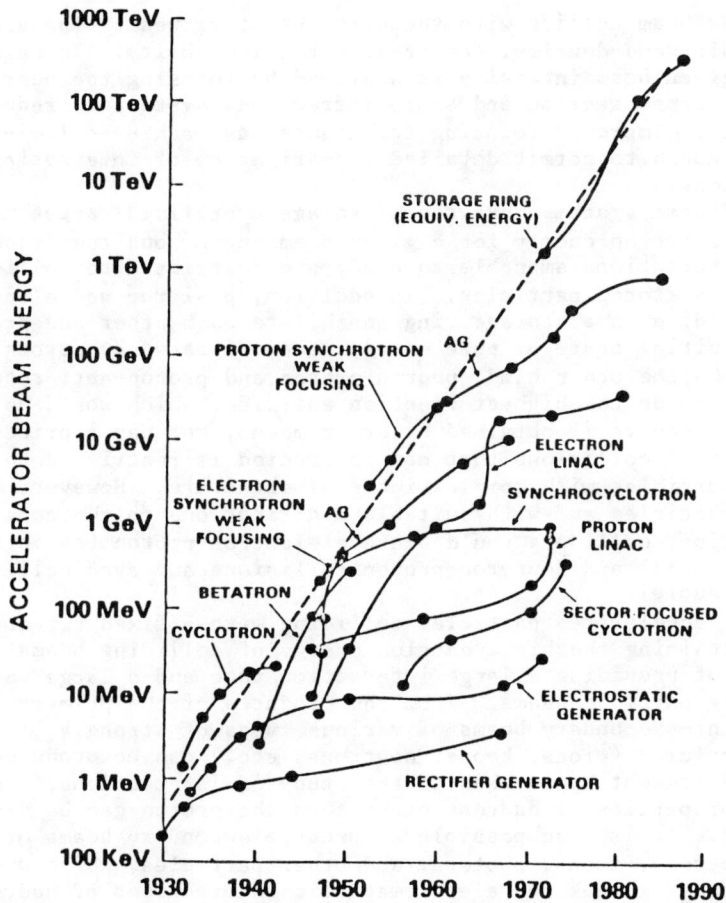

Figure 6. Energy growth of accelerators and storage rings.

trons powerful magnets keep the particles in the circular path and focus them into a narrow beam.

During the 1960s and 1970s major advances in accelerator technology (the accumulation and long-time storage of intense beams of hadrons and leptons) made possible the head-on collision of high energy particle beams in circular accelerators. This dramatically increased the energy available for initiating new forms of elementary processes and probing smaller subnuclear distances. For example, in order to reach the available energy produced in a head-on collision between two 1000-GeV protons, a proton colliding with a fixed target proton would have to have an energy of 2 million GeV, requiring an accelerator with a diameter of ∼8000 miles.

In addition to particle beam energy, another basic accelerator parameter is the beam intensity, or the associated collider luminosity. The luminosity of a collider determines the rate at which the particles in one beam collide with those in the other beam. The greater the beam intensity and density, the greater the luminosity. Increased density for a given beam intensity is achieved by focusing the beams to reduce their cross section and by radiofrequency systems to reduce the bunch length. Improved focusing techniques have achieved luminosities high enough to permit detailed investigation of interesting and rare reactions.

Colliding beam systems have the advantage over fixed-target machines of high reaction energy for a given beam energy but the disadvantage that interactions amenable to study are restricted to collisions between the stored particles. In addition, positron and electron beams colliding in a storage ring annihilate each other and provide a simple initial state of pure energy, uncomplicated by strong interactions. On the other hand, proton-proton and proton-antiproton colliders can provide the highest reaction energies, which would be prohibitively expensive if obtained by other means, but pay a price in that the variety of collisions that can be studied is restricted to interactions of particles with complex internal structure. However, by storing other particles and with suitable modifications to the collision configuration, colliders could achieve electron-proton (as in the HERA project at DESY) and deuteron-proton collisions and even collisions of heavy nuclei.

A beam of accelerated particles colliding with a fixed target, although not attaining the high reaction energy of colliding beams, has the virtue of providing a large interaction rate and a large variety of secondary particle beams. From the products of the primary collision, intense secondary beams of various types of strongly interacting particles (pions, kaons, neutrons, etc.) can be produced to allow many different approaches to the study of interactions. In this way, the properties of hadrons other than the proton can be directly explored. It is also possible to produce secondary beams of neutrinos, electrons, muons, photons, and other particles, which are valuable for studying weak and electromagnetic interactions of hadrons and the properties of their internal constituents.

Schematics of these modern high energy physics accelerator facilities, emphasizing their special features, are shown in Figure 7, including fixed target, e^+e^-, pp and $p\bar{p}$ facilities.

2. Detectors

As their energy has increased, particle accelerators and colliders not only have increased in size and complexity but also have come to depend on large and intricate detectors to record the pattern of particles emanating from the collisions, often in time intervals of billionths of a second. These detectors, consisting of a number of devices (bubble chambers, Cerenkov counters, proportional wire chambers, drift chambers, charge coupled detectors, time projection chambers, scintillation counters, etc.) require advanced technology and high precision.

Detectors extract the physics from particle collisions by making "visible" the products of the interactions of the high energy particles. Elementary particles cannot be seen directly; their path

Figure 7A. Proton beam colliding with fixed-target accelerator. Available reaction energy $\backsim \sqrt{2 \times \text{beam energy}}$.

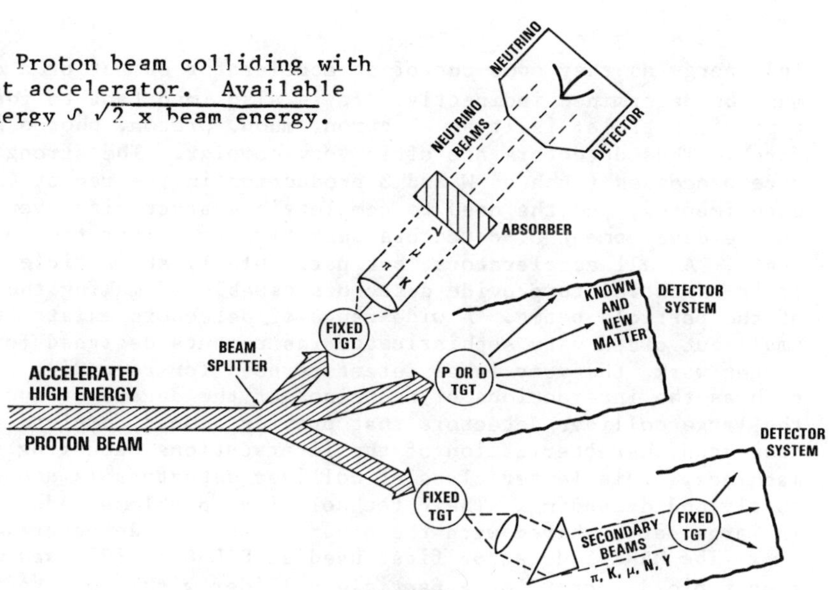

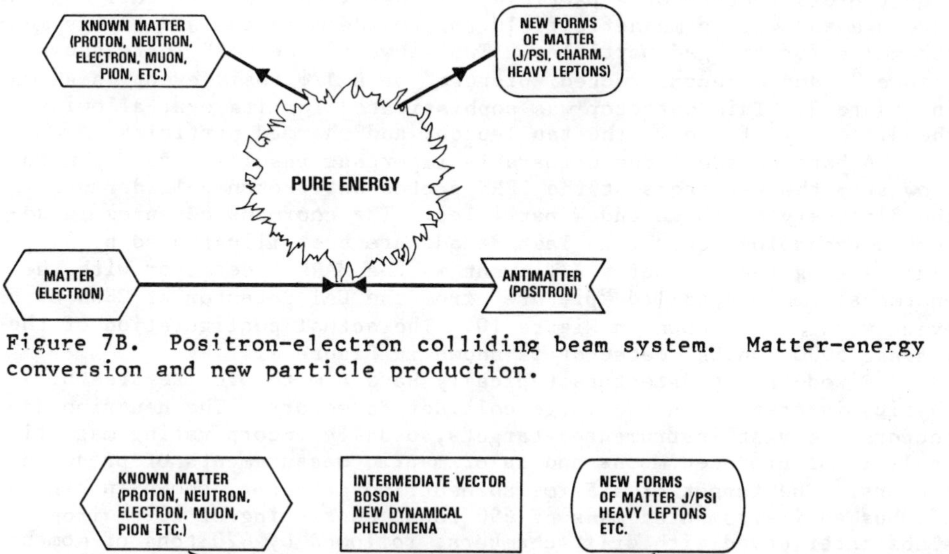

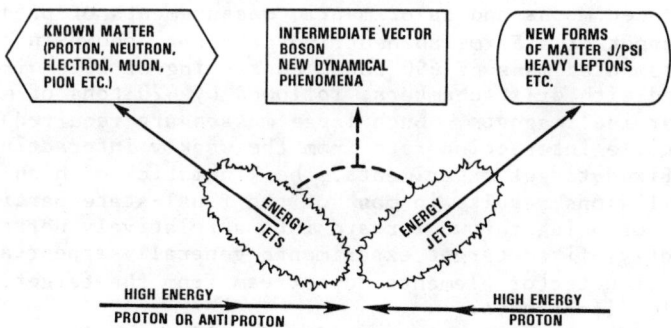

Figure 7B. Positron-electron colliding beam system. Matter-energy conversion and new particle production.

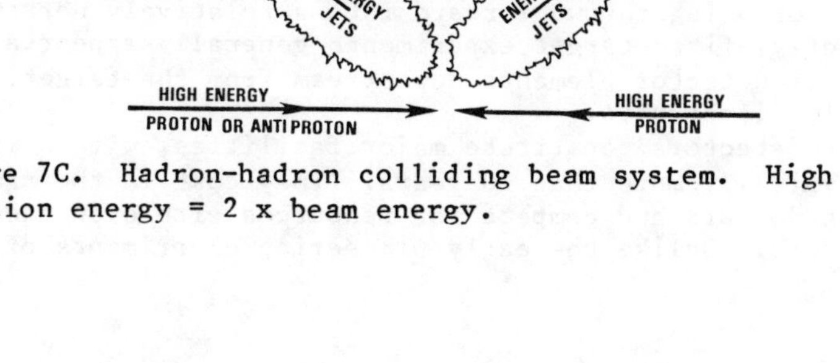

Figure 7C. Hadron-hadron colliding beam system. High available reaction energy = 2 x beam energy.

and energy as they come out of an accelerator or out of a collison must be determined indirectly. It is also important to identify the type of particle: is it an electron, muon, proton, photon and so forth. Thus detectors are often very complex. The strong interest in rare processes (such as W and Z production in the recent CERN experiments), and the need to completely characterize events, have led to the development of detectors sensitive to almost the total solid angle. At all accelerators, but particularly at particle colliders, it is essential to provide detectors capable of making the fullest use of the particle beams. A wide range of detectors exists including the small but often very sophisticated instruments designed for fixed-target work, the very large detectors used for recording rare events such as the interactions of neutrinos or the decays of nucleons, and the large collider detectors that provide almost complete angular coverage and characterization of the interactions occurring in these machines. This latter class of collider detectors is among the most costly and demanding. Their technological problems and solutions are, in large part, shared with the other classes of detectors.

The MARK I detector first used at SPEAR in 1972 was the first electronic detector for a particle collider with close to full angular coverage and with a magnetic field to provide momentum and energy measurements for charged particles. Two views of the MARK I are shown in Figure 8, and a reconstructed "picture" of a J/ψ (psi) event is shown in Figure 9. This detector was sophisticated for its era, allowing the discovery of the ψ, the tau lepton, and charmed particles.

A bare decade later comparably important results have begun to flow from the detectors at the CERN proton-antiproton collider with the discovery of the Z and W particles. The enormous advances in detector technology over this last decade are best illustrated by constrasting the view of a J/ψ event in the MARK I detector with the enormously more detailed "picture" from the UA1 detector at CERN of an event with a Z°, shown in Figure 10. The actual configuration of the immense 5000 ton UA1 detector is shown in Figure 11.

Fixed-target detectors typically have a more precise area of physics interest than the large collider detectors. The neutrino detectors are vast instrumented targets, usually incorporating magnetic analysis of produced muons and calorimetric measurements of produced hadrons. The target of a Fermilab neutrino detector, shown in Figure 12, has an instrumented mass of 690 tons, consisting of 20-cm iron slabs interleaved with drift chambers, followed by 420 tons of momentum analyzing toroidal magnets. Such large masses are required to achieve a reasonable interaction rate from the weakly interacting neutrinos. In fixed-target experiments, the kinematics of high-energy relativistic collisions results in most of the final-state particles from an interaction being thrown forward into a relatively narrow cone. Consequently, fixed-target experiments generally appear as a linear sequence of detector elements downstream from the target, as Figure 12 clearly illustrates.

The larger detectors constitute major facilities, with a lifetime of usage typically more than 10 years. They cost in the range of 10 to 50 million dollars and compete for resources with even the large parent accelerators. Unlike the early pioneering experiments of

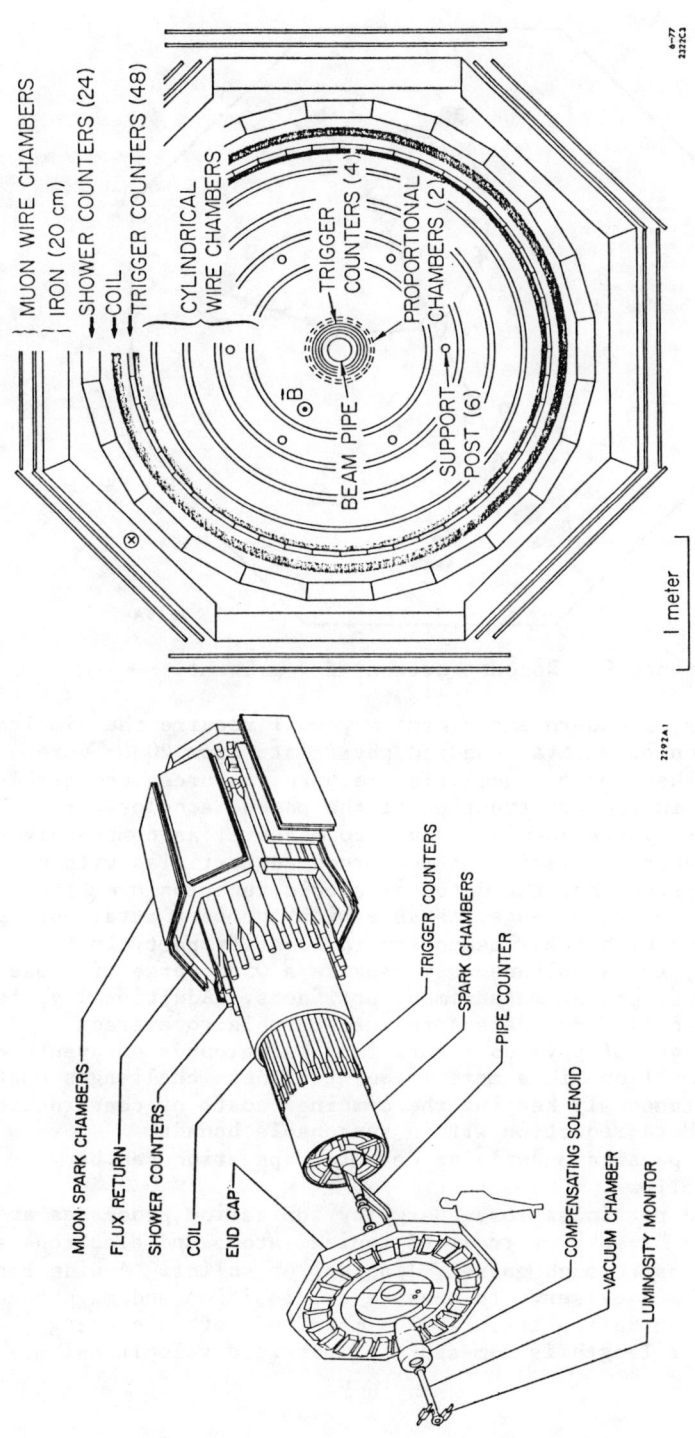

Figure 8. (a) Mark I detector; (b) cross section.

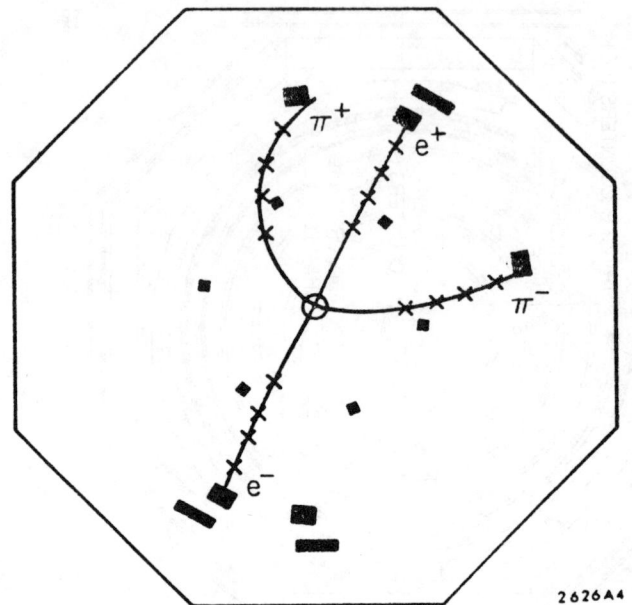

Figure 9. Reconstruction of J/ψ event.

particle physics, a modern experiment may well require the simultaneous collaboration of several hundred physicists from 20 or more institutions. These major facilities require resources comparable with those used in the construction of the parent accelerator.

A detector system should be able to measure, as completely as possible, all the characteristics of the produced particles within an event. This implies that the detector should function over the largest possible angular range, measure with the best attainable precision, be provided with instrumentation to identify particle characteristics, and simultaneously provide a wide range of cross checks to protect against measurement artifacts. Additionally, for use with hadron colliders, detectors must be able to extract interesting classes of physics events from backgrounds of events perhaps a hundred million times more frequent. These challenges must be met while simultaneously keeping the combined costs of construction, operation, and data-reduction within reasonable bounds.

The basic physics underlying detector operation can be summarized as follows:

- Charged particles lose energy by ionization processes and leave a "track" or trail of ionized atoms and electrons as they pass through gasses, liquids, or solids. A wide range of techniques serve to measure the position and magnitude of these ionization trails. The magnitude of this energy loss per unit length is a measure of particle velocities.

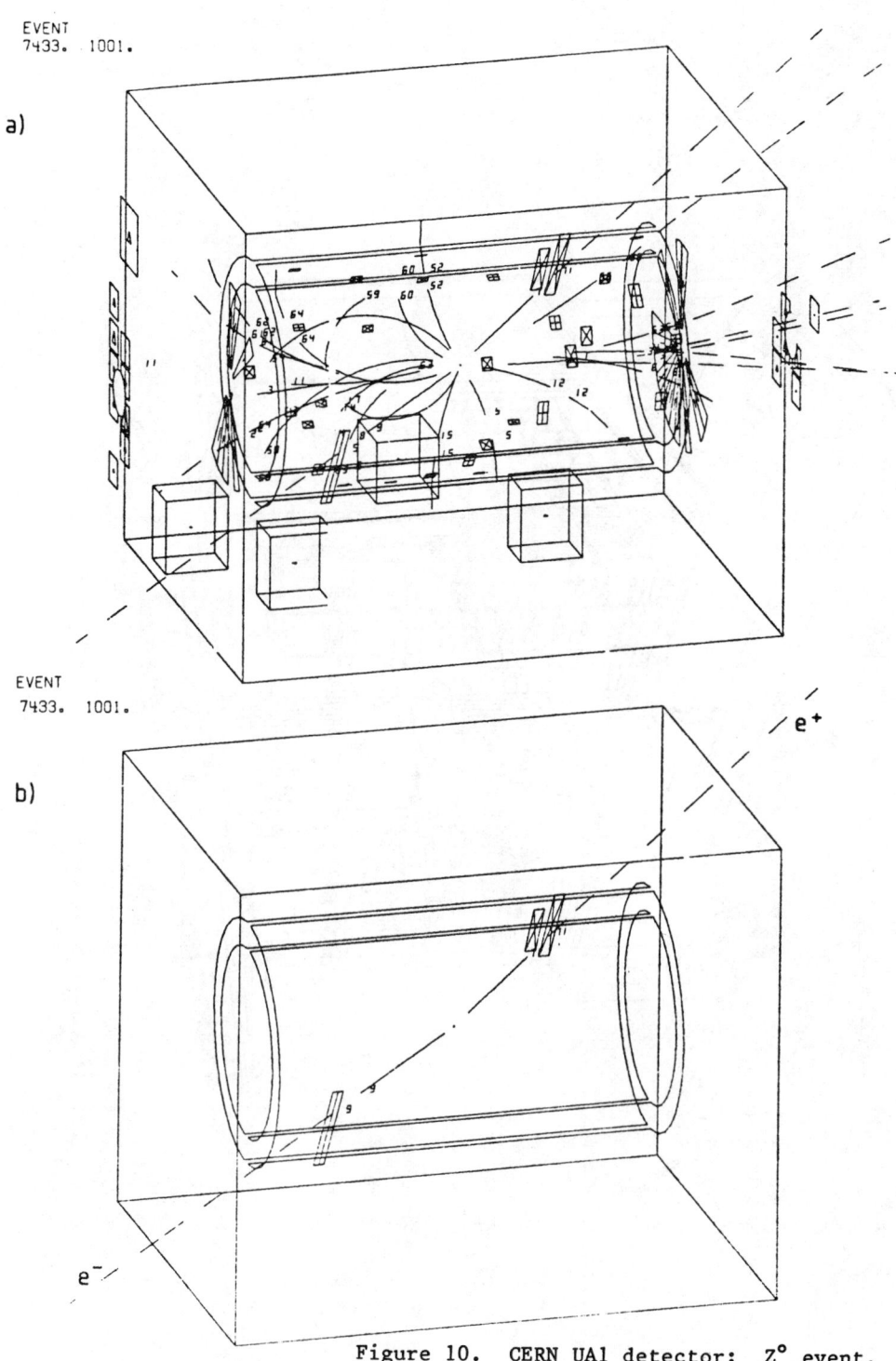

Figure 10. CERN UA1 detector: Z^0 event.

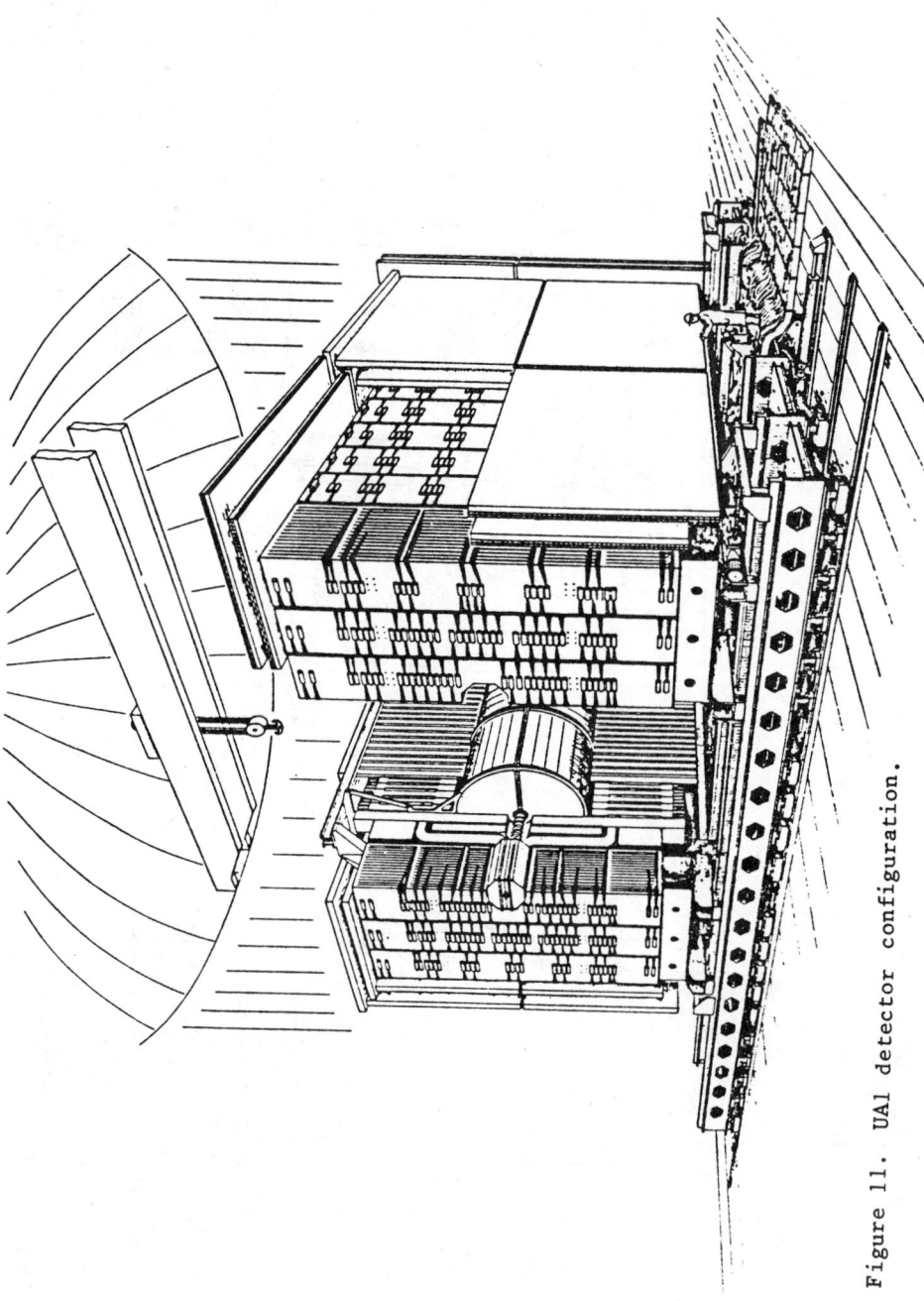

Figure 11. UA1 detector configuration.

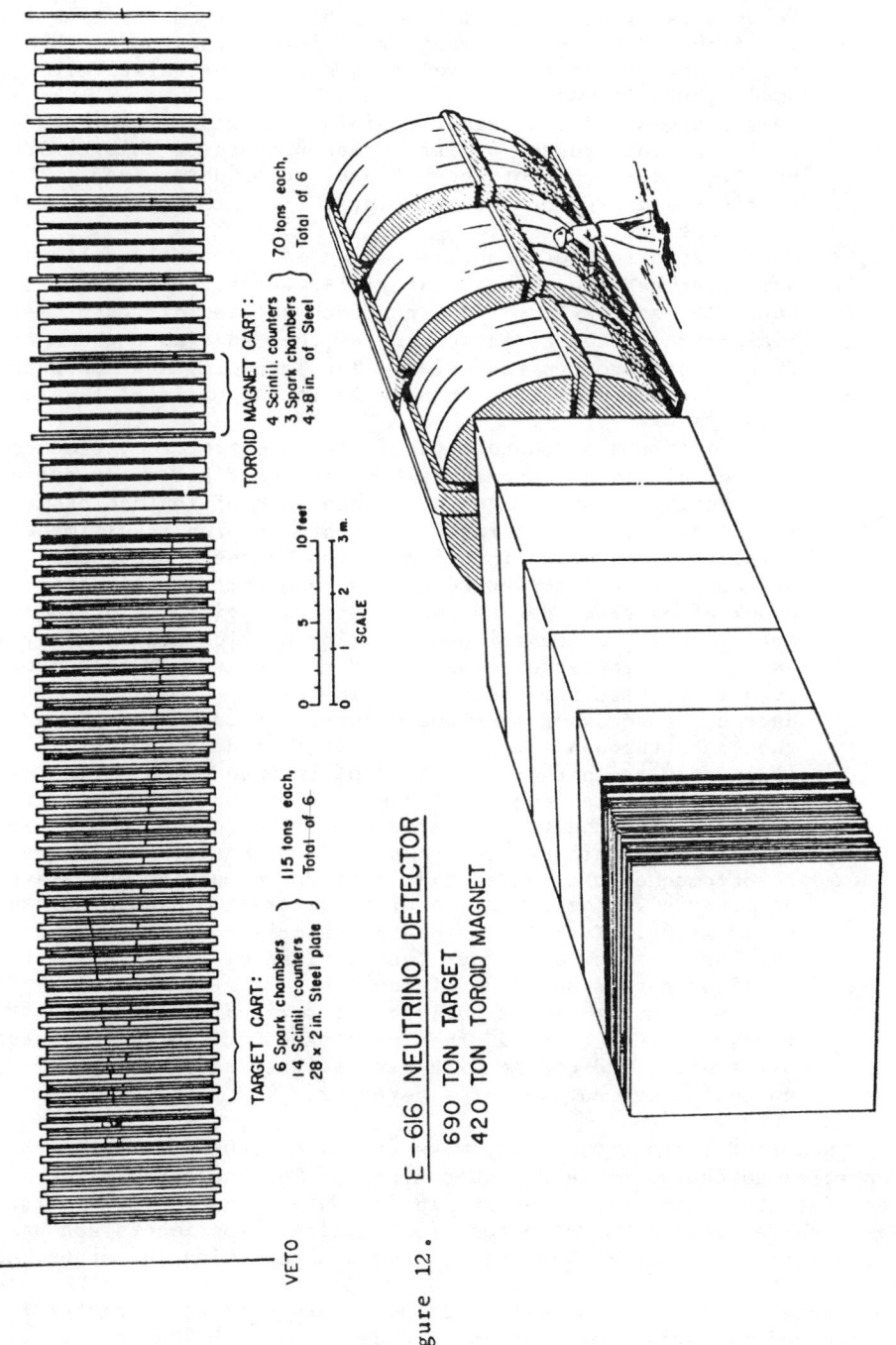

Figure 12. E-616 NEUTRINO DETECTOR
690 TON TARGET
420 TON TOROID MAGNET

- Velocities of particles may be determined from the time interval required to pass between two points. In general this technique differentiates velocities only for relatively small particle energies.
- In the presence of a magnetic field, charged particles are deflected into curved orbits. Measurements of this curvature permits the momenta of these tracks to be determined. The particle's energy can be calculated from its momentum if the mass is known.
- Characteristic, but weak, radiation is coherently emitted by particles passing through material--Cerenkov radiation; or by particles as they cross an interface between different materials--transition radiation; or as they pass through magnetic fields--synchrotron radiation. The intensity and characteristics of these radiations can serve as the basis of a velocity measurement.
- When electrons or photons pass through matter they produce characteristic electromagnetic cascades of secondary radiation which in turn leave an intense core of ionized atoms. The energy of the original electron or photon is ultimately completely converted into ionization by these processes. Measurement of this converted energy in a sufficiently thick block of material determines the total incident electromagnetic energy and constitutes a calorimetric shower-energy measurement. A detector constructed to make use of this property is an electromagnetic calorimeter.
- Hadrons, such as protons and mesons, interact strongly as they pass through matter, producing secondary hadrons. This again results in the production of intense cores of ionized atoms. The total energy of the incident particles can be measured from the total energy deposited in the form of ionization. Hadronic cascades can be differentiated from the electromagnetic cascades that develop in much thinner layers of material. The technique of measurement is known as hadron calorimetry. Typically a hadron calorimeter might require a thickness of three to four feet of instrumented steel with a total weight of hundreds of tons.
- Energetic muons are uniquely characterized by the property that, although charged, they penetrate very large thicknesses of material and emerge with a relatively small change of energy at the outside of a detector.

Modern detectors typically make use of many or all of the above properties to characterize detected events. The characteristics of these detectors have many features in common and share very similar design architectures. The detectors for collider experiments are based on a series of concentric shells or layers, one behind the other, each of which is devoted to some particular aspect or aspects of the detection process. The initial detector layers are used to characterize the charged-particle component and are designed to be "nondestructive"--i.e, to contain very little material so that charged particles will not interact or degrade in energy and thus will maintain their identity while traversing the layers. The outer detector layers deliberately use large amounts of material in order to materialize the neutral particles and to convert the energy carried by the particles into detectable ionization; such a device is known as a calorimeter.

The outer layer of the calorimeter, or an additional detection layer, is frequently used to detect muons. As mentioned earlier, energetic muons are usually the only particles that can reach this outermost layer.

Inevitably, as the layered levels of detection systems are built up, a detector will become very large and correspondingly complex and expensive. A prime objective of detector development is, therefore, to keep detection systems as compact as possible and to combine detection roles whenever possible.

Additional demands are imposed on detector systems associated with hadron colliders by the high ambient radiation levels at the detector, and by the fact that events of interest may be separated only by very short times from uninteresting background events.

The elements or layers constituting a typical modern detector can be seen in the schematic of the CDF detector at the Tevatron collider facility at Fermilab (Figures 13a and 13b).

Detectors for the new colliders face many new challenges. For example, experiments at the Stanford Linear Collider (SLC) with its micron-size beams would benefit from improved spatial resolution if the existing 50 to 100-micron resolution of drift chambers could be improved to a level of about 5 microns. Micro-vertex detectors that can identify the secondary vertex when bottom or charm quarks are produced are being developed. Another important area for R and D is in the use of charge-manipulation structures of silicon or gallium arsenide for particle detection. Silicon strip detectors are being used successfully in particle physics experiments and have demonstrated 5-micron spatial resolution. Two-dimensional Charge-Coupled Devices (CCD) are being developed for television cameras and other optical imaging applications by industry; there is a clear benefit here in a collaborative approach both between laboratories and between laboratories and the micro-electronic industry.

To be effective and productive, the U.S. High Energy Physics Program requires a diversity of particle acccelerator and colliding beam facilities with adequate detector facilities. In the study of fundamental particles, valuable information comes also from sources other than accelerators. The detection of cosmic rays, measurements at nuclear reactors, the search for rare decays in the shielded environment of deep tunnels and mines, and the precise study of atomic transitions have all contributed to high energy physics. The most crucial and fundamental questions, however, can be answered only by probing deeper into the nucleon with increasingly energetic particle collisions.

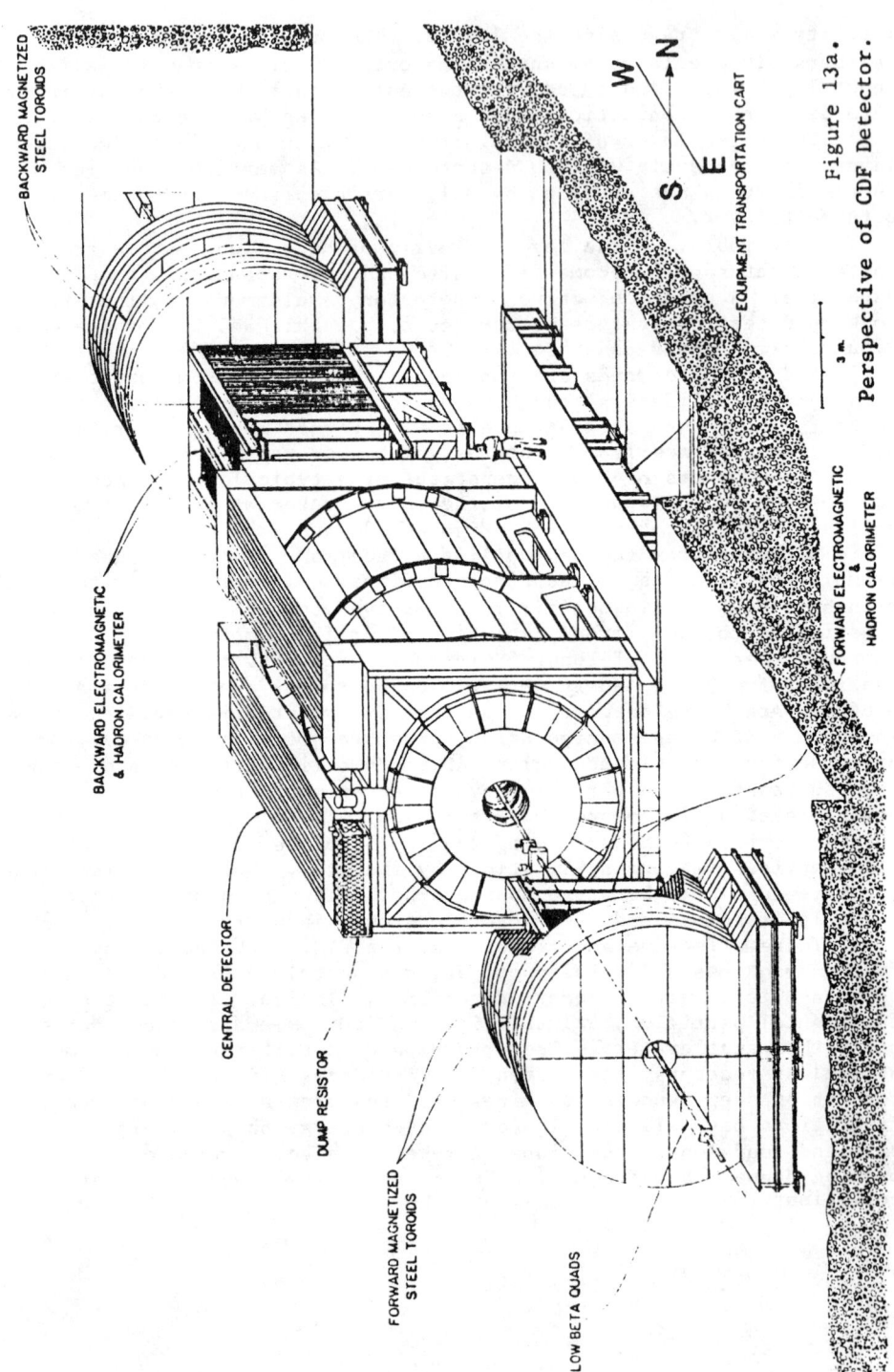

Figure 13a. Perspective of CDF Detector.

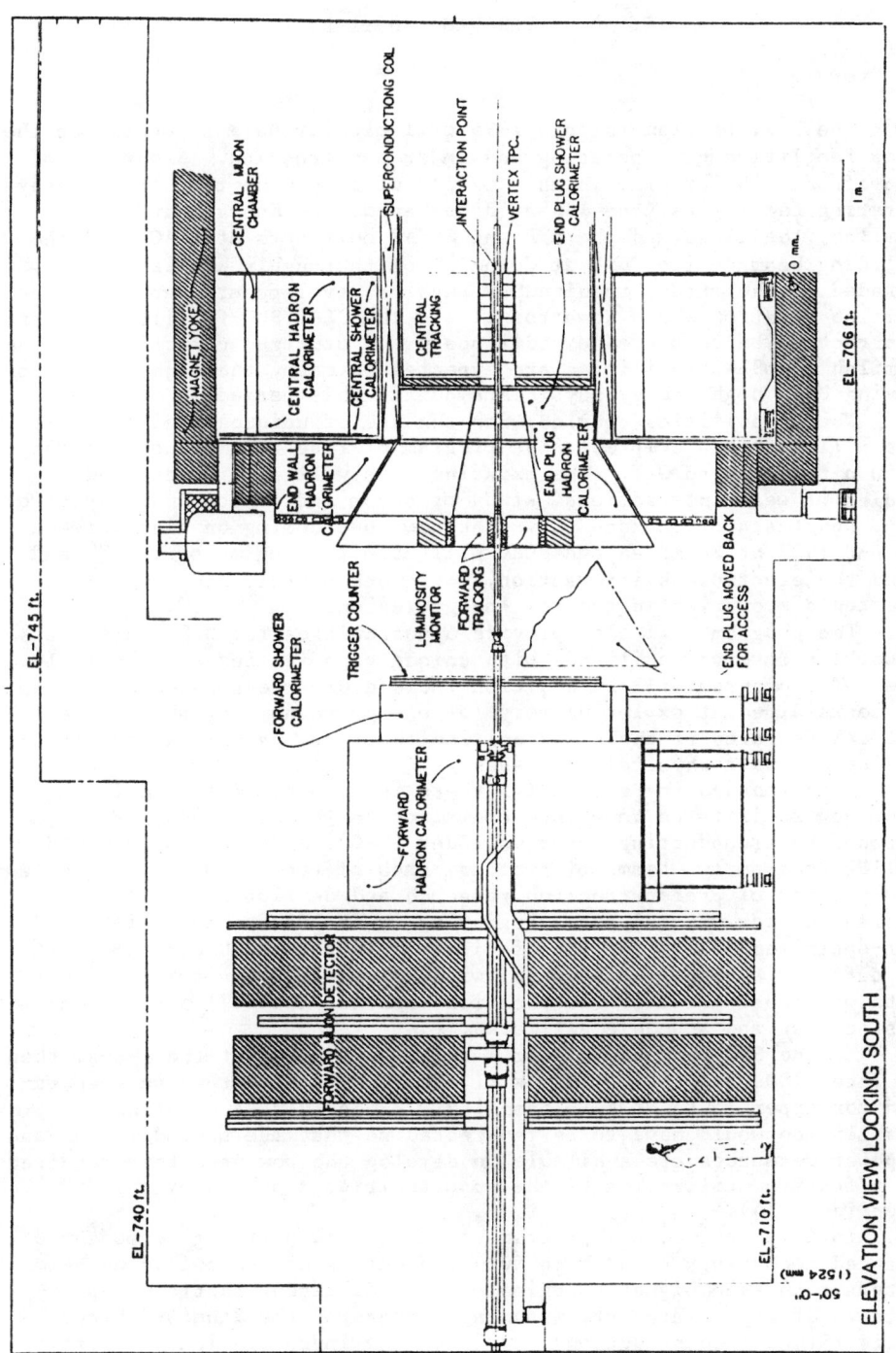

Figure 13b. Cross Section of CDF Detector.

II. U.S. PROGRAM 1985-1995

A. Overview

The U.S. program through 1995 will clearly have to depend on the major facilities now operating and under construction. Assurance of a forefront U.S. program in this period will require that the already operating facilities (the AGS at Brookhaven, the Energy Saver at Fermilab, the linac and the PEP and SPEAR colliders at SLAC, and the colliding beam device CESR at Cornell) be thoroughly utilized and upgraded to the extent required by physics developments, and that the Fermilab Tevatron I and Tevatron II and the SLAC SLC facilities under construction be completed expeditiously and utilized effectively. The Fermilab and SLAC facilities are expected to carry the major burden of serving U.S. high energy physicists during this period.

These facilities coupled with the use of unique foreign facilities will permit a comprehensive program of research in the mass range below a few hundred GeV. The existing ones will permit detailed studies of weak interactions, study of charm and bottom quark spectroscopy, and tests of QED and QCD. The new ones coming on line between now and 1987 at Fermilab and SLAC will open the regime of the Z^0 and W and the electroweak interaction, top quark states, and any unexpected discoveries in the new energy regions.

The program will also provide opportunities for U.S. scientists to work at foreign facilities with unique capabilities not available here. The program will in addition include diverse nonaccelerator experiments aimed at exploring very low energy phenomena, which reveal critical features of the grand unified theories, as well as cosmic-ray physics and astrophysical processes.

As discussed in Section IV, there is an essential need for a major new facility to come into operation in the mid-1990s, the proposed Superconducting Super Collider (SSC), a device which would collide head-on two beams of protons, each of energy 20 TeV. An extensive program of preconstruction research and development will be carried out for several years beginning in FY 1984 to establish a cost-optimized design for this facility. This R and D will provide the data for a conceptual design, will integrate technical requirements with cost-effective fabrication methods, and will provide realistic cost and schedule estimates.

If the SSC is approved for a start at the end of the 1980s, then the late 1980s and early 1990s will be a period of extensive preparation for experiments at this facility. Probably some existing capabilities would have to be redirected so that the needed fiscal and manpower resources are available to develop the new detectors required for effective utilization of the opportunities provided by the SSC (see Figure 14).

In this period, the program would include a strong component of advanced technology R and D on future accelerator and colliding beam systems with even higher energies or with different particle capabilities. Of major interest, assuming success of the Stanford Linear Collider (under construction), would be the investigation of energy

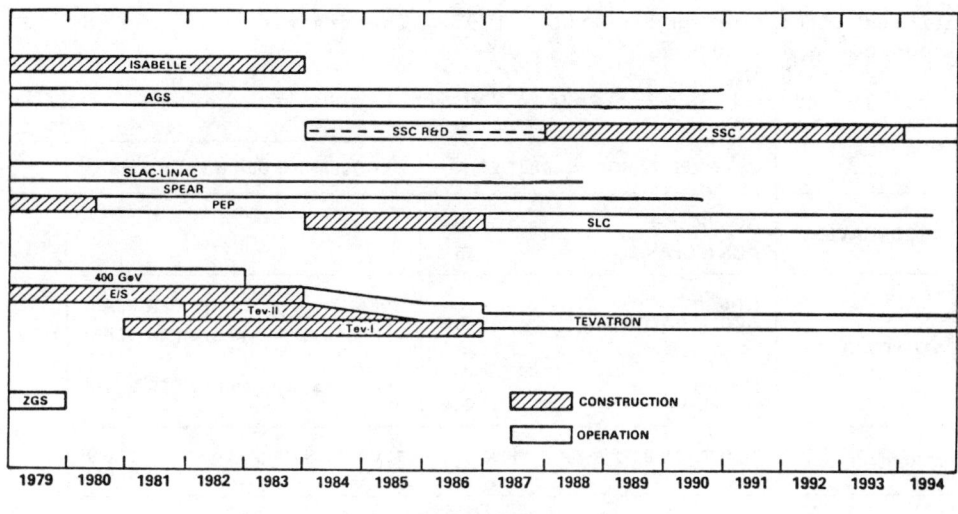

Figure 14. The time line indicates a possible plan for the future utilization of major DOE-funded accelerator facilities.

efficient acceleration structures and power systems for a large electron positron linear collider in the TeV energy range. Advantage would be taken of novel acceleration or detection schemes as the opportunities arise.

B. Details of the Program

The U.S. High Energy Physics Program at present depends primarily on the accelerator facilities at the three DOE-supported national accelerator centers: Brookhaven National Laboratory (BNL), Fermi National Accelerator Laboratory (Fermilab), and Stanford Linear Accelerator Center (SLAC). In addition, there is the CESR colliding beam facility at Cornell University supported by the National Science Foundation. Each has unique features complementary to those of the others, and together they provide the capabilities essential for a forefront U.S. program in the late 1980s and early 1990s. The DOE three-center approach for high energy physics facilities has proven very effective in the past and, though faced with serious challenges from abroad, the U.S. continues to be recognized as a world leader in the field.

This section describes U.S. accelerator facilities (see Table IV); those in other countries are described in Section III. For comparison, Figure 15 shows the energy and luminosity at world-class high energy colliding beam facilities in operation, under construction, or being designed, and Tables V and VI give more details of the U.S. high energy physics research centers and the major foreign ones. The U.S. High Energy Physics Program from 1985 to 1995, with emphasis on the major facilities existing or under construction, will be a vital program for the next decade. The planned SSC, if approved, is anticipated to come on line about 1995. The transition to the SSC is discussed in Section IV.

Table IV Major U.S. High Energy Physics Facilities

	FIXED TARGET FACILITIES		COLLIDING BEAM FACILITIES	
BROOKHAVEN	AGS (P)	30 GeV		
	AGS (POLARIZED P)	26 GeV		
STANFORD	LINEAR ACCELERATOR (e)	33 GeV	SPEAR (e+e−)	4 GeV PER BEAM
			PEP (e+e−)	18 GeV PER BEAM
	LINEAR ACCELERATOR*	50 GeV	SLC* (e+e−)	50 GeV PER BEAM
FERMILAB	TEVATRON-II (P)	1000 GeV	TEVATRON-I** ($\bar{P}$·P)	1000 GeV PER BEAM
CORNELL***			CESR (e+e−)	8 GeV PER BEAM

* LINAC UPGRADE & SLC CONSTRUCTION; INITIAL OPERATION FALL 1986.
** TEVATRON-I CONSTRUCTION; INITIAL OPERATION FALL 1986.
*** NSF SUPPORTED.

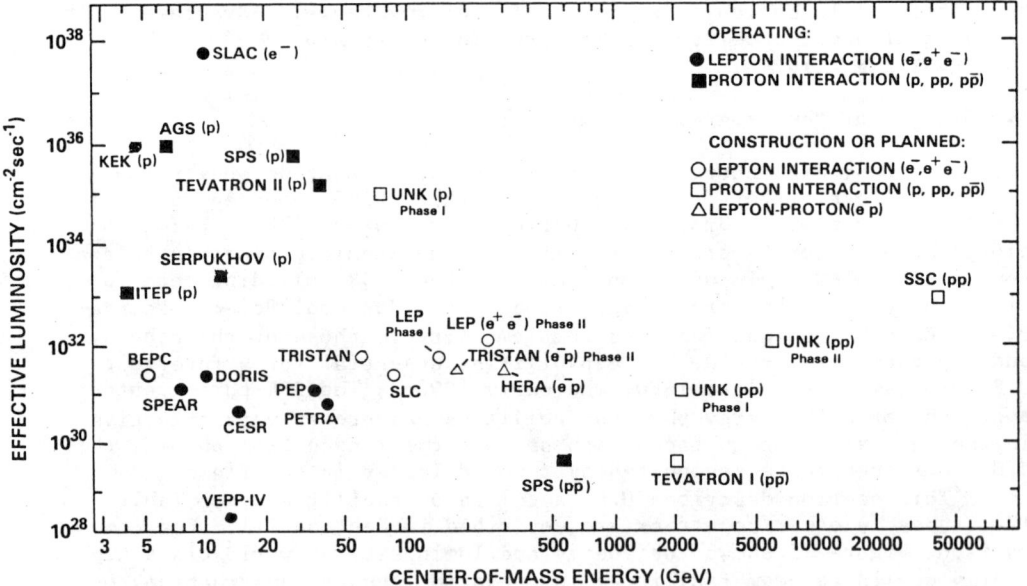

Figure 15. World accelerator and colliding-beam facilities. The major parameters, center-of-mass energy and luminosity, for colliding beam facilities in operation, under construction, or planned. The Superconducting Super Collider (SSC) would reach a center-of-mass energy of 40 TeV at a luminosity of 10^{33} cm^{-2} sec^{-1}, far surpassing the other facilities.

Table V U.S. High Energy Physics Research Centers

LABORATORY (CITY, STATE)	TYPE OF MACHINE	PARTICLES	MAXIMUM ENERGY (GeV)	CENTER OF MASS ENERGY (GeV)	EFFECTIVE LUMINOSITY ($cm^{-2}sec^{-1}$)	FIRST OPERATION IN THIS MODE	STATUS
BROOKHAVEN NATIONAL LABORATORY (UPTON, LI, NY)	SYNCHROTRON (AGS)	p	30	7.5	1.3×10^{36}	JULY, 1960	OPERATING
		p (polarized)	26	7.0	10^{33}	JULY, 1984	OPERATING
CORNELL U. (ITHACA, NY)	STORAGE RING (CESR)	e^+e^-	8×8	16	7×10^{30}	JUNE, 1979	OPERATING
FERMILAB (BATAVIA, IL)	SYNCHROTRON (TeV-II)	p	1,000*	43*	2×10^{35}*	(1983-1986)	OPERATING
	STORAGE RING (TeV-I)	$p\bar{p}$	$1,000 \times 1,000$	2,000*	5×10^{29}*	(1986)	UNDER CONSTR.
SLAC (PALO ALTO, CA)	LINAC (SLED-II)	e^-, e^+	33	7.9	10^{38}	1976	OPERATING
	LINAC (UPGRADE)	e^-, e^+	50*	9.7	7×10^{37}*	(1986)	UNDER CONSTR.
	STORAGE RING (SPEAR)	e^+e^-	4.2×4.2	8.4	10^{31}	APRIL, 1972	OPERATING
	STORAGE RING (PEP)	e^+e^-	18×18	36	5×10^{30}	MAY, 1980	OPERATING
	LINEAR COLLIDER (SLC)	e^+e^-	50×50*	100*	6×10^{30}*	(1986)	UNDER CONSTR.
UNSPECIFIED	SUPERCONDUCTING SUPER COLLIDER (SSC)	pp	$20,000 \times 20,000$*	40,000*	10^{33}*	(~ 1994)	PROPOSED

DECEMBER 12, 1984

*THE NUMBER GIVEN IS A DESIGN GOAL, NOT A MEASURED VALUE.

Table VI Major Foreign High Energy Physics Research Centers

LOCATION	LABORATORY (CITY)	TYPE OF MACHINE	PARTICLES	MAXIMUM ENERGY (GEV)	CENTER OF MASS ENERGY (GEV)	EFFECTIVE LUMINOSITY (cm^{-2}sec^{-1})	FIRST OPERATION IN THIS MODE	STATUS
WESTERN EUROPE	CERN (GENEVA, SWITZERLAND)	SYNCHROTRON AND STORAGE RING (SPS)	p	450	29	5×10^{36}	MAY, 1976	OPERATING
			$p\bar{p}$	270×270	540	5×10^{29}*	JULY, 1981	OPERATING
		STORAGE RING (LEP)						
		PHASE I	e^+e^-	60×60*	120*	4×10^{31}*	(1988)	UNDER CONSTR.
		PHASE II	e^+e^-	130×130*	260*	$\sim 10^{32}$*		(IN PLANNING)
	DESY (HAMBURG, GERMANY)	DORIS II	e^+e^-	5.6×5.6	11.2	3×10^{31}*	JULY, 1982	OPERATING
		PETRA UPGRADE	e^+e^-	23×23	46	$\sim 5 \times 10^{30}$	OCTOBER, 1983	OPERATING
		STORAGE RINGS (HERA)	e^-p	30×820*	314	4×10^{31}*	(1990)	UNDER CONSTR.
JAPAN	KEK (TSUKUBA)	SYNCHROTRON (KEK)	p	12	5	10^{36}	MARCH, 1976	OPERATING
		STORAGE RINGS (TRISTAN)						
		PHASE I	e^+e^-	30×30*	60*	8×10^{31}*	(1986)	UNDER CONSTR.
		PHASE II	e^-p	25×300*	173*	5×10^{31}*	—	(IN PLANNING)
PRC	IHEP (BEIJING)	SYNCHROTRON (BPS)	p	50*	10*	$\sim 10^{36}$*	—	(PSTPND. INDEF.)
		STORAGE RING (BEPC)	e^+e^-	2.5×2.5*	5*	$\sim 5 \times 10^{31}$*	(1987)	UNDER CONSTR.
USSR	IHEP (SERPUKHOV)	SYNCHROTRON	p	76	12	5×10^{33}	1967	OPERATING
		SYNCHROTRON AND STORAGE RING (UNK)						
		PHASE I	p	3,000*	75*	$\sim 10^{36}$*	(1990, PROJECTED)	UNDER CONSTR.
			$p\bar{p}$	$400 \times 3,000$*	2,190*	$\sim 10^{31}$*	(1990, PROJECTED)	UNDER CONSTR.
		PHASE II	$p\bar{p}$	$3,000 \times 3,000$*	6,000*	10^{32}	—	PLANNED
	INSTITUTE OF NUCL. PHYSICS (NOVOSIBIRSK)	STORAGE RING (VEPP-IV)	e^+e^-	7×7*	14*	3×10^{28}*	1979	OPERATING
		STORAGE RING (VAPP-IV)	$p\bar{p}$	25×25*	50*	(UNKNOWN)	(UNKNOWN)	(UNKNOWN)
	ITEP (MOSCOW)	SYNCHROTRON (U-10)	p	10	4	(UNKNOWN)	(OCTOBER, 1961)	OPERATING

* MEANS THE NUMBER GIVEN IS A DESIGN GOAL NOT A MEASURED VALUE.

DECEMBER 12, 1984

1. Brookhaven National Laboratory (BNL)

The fixed-target experimental program at BNL is based on the Alternating Gradient Synchrotron (AGS), a 30-GeV proton synchrotron that provides intense external beams of high energy protons and also secondary beams of pions and kaons and low energy neutrino beams. A schematic of the present BNL accelerator facility is given in Figure 16. Many major discoveries have been made at the AGS, such as the finding that the neutrino associated with the electron is different from that associated with the muon; the omega minus hyperon, which gave strong evidence for the quark hypothesis; CP violation and, indirectly, the violation of time-reversal invariance, which led to a Nobel Prize for J. Cronin and V. Fitch in 1980; the J/ψ particle, which led to a Nobel Prize for S.C.C. Ting and B. Richter in 1976; and the charmed baryon. Currently the physics program at the AGS focuses on studies of the weak interactions via neutrino interactions and rare K-meson decays; the search for neutrino oscillations; hadronic spectroscopy, including the study of gluonium states; a program of spin physics using the polarized proton beam; QED, via measurements on muonic atoms with helium nuclei; intermediate energy nuclear physics studies on hypernuclei including hyperon-induced atomic x-ray emission; and a search for an exotic six-quark state, termed H.

The AGS is the only proton accelerator in the world operating for physics experiments in the energy range up to 30 GeV. In addition, it provides more than 10^{13} protons per pulse every 1.2 seconds for neutrino physics and, with a one-second flattop every 2.5 seconds, for electronic counter experiments.

Vigorous work is in progress to improve the AGS performance so that more sensitive experiments can be done in the very interesting areas of rare K decays and neutrino oscillations. In addition, a more intense polarized proton source is being developed and will enhance this unique BNL capability. Soon it will be possible to accelerate heavy ions in the AGS with initially the availability of ions up to mass 32 and energies up to 15 GeV per nucleon. A booster synchrotron

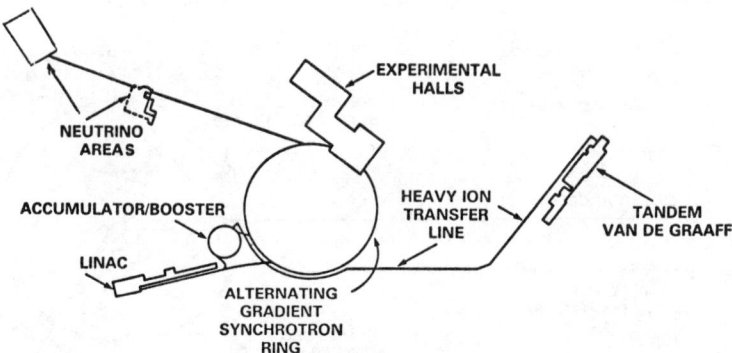

Figure 16. BNL high energy accelerator facilities.

is being considered as an addition to the AGS in the late 1980s to
allow operation with significantly increased proton and polarized proton beam intensities, to permit sumultaneous operation of the Fast External Beam and the Slow External Beam as well as to extnd the mass
range of accelerated heavy ions. In the future, it may be possible
that these facilities could be used to inject heavy ions into a 100 on
100-GeV/nucleon collider located in the now unused CBA tunnel. This
proposed Relativistic Heavy Ion Collider (RHIC) could allow the study
of the high baryon density phase transition and quark-gluon plasmas.

2. Stanford Linear Accelerator Center (SLAC)

The SLAC linear accelerator is the highest energy and highest
intensity electron accelerator in the world. It currently provides
beams of 33-GeV electrons and 22-GeV positrons and serves as injector
for the SPEAR and PEP colliding beam facilities. Its energy is being
upgraded to 50 GeV so that it can serve as the central element in the
Stanford Linear Collider (SLC). A schematic is given in Figure 17.

a. *SLAC Fixed-Target Program.* Since its start in 1966, the
fixed-target experimental program at SLAC has contributed greatly to
the elucidation of the laws of physics. It gave rise to such pioneer-ring experiments as electron-nucleon inelastic scattering, which
gave the first dynamic evidence for the quark and gluon composition of
the nucleons, and the parity violation in polarized electron-deuteron
scattering, which helped to establish the validity of the Glashow-
Weinberg-Salam model of the unified electroweak force. The program
also included extensive photoproduction experiments and a spectrum of
experiments with secondary hadron beams to study the hadron spectroscopy and weak interactions in K-meson decays.

In recent years, the linear accelerator has been used mainly as
an injector for the SPEAR and PEP storage rings, and the fixed-target
program has been winding down. Recent experiments have included an
axion search and a charmed particle lifetime measurement in the SLAC
Hybrid Bubble Chamber facility. This has been completed, and no further bubble chamber experiments are planned. Another experiment,
designed to study the scattering of electrons on nuclei and to eluci-

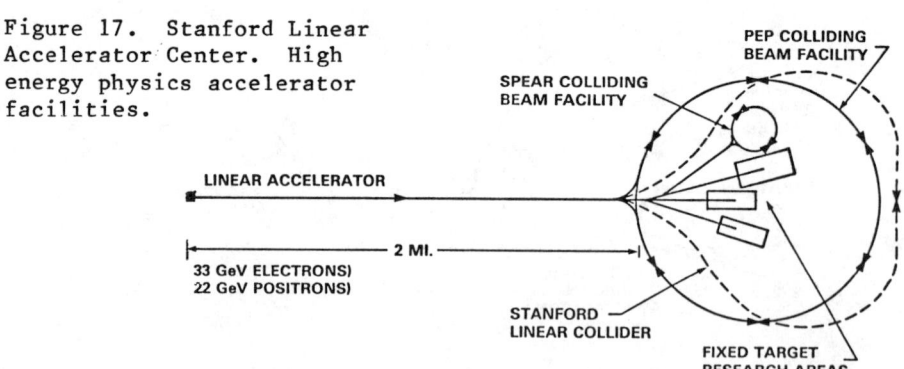

Figure 17. Stanford Linear Accelerator Center. High energy physics accelerator facilities.

date some of the recently observed anomalies in these processes, has also been completed. An off-axis injector has been constructed to give a relatively cheap way of generating 3- to 4-GeV electrons. This improvement is motivated mainly by the approved nuclear physics program but will also provide an economical way of filling SPEAR.

b. <u>PEP Program at SLAC</u>. PEP is a high energy electron-positron storage ring, which can provide collisions at reaction energies between 8 and 30 GeV. Since it started operation in 1980, its luminosity has been steadily improved and has recently reached 3×10^{31} cm^{-2} sec^{-1}. The parameters of PEP are very similar to those of the PETRA storage ring at DESY in Hamburg, West Germany, and during their initial operation the experiments at these two storage rings addressed the same basic physics questions. Subsequently their physics goals have diverged somewhat. The staff at PETRA has pursued added rf capability to reach the highest energies possible ($\sim$45 GeV total center-of-mass energy), hoping to exceed the threshold for production of the postulated top quark; while workers at PEP have been capitalizing on its high luminosity to obtain very accurate measurements at a single energy (29 GeV total energy).

Planned future upgrades of PEP include further increases in luminosity by micro-beta beam focusing insertions as well as corresponding improvements to the detectors. Five general purpose detectors have been operating at PEP: Mark II, a high resolution spectrometer (HRS), a detector oriented around a time projection chamber (TPC), a magnetic calorimeter detector (MAC), and a detector oriented around a large threshold Cerenkov counter (DELCO). Recently, a specialized detector to look for anomalous single photons (ASP), covering full 4π solid angle, has been operating in the sixth interaction region to look for missing energy and transverse momentum in the search for possible long-lived non-interacting neutral particles postulated in supersymmetric theories. Two specialized experiments have already been completed: a free quark search and a monopole search, both with negative results.

These diverse detectors allow a correspondingly diverse physics program at PEP. In hadronic physics, the emphasis is on detailed QCD tests looking at quark and gluon jets, particle correlations in multi-jet events, and studies of the hadronization process for both the light and heavy quarks. The electroweak sector includes measurements of the neutral weak coupling constants of quarks and leptons by studying forward-backward asymmetries and determinations of the weak mixing angles from B-meson decays. Recent upgrading of several detectors with precision vertex chambers has allowed precise determination of lifetimes of the tau leptons, charmed states, and B-quark states. An extensive two-photon physics program is under way, and a dedicated experiment to search for the photino, the photon's partner in supersymmetric theories, is now in progress.

The recent planned improvements in luminosity will allow several years of productive research with high statistics experiments.

c. <u>SPEAR Program at SLAC</u>. SPEAR is a medium energy electron-positron storage ring with reaction energies between 2.5 and 8 GeV.

It has been operating for more than a decade, but its physics program continues to be very productive. Its main value is that it is the unique e^+e^- facility in the world in the 3- to 5-GeV total energy region--the region that allows study not only of charmonium spectroscopy but also of the production and decay of various charmed states in relatively clean circumstances. Because different studies (ψ decay modes, D mesons, F mesons, charmed baryons) require different energies, they cannot be performed concurrently.

SPEAR has two interaction regions, but, since the Crystal Ball detector was moved to DORIS at DESY, only one has been occupied. It serves a general purpose detector known as Mark III. This detector has been optimized for study of the SPEAR energy region and features good efficiency for low energy gamma rays, a precision time-of-flight system to separate pions from kaons and protons over most of the kinematic range, and good charged particle momentum measurement. It has large solid-angle coverage and can be used to reconstruct a large fraction of charmed particle decays.

Recently SPEAR has been running half time as a dedicated synchrotron radiation source for the Stanford Synchrotron Radiation Laboratory, and it will continue to do so in the forseeable future. The Mark III program is expected to remain productive at least until 1988.

d. <u>SLAC Linear Collider (SLC)</u>. Construction of the SLC began in October 1983. Its justification is twofold. First, it provides for the development of techniques essential for establishing the technical feasibility of linear electron-positron colliders, whose energies could reach well beyond those feasible with circular colliders in view of projections that the cost of linear colliders rises linearly with the e^+e^- center-of-mass energy whereas that of storage ring colliders rises approximately as the square of this energy. Second, the SLC is designed with a reaction energy high enough to copiously produce directly the neutral intermediate boson, the Z^0, recently discovered at CERN. The scheme (Figure 18) will bring into collision high-intensity micron-sized bunches of electrons and positrons, each beam with an energy up to 50 GeV for a total center-of-mass energy of up to 100 GeV in the e^+e^- system. These bunch pairs arrive at the collision point from opposite directions, at the pulse rate of the SLAC linear accelerator (120 per second, eventually to be increased to 180). After the beams collide, the residual electrons and positrons are disposed of in beam dumps. SLC will also be capable of collisions with polarized beams, and its very small beam size will allow increased opportunities for close-in vertex detectors and hence good conditions for studies of short-lived particles. It is scheduled to be completed in the second quarter of FY 1987.

The topics to be studied at the SLC include Z^0 production and decays (mass, width, decay modes and branching ratios, number of flavors, number of generations of neutrinos); study of top quarks by directly producing the vector $t\bar{t}$ bound state (toponium); search for new particles such as the Higgs, particles predicted by supersymmetry or technicolor, heavy leptons, and other new heavy quarks; QCD tests by studying quark and gluon jets; and study of the mixing between the

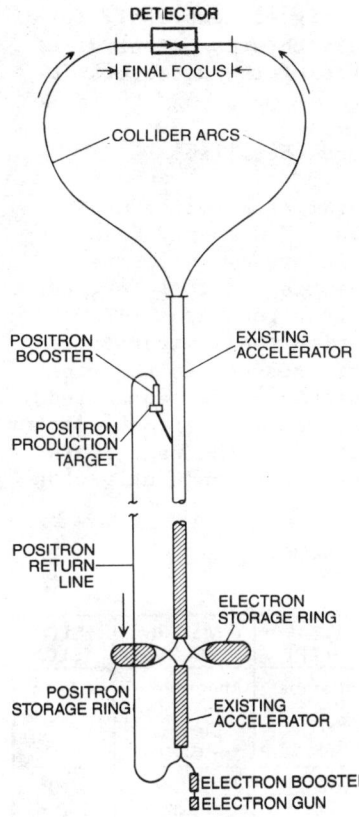

Figure 18. General layout of the SLAC Linear Collider (SLC): the injection system (electron gun and boosters), storage (damping) rings which serve to reduce the size and energy spread of the electron and positron bunches by radiation damping, the existing linac which accelerates bunches of electrons and positrons to 50 GeV, and the transport and final focusing systems which bring micron-sized bunches of electrons and positrons into head-on collisions. The positron target and booster use electron bunches to produce positrons for injection into the front end of the linac.

different quark flavors (K-M mixing angles). These objectives are similar to those of LEP at CERN. The SLC, although lower in energy than LEP, has a design luminosity (16×10^{30} cm^{-2} sec^{-1}) that is comparable with that of LEP, but it is due to have its first physics run about two years earlier.

The SLC has two technical advantages over storage rings. First, detectors can be placed closer to the interaction region than in a storage ring. This allows the position of secondary vertices to be measured with high accuracy, improving the measurement accuracy of short lifetimes and increasing the efficiency for tagging heavy quark decays. The second advantage is the ability to produce longitudinally polarized electron beams. This allows the Z°s to be produced with a tunable polarization, aiding in the measurement of electroweak interaction parameters and the determination of new particle properties. The ability to produce a polarized electron beam and to measure its polarization will probably not be available until 1989.

During SLC commissioning and initial stages of operation, the Mark II detector (suitably upgraded) will be used in the single interaction region. This has the advantage that physics results can be obtained relatively quickly with a detector that is well understood because of its previous operation at SPEAR and PEP. Fabrication of a general purpose detector, designated SLD, designed expressly for this energy range, is underway so that the full physics potential of the SLC will be exploited. The characteristics of the detectors for LEP and SLC are compared in Table VII.

3. Fermi National Accelerator Laboratory (Fermilab)

Until the fall of 1983 the physics program at Fermilab used beams produced by a 400-GeV proton synchrotron. The Energy Saver superconducting magnet ring (also called the Tevatron), which is expected eventually to reach particle beam energies up to 1 TeV, has recently come into operation. (Although the Tevatron has a 1-TeV capability, it is being operated at 800 GeV, with the limitation on energy being due to cryogenic capacity and the presence in the ring lattice of a small number of field-limited magnets.) The so-called Main Ring synchrotron, now operated at 150 GeV, serves as the injector for the Energy Saver. The 1-TeV beam capability will be used in two different modes of operation: colliding 1-TeV proton and antiproton

TABLE VII
Summary of the New e^+e^- Detectors

	ALEPH (LEP)	DELPHI (LEP)	L3 (LEP)	OPAL (LEP)	Mark II (SLC)	SLD (SLC)
Vertex Detector	chamber + μ-strips	chamber + μ-strips	TEC	chamber	chamber + possible μ-strips or CCD's	CCD's
Tracking Device	TPC 1 atm r = 1.8 m n = 300	TPC 1 atm r = 1.2 m + outer	—	DC 4 atm r = 1.6 m n = 160	DC 1 atm r = 1.5 m n = 72	DC 1 atm r = 1.0 m n = 80
Coil	s.c. 1.5 T	s.c. 1.2 T	warm 0.5 T	warm 0.4 T	warm 0.5 T	warm 0.6 T
Tracking Resolution	0.13% p	0.15% p	0.03% p μ only	0.15% p	0.12% p	0.13% p
Particle Id	dE/dx 4.5% σ	dE/dx + RICH	—	dE/dx 3.5% σ	dE/dx 7% σ	RICH
EM calor.	Pb-gas	HDPC	BGO	Pb glass	Pb-LA	U-LA
EM resolution	$18\%/\sqrt{E}$	$18 - 20\%/\sqrt{E}$	1%	$6.5\%/\sqrt{E}$	$12\%/\sqrt{E}$	$8\%/\sqrt{E}$
Hadron cal. Resolution	$100\%/\sqrt{E}$	$100\%/\sqrt{E}$	$50\%/\sqrt{E}$ + 10%	$100\%/\sqrt{E}$	—	$45\%/\sqrt{E}$
Muon Det.	full	full	high res	full	55%	full

Key: TEC — Time Expansion Chamber
CCD — Charged Coupled Device
TPC — Time Projection Chamber
DC — Drift Chamber
RICH — Ring Imaging Cherenkov Detector
HDPC — High Density Projection Chamber
LA — Liquid Argon
(p in GeV)

beams (Tevatron I project), and fixed-target experiments with 1-TeV
protons (Tevatron II project).

    a. <u>Tevatron II Program at Fermilab</u>. The Tevatron II fixed-target
program (Figure 19) provides for extraction of a 1-TeV beam from the
Energy Saver ring; upgrading the switchyard, external proton beam lines,
target stations, and existing experimental halls for 1-TeV operation;
and four new secondary beams including a high flux muon beam, a wide-
band photon beam, a polarized proton beam, and a meson-west pion beam.
The proton beam extraction and proton beam lines in the switchyard
are now in operation, and the entire project is expected to be completed
in 1986. The increased proton beam energy will provide secondary particle beams with higher energy and increased intensity per pulse; however
with reduced cycle rate. In addition, the use of superconducting magnets
substantially improves the operating performance, in particular the
beam spill time. The result is that the duty cycle (i.e. beam-on-target
vs. cycle time) is as high as 30%. This is a substantial improvement
over the non-superconducting machine (Main Ring), where the duty cycle
was 8 to 10%.

    The fixed-target experimental program covers a broad spectrum.
Measurements of leptons produced at large transverse momentum have
resulted in interesting comparisons with quark model predictions as
well as the discovery of the upsilon particle. These programs will
continue to higher energies and will be broadened to include beam particles other than protons and final state particles other than
leptons. Studies of hadron jets and direct photons are also being
emphasized. Measurements with a new polarized proton beam and with
polarized targets will be possible.

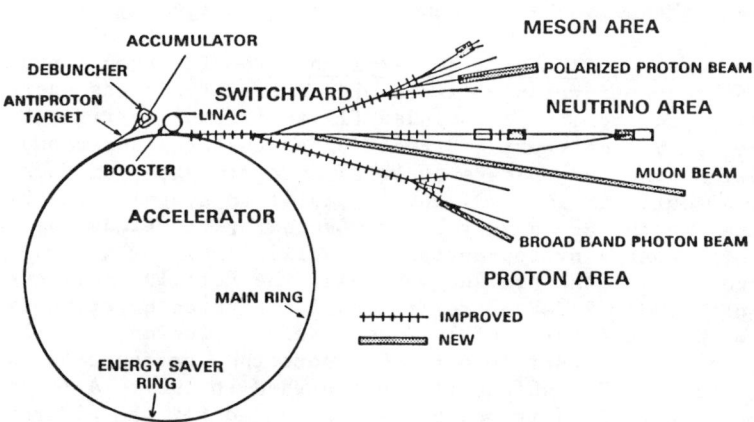

Figure 19. The Tevatron facility at Fermilab: the 6-km-circumference
Main Ring and superconducting (Energy Saver) ring; the switchyard and
external beamlines for fixed-target operation (TeV II); the antiproton
source (antiproton target, accumulator, and debuncher) for the colliding
p$\bar{\text{p}}$ operating mode (TeV I); and the Main Ring injector complex (linac
and booster).

Several experiments on studies of charm and bottom production will use various novel techniques in different secondary beams. Of special interest are efforts to develop vertex detectors and fast on-line processors to extract the events containing decays of heavy quarks, providing measurements of the Kobayashi-Maskawa (K-M) mixing angles.

A series of experiments using neutrino and muon beams will extend the measurements of nucleon structure functions by about a factor of three in momentum transfer (Q^2). This may turn out to be important because of the present confusion in the Q^2 evolution of structure functions (the value of Λ_{QCD} is smaller when it is obtained from data at higher Q^2). The neutrino experiments will be capable of more precise measurements of neutral current couplings, which can be compared with measurements from e^+e^- colliders operating at the Z^0 mass. The muon experiments will study jet fragmentation and atomic number dependence at very high Q^2. The latter has recently been of considerable interest. Study of rare phenomena, like multi-lepton production by both muons and neutrinos, with much higher event rates at higher energies will improve understanding of these processes. Holography will be used in the 15-ft bubble chamber to improve track resolution and thus permit easier observation of heavy quark decays.

Some of the many phenomena from lower energies that are still not understood will be subjects of Tevatron II experimentation. A high statistics measurement of the electron asymmetry parameter in the β-decay of Σ^- should corroborate or remove the present large discrepancy between this quantity as inferred from existing data and the prediction of the Cabibbo model. Precision measurements of CP violation parameters will be made at levels where effects predicted by certain classes of theories should become manifest. CP violation may be observed in channels where it has not yet been seen.

b. <u>Tevatron I Program at Fermilab</u>. The Tevatron I project provides a proton-antiproton collider with a center-of-mass energy of up to 2 TeV. Specifically it includes (1) additional refrigeration and radio-frequency accelerating structures so that the superconducting Energy Saver ring can operate at 1 TeV in a storage ring mode; (2) an antiproton source (Figure 20) and accumulation system; and (3) two large experimental areas suitable for detectors to study the results of very high energy proton-antiproton collisions.

According to the present schedule, the Fermilab collider program will furnish up to 2-TeV $p\bar{p}$ collisions for physics experiments by 1986-87 with a luminosity of 10^{30} cm^{-2} sec^{-1}. Two major collision regions are being planned: one at B0, where the construction is complete, and one at D0, which will be finished in 1987. A large general purpose detector (CDF) is now being fabricated for the B0 area. Fabrication of a complementary detector for the D0 area is under way. The first round of experiments is expected to run for four to five years with an integrated luminosity of up to 10^{37} cm^{-2} or 10^4 nb^{-1} per year of running. The Tevatron I program will divide the available time with the Tevatron II fixed-target program, in a manner similar to CERN operation.

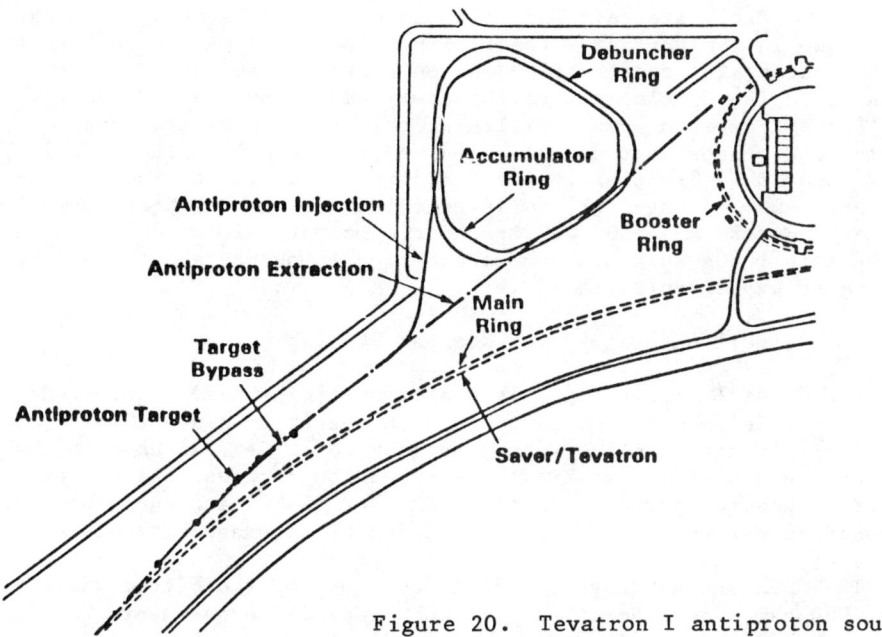

Figure 20. Tevatron I antiproton source.

The CDF detector employs a large solenoidal magnet which contains track chambers surrounding the beam pipe. The superconducting solenoid magnet is enclosed within a lead-scintillator electromagnetic calorimeter which, in turn, is enclosed in an iron-scintillator hadron calorimeter. The calorimeters utilize a highly segmented tower geometry in order to define the energy and direction of a jet. It will be possible to detect electrons and hadrons emerging from the interaction point at an angle of 1° or greater. Muon detection is provided over part of the solid angle. The construction of the detector will be complete in late 1986. The partially completed detector will be placed in the Tevatron Collider during a September 1985 beam test, and it will be used again in another colliding beams test planned for 1986.

Unlike CDF, the D0 detector has no central magnetic field, but emphasizes optimum muon and electron identification and minimizes missing transverse energy resolution. Its designers have chosen highly segmented uranium--liquid argon calorimetry, utilizing projective geometry. The calorimeters will measure the energy of particles which emerge from the interaction point at angles $>1°$ with respect to the beam axis. A special effort is being made to reduce both the statistical errors and the systematic errors in the measurement of electron and hadron energies. Electron identification is achieved through measuring shower development in the calorimeter and the ionization deposited in a transition radiation detector. The detector also has a magnetized iron muon detector covering nearly 4 steradians. On the basis of the current plan the detector can be completed in 1990, with a staged turn-on beginning in 1989.

The Tevatron I program has two main features providing physics opportunities. The primary one is the 2 TeV available for producing new heavy particle states. This high energy gives Tevatron I a position unique in the world for the 1986-92 period. An added feature is that the successful construction and operation of an antiproton source will enhance the development of that complex technology in the U.S.

The anticipated annual yields at the Tevatron of standard model intermediate bosons that can be exploited for detailed study are 2×10^5 $W^{\pm}$ and 6×10^4 Z^0. Examples of the discovery potential of the new machine are $\sim$400-GeV mass for new intermediate bosons, $\sim$100-GeV mass for new heavy quarks, and $\sim$300-GeV transverse momentum for QCD jets. During the late 1980s this program is expected to occupy about 400 U.S. high energy experimental physicists.

4. Cornell Laboratory for Nuclear Studies

The Cornell electron-positron storage ring (CESR), supported by the National Science Foundation, has been operating since 1979, principally in the reaction energy range 9 to 12 GeV, although higher energies are possible. It has two interaction regions, one occupied by a large general purpose magnetic detector (CLEO) and the other by a nonmagnetic detector for high resolution electromagnetic calorimetry (CUSB).

The CESR energy range is ideal for studying the bottom quark. Since CESR came into operation, the CLEO and CUSB experiments have studied the first six ^3S states and the first two triplets of ^3P states in the $b\bar{b}$ quarkonium system. They have measured the dilepton and hadron decay widths of the first three ^3S upsilons as well as the branching ratios for $\pi\pi$ transitions among them and the radiative transitions involving the ^3P states. The fourth upsilon has served as a "B-meson factory," allowing studies of the weak decays of the b quark: semileptonic branching ratios, multiplicities, inclusive K, D, p, and lepton yields, and exclusive modes. The results have confirmed the standard six-quark model and have established the $b \to c \to s$ cascade as the dominant decay mode.

An important gap in the $b\bar{b}$ bound state spectroscopy is the ^1S η_b state which should be accessible from the upsilon states by a sequence of hadronic and radiative transitions. The systematic reconstruction of upsilon decay modes permits possible discoveries of glueballs, Higgs, axions, supersymmetric partners, and perhaps other surprises. A thorough study of B decays will be required to measure the generalized Cabibbo weak mixing angles, $B^0\bar{B}^0$ mixing, CP violation, and rare modes such as $B \to \tau\nu$. Of the higher mass b-flavored hadrons, B^*, B_s, Λ_b, etc., only the B^* has been observed.

The studies of the b quark at CESR are important for progress in understanding both strong and weak interactions. Because the b quark is so massive, the $b\bar{b}$ system is especially amenable to quantitative QCD predictions. The weak decays of the b quark are unique in that they permit measurement of branching ratios which depend significantly on the weak mixing angles other than the original Cabibbo angle; the $B^0\bar{B}^0$ system (or $B_s^0\bar{B}_s^0$) may provide the only chance to observe CP violation outside the $K^0\bar{K}^0$ system. One can hope that this work will lead to insights as to why there are three generations of quarks and leptons.

The future CESR physics program requires significant increases in quantity and quality of data for the following reasons. Relative to cross sections in the charmonium region, $b\bar{b}$ cross sections are suppressed because of quark charge and because of increased energy. Moreover, in the $b\bar{b}$ case there are more states and more decay modes and, therefore, smaller branching ratios. The CESR luminosity improvement program is needed to provide at least a tenfold increase in rates. In addition, an extensive upgrade of the CLEO detector is under way and will provide significantly improved charged particle tracking and particle identification as well as photon energy resolution. The present CUSB experiment has been upgraded by improving the photon energy resolution.

5. Non-Accelerator Program

The U.S. High Energy Physics Program has traditionally supported a modest but diversified program of non-accelerator physics experiments. Recent predictions from grand unified theories have rekindled interest in searches for nucleon instability, heavy primordial magnetic monopoles, neutron oscillations, and other new phenomena. In fact, a stimulating merging of the interests of traditional high energy physicists with those of cosmic-ray physicists, astrophysicists, astronomers, and cosmologists is taking place. Non-accelerator experiments in these areas include searches for gravity radiation, neutrinoless double beta decay, neutrino mass, neutrinos from gravitational collapse, proton decay, and magnetic monopoles; studies of atmospheric neutrinos and neutrino oscillations, extraterrestrial and extragalactic cosmic rays, and gamma-ray and neutrino astronomy up to several TeV in energy; and solar neutrino measurements.

It is important to maintain U.S. effort in these areas. These experiments add significantly to the diversity of the U.S. high energy physics program and are especially desirable in the present theoretical climate where seeing what lies beyond the standard model is of prime importance. Most have direct relevance to the studies at high energy accelerators, and some explore physics on an energy scale beyond that attainable with accelerators. Moreover, constraints on new theoretical ideas often come from experiments done for an entirely different purpose. For example, measurements in x-ray astronomy put constraints on monopole catalysis of nucleon decay, and solar neutrino flux measurements put limits on the rate of monopole catalysis in the sun.

During the last few years, the proton decay program has been especially strong and diversified. Recent results from the large water Cerenkov experiment (Irvine-Michigan-Brookhaven collaboration) have put into question the simplest grand unified theories, and refined measurements continue to come from new data. The future tracking calorimeter experiment (Soudan II) should provide information that is complementary to that from the water Cerenkov detectors. Reactor neutrino experiments provide a unique means of studying neutrino masses and lepton mixing, because of the high fluxes and low energies of reactor neutrinos.

During the next decade the non-accelerator elementary particle program should continue to play a vital role, complementing accelerator-based research. It not only provides a means of glimpsing an energy scale inaccessible to currently conceivable accelerators, but also connects the field of high energy physics with many other branches of physics.

C. University-Based Research

Modern high energy physics research started on the campuses of U.S. universities soon after the end of World War II, when a large group of scientists became free to pursue basic research. These scientists, with funding from the Federal Government, enabled major universities, from coast to coast, to establish particle accelerators on campuses. As the machines reached the GeV energy range and their costs multiplied, a growing number of universities could not afford their own particle accelerator. Nevertheless, this field of research required higher energy accelerator facilities. The result was the arrangement now in effect: accelerators are built and operated at a few large national laboratories, (managed by individual or groups of universities), and the experiments at them are done primarily by university-based scientists--in the user mode (see Figure 21 and Table VIII).

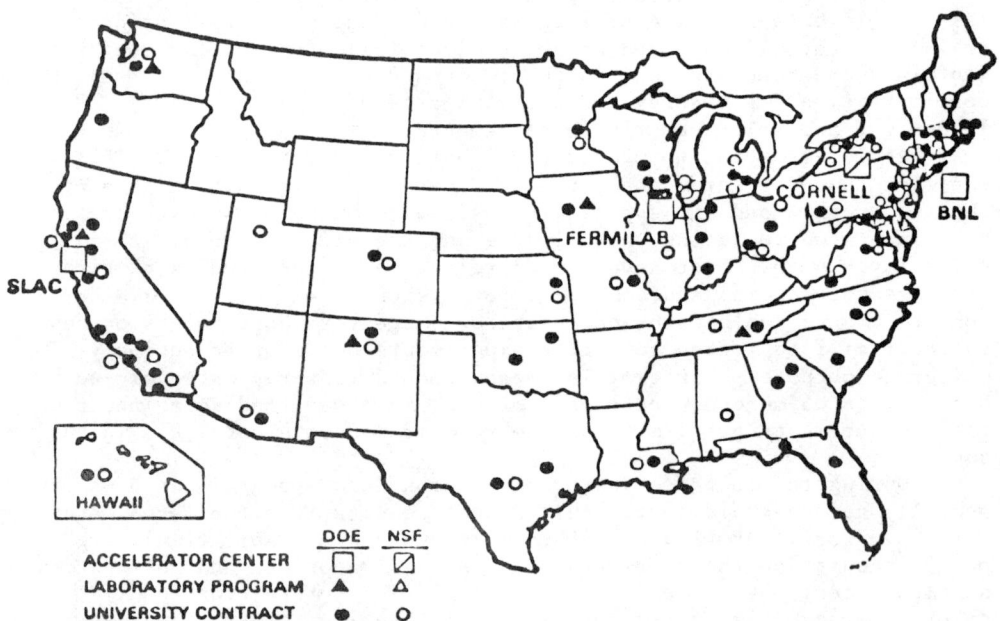

Figure 21. The location of participants in the U.S. High Energy Physics Program as of 1983. Accelerator centers, national laboratories, and universities supported by DOE and NSF are indicated.

Table VIII User group mode

- USER MAKES RESEARCH PROPOSAL TO LABORATORY. SCIENTIFIC REVIEW BY PROGRAM ADVISORY COMMITTEE
- LABORATORY FITS APPROVED EXPERIMENTS INTO ACCELERATOR SCHEDULE
- LABORATORY PROVIDES BEAM AND TECHNICAL SUPPORT
- USER ORGANIZES, SETS UP AND CARRIES OUT EXPERIMENTS AT LABORATORY
- USER ANALYZES DATA AND PUBLISHES RESULTS FROM HOME BASE
- LARGE EXPERIMENTS OFTEN REQUIRE COLLABORATION OF SEVERAL GROUPS
- FUNDING SUPPORT OBTAINED THROUGH SEPARATE RESEARCH SUPPORT PROPOSAL TO FUNDING AGENCIES

This arrangement has been extremely successful, and Western Europe has also adopted close coupling between large accelerator laboratories and universities.

The user mode allows the construction of facilities far larger than could be supported by any one university, and yet maintains the major advantages of university-based research. Because high energy physics is recognized as being of fundamental importance by the general public as well as by scientists, universities have made an effort to attract and hold the best scientists (both from the U.S. and from abroad) and have thus remained the principal source of advances in both theory and experimental work.

A great advantage of keeping scientists at universities while providing them with access to forefront research centers is that they maintain their association not only with the university staff but also with the students, which is vital to the continuation of scientific advance. High energy physics needs new entrants, but even students who do not stay in this field benefit from contact with it.

An important interaction on campus is that between high energy theorists and experimentalists: the former learn which ideas might be explored by experiment, and the latter learn which experiments are needed for theoretical progress. Also important is the flow of ideas to and from other disciplines. High energy physics results have found their way into solid state theory, molecular and neurobiology, medical physics, plasma physics, synchrotron radiation facilities, cosmology, and astrophysics. In turn, computer science, solid state physics, cosmology, laser physics, the physics of gases, and electronic engineering have advanced high energy physics.

In a university environment, scientists are free to change the emphasis of their research. New opportunities quickly find energetic proponents, and this fosters the timeliness of research trends. Since peer review panels and program advisory committees at the accelerator laboratories include many university scientists, the high energy physics research community can adapt quickly to new ideas.

During the academic year 1982-83, a Technical Advisory Committee on University Programs (TACUP) appointed by the High Energy Physics Division of DOE carried out a comprehensive review of the entire DOE-supported university-based program in high energy physics in order to assess the quality of the individual university contracts funded by DOE. TACUP was also asked to comment on the balance between the various components of the DOE High Energy Physics Program and to report opinions based on its review of the separate contracts and on comments solicited from the high energy community.

TACUP was asked to review 82 contracts comprising 64 separately funded theory tasks at 53 institutions, and 107 experimental tasks at 58 institutions. Eight panels, three for theory and five for experimental work, were appointed, and all used similar quality criteria. The evaluation was based on (1) a dossier submitted by the task leader, and (2) a conference between the panel and the task leader, with a site visit for larger contracts. Final reports, providing a quality evaluation for each of the 171 tasks, were submitted to DOE in May 1983.

Such a comprehensive review of all university-based DOE contracts simultaneously has not been carried out before, and the TACUP review provides a unique overall picture. The main conclusion from the individual task evaluations is that, with few exceptions, the university-based contracts produce research of very high quality. The high energy physics community is proud of the scientific achievements of the last decade and eager to deal with the new and deeper questions which these successes have suggested. The individual research programs, even though they address a bewildering array of topics, are well focused on issues of current scientific importance, and competition to obtain significant results is keen.

The High Energy Physics Program has subdivisions beyond theory and experiment. Experiments can be speculative or can be designed to add to knowledge in a predictable way; they can depend on accelerators--U.S. or foreign--or be independent of them. Given the totality of funds available, TACUP concluded that the allocation of resources to these various program components was reasonably matched to the opportunities in each area.

Enthusiasm in the high energy community continues to be high, but the TACUP panels reported a significant level of concern. The number of U.S. accelerator facilities that can support front-rank experiments has been decreasing steadily. For example, only three fully instrumented collider interaction regions, two at Tevatron I and the other at SLC, will permit access to new energy regimes at the end of the decade. The number of U.S. research teams planning experiments in Western Europe is increasing. Many university scientists are concerned that the shrinking opportunities foreseen in the 1990s will discourage young researchers from committing their careers to the field. TACUP found the question of the development of frontier facilities in the U.S. in the 1990s with adequate opportunities for university research groups to be a matter of great urgency.

Over the years, funding limitations have not allowed the full exploitation of the existing U.S. accelerators, resulting in an increase in the already long time spans between initiation and completion of experiments and in a reduction in the number of experiments performed. TACUP recommended, as have many others, that operating funds earmarked for accelerator operation be increased until these large facilities are fully exploited.

Universities have opportunities for input into the decision-making process, but there is probably a need to develop further effective channels for participation in shaping the High Energy Physics Program. The Snowmass meeting of 1982 was useful in this respect, and workshops under APS Division of Particles and Fields sponsorship should be continued, with broad university participation.

TACUP did not study the program funded by the Advanced Technology R and D Branch of DOE, but many of the task leaders told the panels that research on advanced accelerator concepts is not being pushed as vigorously as it should. This would be an appropriate activity for universities since accelerator laboratories must be concerned also with more immediate improvements. The advisability of additional funding for universities for advanced accelerator research should be determined.

TACUP found that opportunities for young physicists to enter faculty positions has somewhat improved. The staffs of many large and medium-size research tasks now include a tenure-track faculty member hired in the last six years. The DOE Outstanding Junior Investigator (OJI) program has contributed significantly to this improvement. Since its inception in 1978, more than 50 theorists and experimentalists have been OJIs.

Since 1974, DOE support of its university-based physics program has been essentially constant at $53 million per year (in 1982 dollars), but indirect costs have increased by about 10%, and the experiments have become much more complex and costly. The TACUP panels found most tasks to be handicapped by serious funding constraints. On the theory side, funding shortages impede the training of students and postdoctorals and restrict travel and visitor exchanges. On the experimental side, universities have infrastructures that are deteriorating and apparatus that is no longer up-to-date, and few of them can maintain a stable group of engineers and technicians. These trends should be reversed by additional funds, especially for equipment modernization and for support of R and D projects not directly tied to approved experiments.

The use of computers by theorists is burgeoning. A major new development is their application to clarifying basic mathematical ideas such as lattice gauge theories. This requires computers that not only are large but also have special characteristics designed to handle such calculations efficiently. The pressing needs in this area cannot be met by a small fractional increase in the theory budget, and DOE should set up a panel to study computer needs and estimate budget requirements.

The assignment of VAX (or eqivalent) computers to a number of experimental tasks has been most welcome. The need for more computers—of higher power and greater variety—will continue and should be given high priority, with at least one VAX (or equivalent) planned per medium-size research team, plus one located at the experiment and shared by the collaborators. The panels also found a large unmet need for "networking," i.e., communication between the computers of a collaboration situated on various compuses, even though a few large collaborations are developing their own systems. DOE has convened a panel to evaluate various networking approaches, and recommendations to the funding agencies are continuing.

TACUP emphasized the importance of graduate and postdoctoral training in the High Energy Physics Program. Summer study institutes have in fact been initiated for the training of postdoctoral theorists and experimentalists. Such institutes provide discussions of important current work, produce relevant pedagogical literature, and enhance interactions between students. Many graduate students doing experimental work at accelerator laboratories must interrupt or discontinue their formal education, and this is a growing problem as the time needed at the accelerator site lengthens. It would be helpful if the accelerator laboratories greatly increased the number of mini-courses and lecture series to compensate for the formal courses on campus.

TACUP found the relations between the university users of accelerators and the laboratory staffs to be excellent and mutually supportive but suggested that both groups would benefit from more frequent exchanges of personnel than at present and encouraged DOE support for such exchanges.

As a result of this comprehensive review, corrective actions were taken with respect to some 10 to 15% of the DOE-supported university contracts, including some terminations.

III. INTERNATIONAL COOPERATION IN HIGH ENERGY PHYSICS

A. General

International collaboration is exceptionally well-suited to high energy physics and has long been effective and mutually beneficial. Shared use provides access to unique facilities in one nation which are often complementary to those in another. It provides research opportunities that would otherwise not be available because of the worldwide distribution of unique facilities. Shared use has proved to be mutually beneficial, and the cash flow, averaged over the years, is well balanced between cooperating nations and entities having major facilities.

Coordination of plans for constructing major new facilities, and international participation in research, including joint fabrication of large-scale equipment, is useful for progress on an international scale. However, no formal joint decision-making mechanism has been set up, and no joint construction venture (if CERN is considered a single participant) has been undertaken. In another area, some non-accelerator experiments require unusual geophysical conditions such as very deep mines or a specific geomagnetic latitude, and these depend on international access, the history of which has generally been positive.

In the late 1950s and early 1960s, when the U.S. facilities (and their support) were better than those elsewhere, individual and institutional participation by foreign scientists in the U.S. was high, but few U.S. physicists worked abroad. The balance improved in the late 1960s and early 1970s with the growth of CERN and DESY in Europe. The past ten years have seen extensive U.S. institutional involvement in both the ISR at CERN and PETRA at DESY. There is now some U.S. participation in the p$\bar{\text{p}}$ program at the CERN collider. Many U.S. groups have made a strong commitment to work at LEP and TRISTAN, and others have expressed significant interest in doing experiments at HERA located at DESY. There is also U.S. participation at the DORIS storage ring at DESY and at LEAR, the low energy p$\bar{\text{p}}$ ring at CERN.

The fraction of the university high energy budget spent for work using unique facilities outside the U.S. has recently risen to >10% with the increasing expenditures for the LEP detectors. A formal policy, initiated by the International Committee on Future Accelerators (ICFA) and adopted by all worldwide participating laboratories, specifies that scientific merit and feasibility of experiments shall be the sole criterion for acceptance of programs and experiments. This provides the basis for a reasonable "trade balance."

Whereas the extensive relations with Western Europe are largely informal, international callaborations with other nations are governed in part by formal bilateral agreements. The agreement with Japan appears to be very successful. The agreement with the People's Republic of China is working out well as they struggle to establish a world identity in high energy physics. Joint activities with the Soviet Union have decreased somewhat in recent years. The consensus remains that U.S.-Soviet exchanges are useful and should be maintained.

U.S.-Soviet Union cooperation in high energy physics is governed by the Bilateral Agreement on Cooperation in the Fundamental Properties of Matter, which operates under the Agreement on Cooperation in Atomic Energy, which was extended in 1983. Agreement on a yearly work program is determined by annual meetings of the Joint Coordinating Committee on Reseach on the Fundamental Properties of Matter (JCC/FPM). There have been some benefits under the JCC/FPM exchanges. More Soviet experimenters have participated in work at American accelerators than have Americans at Soviet installations, but the Soviets have contributed novel instrumentation techniques that have been applied at American laboratories. Among these are the gas-jet target for 400-GeV experiments at Fermilab and the lithium high-current magnet lens for the Tevatron I antiproton source. A test facility for the electron cooling of proton beams, a technique invented in the Soviet Union at Novosibirsk, was designed and built at Fermilab a few years ago. These comprise significant "technology transfer" from the U.S.S.R. to the U.S.

Cooperation in high energy physics with Japan operates in part under a 10-year formal bilateral agreement initiated in 1978. It has consisted primarily of Japanese physicists working on U.S. programs, and equipment contributions, in exchange for access to U.S. facilities. With TRISTAN being brought into operation there is increasing interest on the part of U.S. physicists in doing experiments in this new facility. Currently an American-led group is assembling a major detector (AMY-CHAN) for use at one of the four TRISTAN intersection regions. There is also substantial U.S.-Japanese cooperation outside the formal agreement.

Cooperation between the U.S. and People's Republic of China in high energy physics is also governed in part by a bilateral agreement. Most of the collaborative efforts have been dominated by U.S. assistance in designing installations and instruments. The agreement is such that this assistance is without significant financial cost to the U.S. programs.

The High Energy Commission under the aegis of the International Union of Pure and Applied Physics (IUPAP) has sponsored a regular cycle of international conferences in high energy physics, rotated among the Soviet Bloc, Western Europe, and the Unites States, and once in Japan. These have been eminently successful in furthering international communication, and in some cases have provided the first informal contact among scientists previously kept apart by political conditions.

Following the Economic Summit Meetings at Versailles and Williamsburg, a Summit Working Group on High Energy Physics was set up to facilitate international cooperation in research and planning. The initial conclusions of the Working Group were the following:

i. Additional studies should be done on the possibility of expanded cooperation in accelerator technology development.
ii. The present intergovernmental working group should continue to meet periodically to ensure maximum collaboration over the coming years, regarding both existing machines and the planning of new ones.
iii. To benefit from international cooperation, both countries with major high energy physics programs and those with minor ones should have a strong home base program.
iv. If a large international project is agreed upon, equity of benefits cannot be assured, but it can be approached through intensive negotiation on a case-by-case basis. Each participant should contribute in a way that leads to a balance of cooperation over the course of time.

The Working Group set up three new subpanels to facilitate the advance of high energy physics:

i. <u>Subpanel for Long-Term Planning</u> will receive from the Summit nations their plans and proposals for high energy physics, with progress reports on existing major facilities, and will also take account of plans for high energy physics worldwide.

ii. <u>Subpanel on Technical Collaboration</u> will review activities that are underway or planned, to carry out the technical recommendations of the Subpanel on Improving International Cooperation in High Energy Physics.

iii. <u>Subpanel on Administrative Issues</u> will review and make recommendations with respect to customs practices, data communications, and personnel exchanges.

Meetings of the Working Group occurred in Washington in October 1983, in Brussels in July 1984 and in Cadarache, France, January 1985. The Group's activities are reviewed in "Annual Report to the Working Group on Technology, Growth, and Employment by the Summit Working Group on High Energy Physics," April 1985, issued by the U.S. DOE.

Several conclusions on long-range planning were reached by the Working group. The present set of new high energy physics facilities under construction, Tevatron and SLC in the U.S.; TRISTAN in Japan; LEP at CERN; and HERA in the Federal Republic of Germany, are complementary and not duplicative. A clear need not met by the present generation of accelerators is the requirement to extend the energy range of hadron colliders. The SSC is the most advanced plan in this regard and the Europeans are considering options in the LEP tunnel. Other open questions are a linear collider for electron positron collisions and an electron-proton collider which could also make use of the LEP tunnel. The required new and advanced facilities can be built and operated within broadly constant worldwide budgets, with some fluctuations during peak capital expenditure years. While further concentration of facilities is perhaps inevitable, more than one region with a forefront accelerator capability working effectively in high energy physics is essential. It is also essential that the limited number of unique facilities remain open to competent scientists from all over the world. Thus, it is considered of great importance to continue the discussion and planning on an intergovernmental level.

With progress in high energy physics critically dependent on the availability of progressively more complex particle accelerators and versatile detectors, it has been necessary to develop new technology to ensure that these new facilities are achievable in the most economically feasible manner. Consequently, the technical cooperation subpanel recommended that the already active international collaboration in accelerator and detector technology be further encouraged; that common international standards be established to reduce costs and ensure compatibility; and that the existing high energy physics laboratories which are centers of advanced technology be used in extended ways for training in science and technology for industry, university, and other institutions.

It is clear that the removal of certain administrative obstacles would facilitate international collaboration via cost sharing and joint exploitation of regional facilities. Tariff and tax exemptions (e.g., on experimental equipment) are required for extended periods rather than the present short intervals. The freer exchange of scientific and technical staff requires simplified administrative country entry formalities, facilitated integration of the research worker's family in the host country, and a guarantee of adequate social coverage (e.g., health and accident insurance). Lastly, a productive international collaboration requires efficient international data transmission. To this end, it is necessary to review the changing policies for scientific data transmission across national borders and to promote effective data communication standards between nations.

In addition, the U.S. is actively pursuing opportunities for international cooperation in the Superconducting Super Collider (SSC) project, for example, through the Economic Summit process. The U.S.

representative and Chairman of the High Energy Physics Working Group has recently formally sent his Summit colleagues an open invitation for scientists from their countries to participate in the SSC effort. With the SSC being internationally available to collaborative experiments upon completion, it would be desirable to have some participation in the early R and D effort by scientists and engineers from those countries that are likely to be involved in the use of the facility. By such participation they would be better informed as to what international collaboration might best suit the future interests of their colleagues. In turn, the U.S. would benefit from their contribution to its R and D efforts to develop a plan for the best possible facility at the lowest possible cost.

Cooperation can take various forms extending from participation in the planning and design, to that of R and D on various problems, to the building of prototype systems, to construction, and to use. Early collaboration does involve the participation of capable and trained people where the contributions can be most significant and influential on the entire future success of the project. The summit offers an opportunity for greatly improving the international cooperation process.

B. The European Program

The European program in experimental high energy physics is centered at the European Organization for Nuclear Research Laboratory (CERN) in Geneva, Switzerland, and at the Deutches Electronen Synchrotron Laboratory (DESY) in Hamburg, West Germany. CERN has 3600 staff members and an annual budget of ∽600 MSF; about 1800 high energy physicists participate in the program. DESY has a staff of 1100 and an annual budget of ∽130 MDM. Although DESY is a national laboratory, it is utilized internationally, with groups from 11 nations and a total of 400 physicists participating in the program. These facilities are described below. For comparison, the 1984 worldwide budgets for high energy physics are shown in Table IX.

1. CERN

a. <u>Fixed-Target Operation of the Super Proton Synchrotron (SPS)</u>. The SPS is the mainstay of the experimental program at CERN. Its maximum energy has been upgraded to 450 GeV. The beam intensity achieved is now about 2×10^{13} protons every 10 seconds. The West Experimental Area will be upgraded to provide two high energy, high quality secondary beams in addition to several test beams. New neutrino beams for beam dump experiments and for neutrino oscillation experiments are being constructed. The fixed-target program using the SPS is expected to remain strong during the second half of the 1980s.

b. <u>The SPS Proton-Antiproton Collider</u>. This facility produces proton-antiproton collisions at a center-of-mass energy which has now reached 630 GeV and with a peak luminosity of 4×10^{29} cm^{-2} sec^{-1} in each of two intersection regions. The integrated luminosity accumulated since 1981 is 0.6 pb^{-1}. The S$\bar{p}$pS collider operates as follows: 26-GeV/c protons from the PS are used to produce 3.5-GeV/c

Table IX World Budgets in High Energy Physics 1984

U.S. —		DOE	480 M$
		NSF ~	40 M$
		TOTAL	520 M$
WESTERN EUROPE —			
	CERN 701 MSF × .4875 $/SF (3/7/84)		342 M$
	REMAINDER OF W. EUROPE ≳ SAME		360 M$
		TOTAL	700 M$
USSR —			
	ESTIMATE, COMPARABLE OR SOMEWHAT LESS THAN THE MANPOWER/FUNDING EFFORT OF THE U.S. PROGRAM IF ONE CONSIDERS TOTAL ACTIVITY.		500 M$
JAPAN —			
	EXTRAPOLATED FROM HEP/GNP GRAPH RECEIVED FROM DELEGATION OF THE COMMISSION OF THE EUROPEAN COMMUNITIES.	~	150 M$
CHINA —			
	ESTIMATE		60-100 M$

antiprotons, which are stored and stochastically cooled to form a dense beam in an accumulator ring (the AA). After enough antiprotons have accumulated and cooled, they are reinjected into the PS, accelerated to 26 GeV, and transferred to the SPS, where they are accelerated to 315 GeV and caused to collide with counter-rotating bunches of 315-GeV protons. The spectacular discoveries of the W and Z^0 particles in 1983, and probably of the top quark in 1984, are all results of the first experiments with this facility. Improvements are underway to raise the luminosity to 3×10^{30} cm^{-2} sec^{-1} by 1987, with an added 3-GeV ring to enhance the accumulation of antiprotons (ACOL), (leading to an expected integrated luminosity of 10 pb^{-1} during 1988), and to upgrade the UA1 and UA2 detectors.

The UA-1 detector is being upgraded by improving the calorimeter energy resolution, by adding a high resolution vertex detector, and by extending the muon detection system to a larger solid angle. In addition to these detector improvements, the data acquisition system will be upgraded to match the increase in luminosity. The muon detection system improvements will be completed in 1985, and the remainder of the improvements will be completed in 1987. These improvements will lead to better energy resolution for electrons, muons, and jets and will, thus, enhance the detectors' ability to determine the transverse energy carried away by particles which do not interact in the detector, such as neutrinos. With 23 U.S.-supported physicists among the 178 UA-1 collaborators, this detector represents a significant effort for the U.S. as well as for Europe.

The UA-2 detector will be upgraded by extending the hadron calorimeter coverage from 40° to 5° with respect to the beam axis. This will be accomplished by replacing the toroidal magnets with calorimeters, thereby making the detector a totally non-magnetic detector. Other changes will be made to improve the electron identification and tracking in the central calorimeter.

Table X Comparison of the S$\bar{p}$pS and Tevatron Colliders *

	S$\bar{p}$pS (upgraded/1988)	Tevatron I (1987)	Tevatron I (upgraded/1989)
Center-of-Mass Energy (GeV)	630	1800	2000
Peak Luminosity (cm^{-2}sec^{-1})	3×10^{30}	10^{30}	10^{31}
Events of $W^{\pm} \to e^{\pm} \nu$	1.0×10^4	1.2×10^4	1.4×10^5
Events of $Z^0 \to e^+ e^-$	3.6×10^2	9.0×10^2	1.1×10^4
Number of t and $\bar{t}$ (40 GeV/c)	6.0×10^4	1.8×10^5	2.2×10^6
Mass Limits for Heavy V (GeV/c^2)**	240	500	700
Mass Limits for Heavy Q (GeV/c^2)**	90	130	220

* The yields assume a run of 10^7 seconds at peak luminosity.

**The limits are based on the production of 100 produced particles in a run of 10^7 seconds at peak luminosity.

A relative comparison of the discovery potential of the S$\bar{p}$pS and the Tevatron can be made by comparing the production of the standard electroweak bosons, more massive gauge bosons (V), and heavy quarks (Q) in a run of 10^7 effective seconds. The results are given in Table X, which considers planned luminosity improvement programs at both Colliders.

c. Large Electron-Positron Facility (LEP). LEP is an electron-positron storage ring with a circumference of 27 km and eight straight sections. Phase 1 of LEP, aimed at providing 60 GeV x 60 GeV electron-positron collisions in four intersection regions was approved by the CERN Council at its October 1981 meeting. The predicted peak luminosity is 10^{31} cm^{-2} sec^{-1}. The CERN schedule calls for the first colliding beams by the start of 1989. In Phase 1, only 1/6 of the available space along the ring will be equipped with conventional radiofrequency accelerating cavities. With the full complement of conventional cavities, LEP will be able to reach 85 GeV x 85 GeV. If the superconducting accelerating cavities are developed successfully and employed in LEP, an energy upgrade of LEP as high as 120 GeV x 120 GeV will be possible. A capsule of the project is shown in Table XI. The four planned detectors for LEP (ALEPH, DELPHI, OPAL, and L3) are composed with those for SLC (MARK II and SLD) in Section II.B.2.d.

At center-of-mass energies of about 200 GeV, LEP will permit searches for new types of leptons and quarks up to masses of 100 GeV. It will permit study of the production of $Z^0 Z^0$ pairs and $W^+ W^-$ pairs

Table XI LEP Project

- e^+e^- STORAGE RING — 60 – 120 GeV/BEAM
  — 120 – 240 GeV CENTER-OF-MASS
- CIRCUMFERENCE — 27 km (17 mi)
- LOCATION — EUROPEAN CENTER FOR PARTICLE PHYSICS RESEARCH, CERN, GENEVA, SWITZERLAND
- AUTHORIZED FOR CONSTRUCTION — DECEMBER 1981
- COST — 910 MSF (1981), EXCLUDING LABOR
  — ESTIMATE ∼ 2000 MSF WITH LABOR
- SCHEDULE — INITIAL OPERATION: START OF 1989
- DETECTORS — 4 OF 6 INTERACTION REGIONS INITIALLY

and searches for heavier vector bosons and other particles. LEP will also be able to search for Higgs particles up to a mass of 70 GeV. These experiments address some of the most fundamental questions with respect to our understanding of electroweak processes, including the origin of symmetry breaking and the origin of masses.

   d. The Next Step at CERN. The European high energy physics community is considering the installation in the LEP tunnel in the mid-1990s of a superconducting magnet ring suitable for storing 5-and perhaps up to 10-TeV protons. This has already been the subject of several European workshops and was discussed at the 1984 meeting in Japan of the International Committee on Future Accelerators (ICFA). Such a facility, designated the large hadron collider (LHC), would probe near the 1-TeV mass range but with capabilities significantly lower than those of the SSC facility being considered in the U.S. A decision may be reached in 1987.

   2. DESY

   a. DORIS. The electron-positron collider ring DORIS has been rebuilt with more accelerating capability added to raise the maximum reaction energy to 12.0 GeV. Both interaction regions are equipped with mini-beta beam-focusing optics to allow high luminosity operation (∼300 to 400 nb^{-1} per day). DORIS is operating with two complementary detectors: a large solenoid detector, ARGUS, and the Crystal Ball detector relocated from SPEAR. These will be used in a detailed study of bound bottom quark states and of decays of particles containing bottom quarks.

   b. PETRA. The electron-positron colliding ring PETRA has so far operated at reaction energies between 10 and 45 GeV. The installation of a mini-beta beam-focusing system in all four interacting regions has resulted in a peak luminosity above 10^{31} cm^{-2} sec^{-1} and an average integrated luminosity between 400 and 600 nb^{-1} per day per interaction region. A program has been implemented recently to increase the maximum reaction energy to about 45 GeV in two steps. An increase from 36.7 to 40.5 GeV was achieved with a doubling of the rf power from 4 to 8 MW. With the installation of rf cavities in the two remaining long straight sections the energy was increased from 40.5 to about 45 GeV in early 1983. (Actually, a peak energy of 47.6 GeV has been reached since that time.) Superconducting rf cavities would be required to increase the energy beyond 45 GeV up to a maximum of 60 to 70 GeV. A superconducting cavity built by Karlsruhe was successfully tested with high rf power and was installed at PETRA in 1982. DESY, in collaboration with CERN and Wuppertal, has a program to build and test superconducting cavities, but it is unlikely that such an energy upgrade for PETRA will be pursued.

c. <u>HERA</u>. The Deutsches Elektronen-Synchrotron (DESY) has begun the construction of a large electron-proton colliding beam facility, HERA, designed to collide either electrons or positrons of 30-GeV nominal energy with protons of energies up to 820 GeV, yielding 314-GeV reaction energies and a maximum momentum-transfer-squared, Q^2, of 98,000 GeV^2. This is equivalent to an electron beam of 52 TeV on a stationary target. The design luminosity of 2.5×10^{31} cm^{-2} sec^{-1} will allow the determination of the proton structure function in a substantially new range of momentum transfers, $2000 \ (GeV/c)^2 < Q^2 < 10000 \ (GeV/c)^2$, far beyond anything acheivable anywhere else in the world. (The maximum Q^2 achievable in deeply inelastic muon or neutrino scattering at the Tevatron II is about $400 \ (GeV/c)^2$). These experiments, together with observations of the final state, will allow new tests of the Standard Model, especially of perturbative QCD. The electron and positron beams are expected to be transversely polarized, and DESY has plans to enable rotation of the polarization, to provide longitudinally polarized electrons and positrons, an extremely useful capability.

Construction of HERA began in the spring of 1984, with initial operation scheduled for the last half of 1989. A capsule of the project is shown in Table XII.

C. The Japanese Program

Elementary particle research in Japan has so far been carried out at the 12-GeV proton synchrotron at KEK. The future program involves the construction at KEK of a large storage ring complex, TRISTAN, designed to collide electrons and positrons and, perhaps at a later stage, electrons (or positrons) with protons. The TRISTAN tunnel will have a total circumference of 3 km with four interaction regions. An electron-positron colliding ring capable of exploring reaction energies up to 60 GeV will be housed in this tunnel. TRISTAN covers an energy range not covered by any other existing or planned e^+e^- collider and thus provides an important new opportunity to discover and study any new phenomena whose thresholds lie in this range. Its physics program will probably stress electroweak interference effects, jet formation and fragmentation, and studies of the toponium system if $\bar{t}t$ bound states are within its energy range. A maximum luminosity in the range $(3 \text{ to } 8) \times 10^{31}$ cm^{-2} sec^{-1} is expected.

Table XII HERA Project

- **EP STORAGE RINGS** — 820 GeV p COLLIDING WITH
  30 GeV e
  — 314 GeV CENTER-OF-MASS
- **CIRCUMFERENCE** — 6.3 km (4 mi)
- **LOCATION** — DEUTSCHES ELEKTRONEN SYNCHROTRON
  (DESY), HAMBURG, WEST GERMANY
- **AUTHORIZED FOR CONSTRUCTION** — APRIL 1984
- **COST** — 950 MDM (1984), EXCLUDING LABOR
  — ESTIMATE ~ 2000 MDM WITH LABOR
- **SCHEDULE** — INITIAL OPERATION: LAST HALF OF 1989
- **DETECTORS** — INITIAL COMPLEMENT OF 3 DETECTORS
  PLANNED. LETTERS OF INTENT TO BE SUBMITTED
  BY JUNE 30, 1985.
- **COLLABORATING NATIONS** — CANADA, FRANCE, ISRAEL, ITALY,
  NETHERLANDS, AND UNITED
  KINGDOM

There are three major detectors at TRISTAN; two of them, TOPAZ and VENUS, are general purpose detectors. VENUS uses a drift chamber in a solenoidal magnetic field for charged particle tracking and a lead glass array for photon detection. TOPAZ also features lead glass but uses a TPC for charged particle identification and tracking. Both detectors are being constructed by Japanese collaborations. The third detector, AMY, is a compact detector emphasizing lepton and photon detection. To this end, the detector uses a high magnetic field, high resolution tracking chambers, and a photon counter before the magnet coil. This detector has major U.S. collaboration and leadership.

The TRISTAN project has been authorized, and construction is underway. The start of the experimental program is scheduled for late 1986. The installation of superconducting rf could bring the energy to 36 GeV per beam.

D. The U.S.S.R. Program

   1. Institute for High Energy Physics (IHEP), Serpukhov

The major proton accelerator facility in the U.S.S.R. at present is the 76-GeV synchrotron at IHEP. Although its proton beam intensity is low by Fermilab and CERN SPS standards, it supports an active program, similar to that at the AGS at BNL, with emphasis on large, relatively permanent experimental setups. Some unique work on charm and particle production by hadrons is being done in a neutral beam; a neutrino counter experiment has ceased operation until the new booster comes into operation to increase the intensity. Work on K^o regeneration experiments has been in progress for 10 years; a large multi-particle spectrometer facility has been used to study charmed particles. A group using emulsion techniques has provided the cold emulsions for the search for charmed particles in the Fermilab neutrino beam using the 15-ft Fermilab bubble chamber. SKAT, the large propane chamber for neutrino physics, is still in operation as is MIRABELLE, a large bubble chamber for hadron physics. The bubble chamber scanning and measuring facilities at both Dubna and Serpukhov are extensive and have recently been upgraded.

In the future the 76-GeV synchrotron will become the injector for a multi-TeV accelerator/storage ring, UNK. Phase I of UNK, which was authorized for construction in 1981, includes a 400-GeV booster synchrotron using conventional magnet technology and a 3000-GeV (3-TeV) synchrotron using superconducting magnets whose design is similar to Fermilab's. Phase I will provide fixed-target beams to 3000-GeV maximum incident energy (75-GeV reaction energy) and proton-proton colliding beams of 400 x 3000 GeV (2190-GeV reaction energy): $\bar{p}p$ collisions at 2.2-TeV center-of-mass energy would be possible at this stage. Phase II, which is still in the planning stage, could provide 3000 x 3000-GeV proton-antiproton colliding beams and, with the addition of a second superconducting magnet ring, 3000 x 3000-GeV proton-proton collisions. If both phases of UNK are successfully completed and operated, the U.S.S.R. will have a high energy physics research capability in the 1990s far beyond that provided by any facility currently authorized. However, either the SSC in the U.S. or the LHC at CERN would surpass the capabilities envisioned for UNK.

2. Institute for Nuclear Physics (INP), Novosibirsk

This Institute is the center of lepton interaction physics in the Soviet Union. The research centers around the VEPP-4 electron-positron storage ring, which has a maximum energy of 5.8 GeV per beam. A luminosity of 3×10^{30} cm^{-2} sec^{-1} has been obtained at 5 GeVx5 GeV (10-GeV reaction energy). This device has been used to make precision measurements of masses of particles containing the charmed quark. An older storage ring, VEPP-3, is used as an accumulator/injector for VEPP-4, allowing the electrons and positrons to polarize before injection. Polarized beams of >50% have been achieved. The main detector facility employs a 400-ton magnet with field perpendicular to the plane of the particle orbits. While the physics research program at INP is considered modest by Western standards, the contributions of INP to the development of accelerator technology have been exceptional. INP is reported to be involved with design of an antiproton source for use with UNK. This effort stems from the pioneering work at INP on electron cooling of proton beams.

3. Joint Institute for Nuclear Research (JINR), Dubna

Although the accelerator facility at Dubna, the 10-GeV synchrophasotron is not competitive by modern standards, its high energy physics program remains active. Its group is allocated about 30% of the research time at Serpukhov and maintains a team of physicists on-site to support this research. At Dubna, work is being done on quark models of the nucleus, using heavy ion beams, as well as some fine polarized jet target research using a very advanced and effective solid state detector array.

E. The Program of the People's Republic of China

The major high energy physics facility under construction in the People's Republic of China is a 2.5 GeV x 2.5 GeV electron-positron collider designated BEPC. It is designed to achieve high luminosity to support an experimental program focused on high precision studies of the particle states containing charmed quarks. The design luminosity is 1.7×10^{31} cm^{-2} sec^{-1} with one bunch per beam. Initial operation is expected in 1988.

IV. NEW FACILITY NEEDS FOR THE U.S. HIGH ENERGY PHYSICS PROGRAM

A. The Next Step

In the past, major scientific breakthroughs in high energy physics have been associated primarily with investigations at higher and higher energies. Recent insights have led to the conclusion that exploration of the TeV mass scale (or equivalently of the 10^{-17} cm distance scale) is the next step toward any major advance.

This could be accomplished with a Superconducting Super Collider (SSC), a device that collides head-on two beams of protons, each of energy 20 TeV. The proposed SSC (Figure 22) would consist of an injector complex of accelerators, of approximately the scale of the Fermilab Tevatron facility, to accelerate protons from rest to an energy of 1 TeV or more for injection into two very large rings of superconducting magnets that confine the counter-rotating bunches of protons in the desired orbits. These bunches of high energy protons are then accelerated to an energy of 20 TeV and caused to collide at six symmetrically placed positions around the rings where facilities for physics experiments can be located. The circumference of the main ring is determined by the strength of the main ring dipole magnets which bend the trajectory of the protons. For the range of dipole magnetic fields considered at this time, the circumference of the main ring is between 90 and 164 km. Figure 23 shows the size of a 90-km SSC main ring compared with the 6-km-circumference Tevatron and the 27-km-circumference LEP facility.

The energy of the SSC and its high design luminosity (10^{33} cm^{-2} sec^{-1}) represent a substantial extension of the major parameters of colliders, proposed or under construction. The SSC will be able to probe to a much higher mass scale or to smaller subnuclear distances than are accessible with any other existing or planned facility. It thus offers a real probability of major advances in our knowledge of elementary particles and the unified forces of nature. Such a device is now within reach of present-day technology because of the success of U.S. investment in superconducting magnet technology and the investment of the world HEP laboratories in high energy beam technology. The SSC could be operational in the mid 1990s.

The 20-TeV proton-proton collider envisioned for the SSC have been discussed for some time. Two workshops sponsored by the International Committee for Future Accelerators (ICFA), at Fermilab in 1978 and at CERN in 1979, considered the technology of such accelerators and other potential frontier instruments for reaching very high

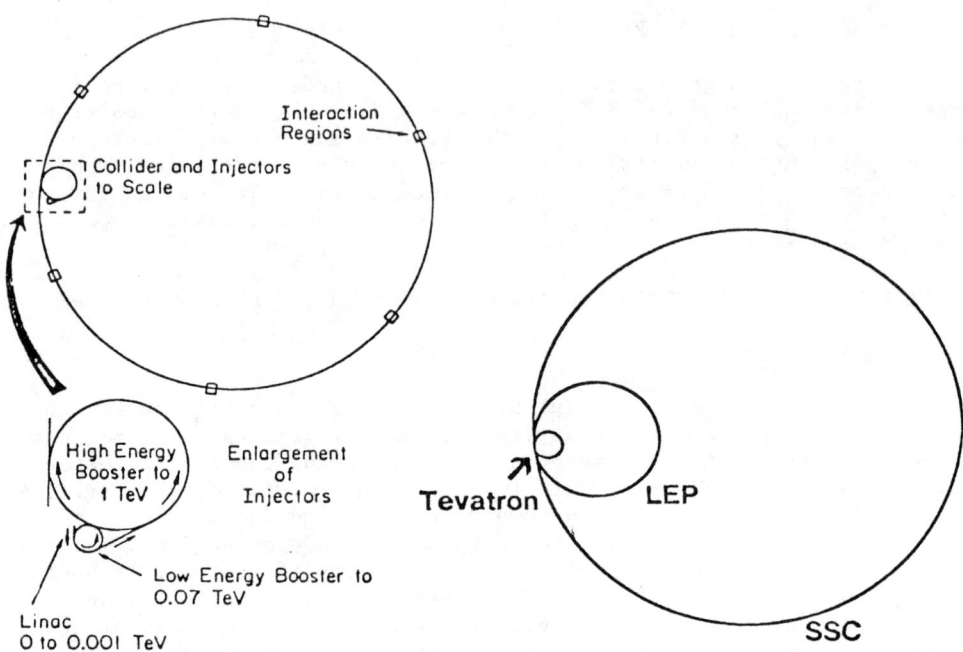

Figure 22 (Left). The layout of the SSC used in the Reference Designs Study, indicating the injector complex and the main ring where protons are accelerated to 20 TeV in counterrotating bunches that collide at six points along the circumference.

Figure 23 (Right). The relative sizes of the Tevatron at Fermilab (6-km circumference), the LEP collider under construction at CERN (27-km circumference), and the SSC main ring (90-km circumference if constructed with 6.5-T bending magnets).

energies but left costs and engineering details to the future. Since then, magnet and colliding beam technologies germane to proton-proton colliders have made substantial advances.

These technical advances and recognition of the enormous scientific potential of a multi-TeV proton-proton collider led U.S. leaders in high energy physics to suggest construction of such a collider as soon as possible. Figure 24 indicates some of the physics motivation for the SSC.

The U.S. high energy physics community started to consider this question at the 1982 Summer Study of the American Physical Society (APS) Division of Particles and Fields at Snowmass, CO (June 28 to July 16). The purposes of the Summer Study were to assess the future of high energy physics, to explore the limits of our technological capabilities, and to consider the nature of future major facilities in the U.S. The 1982 Summer Study led to a widespread recognition of the need for (and importance of) a multi-TEV proton-proton collider, based

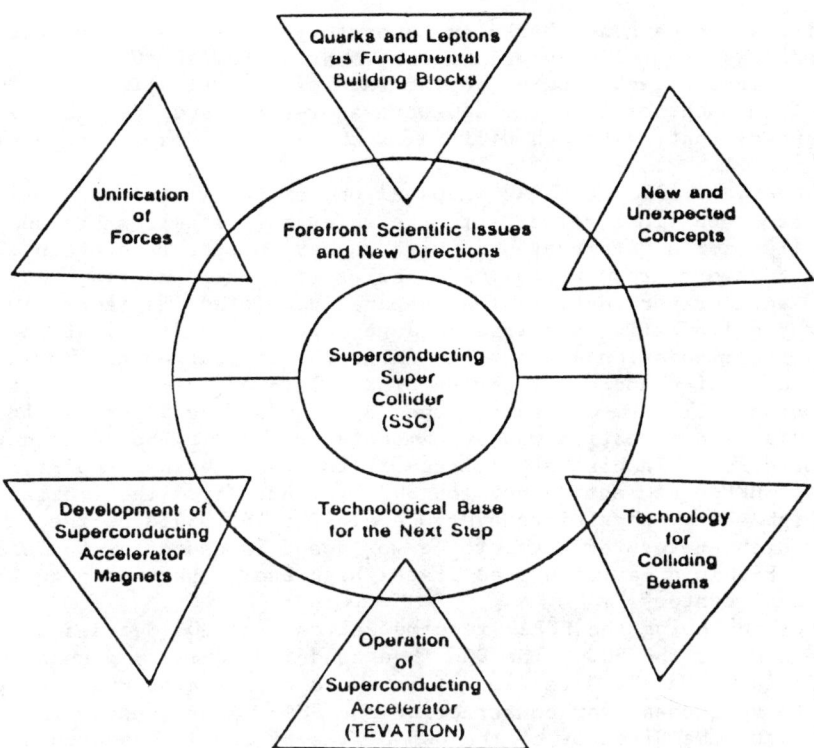

Figure 24. In recent years progress in theoretical and experimental physics has pointed to several forefront scientific issues and new directions for research that must be addressed by a vital U.S. high energy physics program, and technical developments have provided a strong base for the next step in major facility development and construction. The SSC concept represents the response of the U.S. high energy physics community to these factors and opportunities.

in part on recent accomplishments (in both experimental and theoretical research) strongly suggesting that the next major step will require exploration of the energy region where new particles (with new properties and masses of a few TeV) can be produced.

At a 1983 20-TeV Hadron Collider Technical Workshop at Cornell University (March 28 to April 20), 40 U.S. and European experts in accelerator science and accelerator construction technology met to consider further the technical issues. They concluded that the time was ripe to intensify development of candidate magnet systems in order to allow design and cost studies that would narrow cost uncertainties and define construction methodologies more sharply. Assuming a reasonable level of effort, they thought construction could begin within four years.

The APS Division of Particles and Fields sponsored a 1983 Workshop on Hadron Collider Detectors: Present Capabilities and Future

Possibilities, at Lawrence Berkeley Laboratory (February 28 to March 4), attended by about 100 experimenters, mostly from the U.S. The issues of detector technologies in the face of the multiplicities expected from machines like the SSC were addressed, and the general conclusion was that, although difficult, the required experiments will be possible.

In February 1983, a HEPAP Subpanel on New Facilities was formed to make recommendations relative to the scientific requirements and opportunities for a forefront U.S. high energy physics program in the next five to ten years. Following a series of on-site meetings at the three DOE accelerator centers, the Subpanel completed its report in two lengthy deliberative meetings in June-July 1983. Its first and unanimous recommendation was for the immediate initiation of a project aimed at the design and construction of a multi-TeV, high-luminosity proton-proton collider, designated the Superconducting Super Collider (SSC). This recommendation was unanimously endorsed by HEPAP with the highest priority. In his letter transmitting the Subpanel report, the Chairman of HEPAP commented that the SSC "... has fired the imagination of high energy physicists everywhere. The SSC would be the forefront high energy facility of the world and is essential for a strong and highly creative United States high energy physics program into the next century."

After reviewing the HEPAP recommendation, the DOE decided to initiate R and D for the SSC. The DOE then used its advisory mechanism to provide input on the direction of advanced accelerator R and D, preliminary to a proposal for constructing the SSC. As requested in August 1983 by the Director of the DOE Office of Energy Research, HEPAP set up a subpanel to provide advice and recommendations on the content and implementation of an FY 1984 R and D effort, preliminary to the initiation of a formal SSC R and D program. In the fall of 1983, the subpanel reviewed the various requests to carry out relevant R and D and recommended several different approaches to SSC magnet design, including high field (8- to 10-tesla) magnets based on Nb_3Sn superconductor, medium field (5- to 6-tesla) magnets based on Nb-Ti superconductor, and low field (2- to 3-tesla) superferric magnets. It also recommended support for development of Nb_3Sn conductor, accelerator physics activities, and other critical technical R and D efforts. The subpanel's recommendations were endorsed by HEPAP and transmitted to the DOE Office of Energy Research.

In addition to the work by HEPAP, other activities have provided significant input from the high energy physics and accelerator physics communities (as well as the relevant technical and industrial communities) into the nationwide consideration of the scope, design, and use of the SSC and of the R and D required to generate an optimized design. These activities include the following:

i. Ad hoc Physics at the SSC (PSSC) Meetings of the potential SSC user community, held periodically since 1983 at various universities and laboratories, attended by about 200 U.S. physicists, focus on the required scientific capability for the SSC as best determined from recent theoretical and experimental results, and on the anticipated requirements for detector instrumentation.

ii. The 1983 APS Division of Particles and Fields Workshop on Accelerator Issues for the SSC at the University of Michigan, Ann Arbor (December 12-17) was attended by 70 accelerator specialists from many laboratories and universities. Its goals were to identify the accelerator physics issues crucial to demonstrating the feasibility of the SSC and the key R and D needs, and to guide the engineering design of technical components and systems.

iii. The 1984 APS Division of Particles and Fields Cryogenic Workshop at Brookhaven National Laboratory (January 17-19) was attended by about 100 experts in cryogenic systems, superconducting magnet design and fabrication, and accelerator physics, who addressed the cryogenics issues most pertinent to the design of a large-scale accelerator using superconducting magnets.

iv. Implementation was completed by DOE, in early 1984, of an interim management arrangement to develop R and D and management plans for a preconstruction R and D and proposal development phase of the SSC activity. These plans were submitted to DOE in July 1984.

v. The 1984 APS Division of Particles and Fields followup workshop to the 1982 Snowmass workshop, at Snowmass (June 23 to July 13), reaffirmed the SSC as the facility essential to ensure a vigorous and creative U.S. program in high energy physics in the latter part of this century and the early part of the next.

In December 1983 the directors of the U.S. high energy accelerator laboratories with DOE concurrence chartered the National SSC Reference Designs Study to review in detail the technical and economic feasibility of various options for creating the Superconducting Super Collider (SSC) facility as a 20-TeV on 20-TeV proton-proton collider having a luminosity up to 10^{33} cm^{-2} sec^{-1}. The objective was to assess the technical feasibility, develop a cost estimate based on clearly stated and credible assumptions, and to help define how best to proceed with SSC R and D directed toward improving the cost effectiveness of applicable accelerator technology. The study was based on three kinds of super-conducting magnets, with different configurations, and was aimed at sharply decreasing the cost of the needed magnet system below that of existing designs. It addressed three key areas: technical feasibility, economic feasibility, and identification of specific R and D needs, with primary emphasis on the range within which the SSC construction cost can confidently be expected to fall--excluding the cost of research equipment, preconstruction R and D, and site acquisition. The results indicated that the basic design principles used successfully for existing accelerators can be conservatively extended to a proton collider having the SSC primary specifications of energy and luminosity, and that each of the three reference magnets could serve as the foundation for such a collider. Vigorous R and D would be needed to refine the cost estimates for the magnets, to determine their actual performance, reliability, and manufacturability, and to develop cost-effective methods for their assembly and quality assurance. A 6.5-tesla magnet was selected in 1985 as part the R and D program. An important goal will be to produce a significant number of magnets by mass-production methods, and to test them under conditions simulating actual accelerator operation.

The estimated construction costs for an SSC facility based on the three kinds of magnets, range from $2.70 to $3.05 billion (FY 1984 dollars). These include sufficiently conservative contingencies that they represent the current best estimate for an upper bound on the SSC cost.

A DOE committee, including consultants, reviewed the National SSC Reference Designs Study Report in May 1984, and concluded that the Report establishes the technical feasibility, economic credibility, and specific R and D needs of the proposed SSC project. The committee recommended increasing the total estimated construction cost by 0.2 billion dollars, and concluded, on the basis of an analysis of the sensitivity of the cost estimate to various critical assumptions, that an upper limit on the SSC cost is no more than 1.25 times the cost of the most expensive design considered in the Study. The committee judged the proposed six-year construction schedule to be feasible.

A decision to proceed with the preconstruction R and D and proposal development phase of the SSC effort was made by the DOE during the summer of 1984.

In preparation for the SSC, research and development are continuing in FY 1985. A critical milestone of the 1985 program is selection of the magnet type. The efforts in FY 1985 were concentrated on magnet R and D to provide a technical basis for this decision, but they also included work on a siting parameter document. In FY 1986 R and D is expected to focus on a cost-optimized design of the selected magnet, conceptual designs of conventional and technical systems (including cost and schedule estimates), and fabrication of magnet prototypes in preparation for spring tests in 1987.

B. Accelerator R and D in Superconductivity

For large circular colliders, it is imperative to use superconductivity to keep costs down, and to keep power consumption and/ or size within reasonable limits.

In the area of hadron colliders, substantial R and D on superconducting magnets is required to continue the improvement in performance, to lower even further the total cost of such colliders, and to increase magnet reliability. This R and D work (being performed in the U.S. at BNL, Fermilab, LBL, and the Texas Accelerator Center) can conveniently be subdivided into two categories:
1. Development of new conductors and related magnet construction techniques suitable for fields $\geq 8T$ (high fields).
2. Improvement of conductors and techniques for fields in the range 2 T to 6 T (low to medium fields).

The most substantial basic development is required for the first category, while cost considerations are important for both categories in order to allow the final choice to be made on sound economic grounds.

Two lines of development are open in the area of high-field conductors and magnets (≥ 8 T): Nb_3Sn conductor used at $\sim 4.5°K$, and NbTi used at $\sim 2°K$. In principle the most promising material is Nb_3Sn because of its higher critical field and higher temperature (more relaxed cryogenic system). Its drawback is brittleness and fragility, which requires either final reaction of the composite to obtain the superconducting state after winding, or else magnet coils wound with relatively large radii using pre-reacted material. Nb_3Sn has been produced for many years, but the specific requirements for this application of small-bore magnets operating over a wide field range (maximum field about 20 times the injection field) are high current density and superconducting filaments of small diameter. Winding and insulation techniques compatible with a heat treatment of $\sim 700°C$ for a few hours, or methods of avoiding too small bending radii in winding of pre-reacted material, must also be developed.

The alternative line of development toward high fields is to use NbTi conductors at lower temperature ($\sim 2°K$). The advantage is that the material can be wound in the reacted state by means of well established techniques; the disadvantages relate to the cryogenic system – more complicated cryostats and larger power consumption. The maximum field is also more limited. This line is a convenient reserve should the development of Nb_3Sn encounter serious difficulties or lead to excessive costs.

Work based on medium-field magnets (4.5 to 6 T) is a natural continuation of present work. The R and D is directed toward low cost and the development of industrial manufacturing techiques. The conductor development would be essentially the same as that required for the high-field case with NbTi at $2°K$.

Still lower fields, between 2 and 3 T, are also being considered. In this case the field distribution is shaped by iron boundaries, and the role of superconductor is to minimize the power consumption. These magnets are called superferric. The development needed here is directed toward simple, inexpensive design and cost-effective manufacturing techniques in order to counterbalance the increase in cost attributable to the longer tunnel and more spread-out infrastructure that would be required.

International collaboration in these superconducting magnet development programs would be very beneficial. Such work should be aimed at enhancing technology transfer to the industries of the collaborating regions. Specific elements in such a collaboration might be the following:
- Common definition of possible new superconductors in order to minimize industrial investment for development and to enlarge potential markets.
- Joint selection of a small number of potentially interesting techniques and conceptual magnet designs to be tested by means of models.
- Eventually, coordinated fabrication and evaluation of full-scale prototypes.

To assess the possibility of acheiving these goals, the exchange of people between interested laboratories should be encouraged and supported.

It should be emphasized that all of these developments could be beneficial to a number of applications in other fields such as fusion, electrical power transmission, cryogenerators, energy storage and recovery, magnetic separation of minerals, nuclear magnetic resonance for medical and other applications, transportation, etc.

The large superconducting magnet systems described above require very substantial cryogenic systems, distributed over long distances, for the production, transfer, and recovery of He. Simple cryostats, a reliable and efficient liquifier plant, and low-loss transfer lines are very important elements of a satisfactory design, which must be tailored to each specific accelerator project. One point that deserves a careful assessment by experts is the possibility of using superfluid He at $\sim 2°K$ for such applications. An evaluation of the additional costs with respect to normal $4.5°K$ systems, because of their increased cryostat complexity and power consumption, is a necessary first step.

During the last few years superconducting radiofrequency (rf) accelerating structures have matured to the point where their use in large circular accelerators seems feasible and realistic. Multicell niobium structures have reached average accelerating gradients of more than 4 MV/m at negligible rf-power loss. Even considering the fact that these small remaining losses occur at the temperature of liquid helium and require powerful refrigerators, the overall power economy of superconducting resonators is one to two orders of magnitude better than that of conventional copper structures. The fact that the accelerating gradients are more than a factor of four higher than those in copper structures (in cw operation) makes these superconducting resonators very well suited to applications in which large rf accelerating voltages are required in a continuous operation, i.e., in electron and proton storage rings.

These new structures have been studied at a number of different laboratories in the U.S. Germany, France, Japan, Italy, and at the CERN laboratory. Successful and reliable operation at high accelerating gradients has been demonstrated at the electron-positron storage rings CESR (Cornell University) and PETRA (DESY, Hamburg). To make this new technology more economical and attractive for routine operational use, vigorous development programs are now under way at several laboratories. These efforts include the development of copper structures with a superconductive niobium coating, simplifications in the cryogenic technology, special cavity shapes to suppress excitation of higher resonances, and simple techniques for industrial production.

C. Longer Range

For the longer range, beyond 1995, several possibilities are envisioned for new or expanded facilities to allow continued vitality and U.S. leadership in high energy physics. These could include upgrading the SSC, a TeV-range colliding linear accelerator, or some type of facility not yet imaginable.

The concept of colliding beams with linear accelerators as a possible approach to high energies in the next century has been discussed within the high energy physics community and informally with the 1983

HEPAP Subpanel on New Facilities for the U.S. High Energy Physics Program. In this concept two opposing, collinear linear accelerators would accelerate a series of charged bunches toward each other in such a way that collisions between the bunches take place in the space between the linacs, where sophisticated beam optical systems focus the beams to submicron size at multiple collision points displaced laterally from one another. Different bunches collide at different collision points.

Technical success with such a colliding beam system depends on the ability to produce and maintain high phase-space density bunches in the face of strong space-charge and wakefield forces, and on the ability to bring these bunches to a sharp, steady focus of typical dimension 0.1 micron: 10 times smaller than that sought in the SLC. In addition, luminosity multiplication through self-pinching of the opposing beam bunches during collision is crucial for achieving the desired luminosity. Some confidence about understanding these factors will be gained in the course of bringing the SLC into operation, but the order-of-magnitude reduction in beam size and transverse phase-space area for the multi-TeV case should be emphasized.

Economic success with this direct evolutionary approach depends on considerable lowering of unit costs, both capital and operating. Major items in the capital cost are the rf sources, including klystrons and modulators; and the accelerator sections, including accelerating and feed waveguides, vacuum systems, and supports.

Net conversion efficiency of line power to beam power for single-bunch operation of a conventional SLAC-style linac is considerably less than 1%. As average beam power of 10 MW or more will be required to produce useful luminosities at a colliding linear accelerator facility, efficiencies of acceleration must be improved considerably to avoid unacceptable electric power usage.

Assuming the necessary technological developments, colliding electron and positron linear accelerators might provide, at 2-TeV center-of-mass, a luminosity of 1.5×10^{32} cm^{-2} sec^{-1} with an energy spread of 10% at each of six parallel interaction regions. Such linacs might be used also for electron-proton collisions if a suitable cooled proton source (e.g., cooling ring) could be developed.

D. Advanced Technology Research and Development

Research and development in the technologies of particle beam acceleration, beam control, other accelerator-related systems and instrumentation, and detection systems for experimentation are essential for maintaining the forefront nature of the High Energy Physics Program.

New technologies developed in the past have resulted in enormous increases in accelerator energy and decreases in cost per unit energy in the past fifty years, but the present scale of R and D is relatively small--and certainly not commensurate with its importance. One problem is the reluctance of individuals to commit themselves to tasks whose possible fruition seems quite distant. Another is the lack of suitably trained multi-disciplinary experts. A third may be the mechanisms for supporting accelerator physics. Encouragement to universities to expand training in accelerator physics may be needed, and it might be desirable to implement a funding mechanism that would allow laboratories to pursue long-term work with an assurance that such funding was truly an addition to that for more immediate goals, since internal priorities tend to force some curtailment of very long-range activities.

Despite these problems, it is encouraging that many people are working on new ideas and that advanced accelerator workshops are held at regular intervals. Calculations and research are being done in the U.S. and abroad on various concepts for obtaining higher accelerating gradients (energy gain per unit of accelerator length). These indicate that accelerator structures can eventually be built to handle gradients up to 200 GeV per kilometer, ten times those now available. This will require a suitable high efficiency, high power source of short-wavelength electromagnetic radiation that can provide a relatively large amount of energy per unit length. Some of the possibilities are as follows:

i. Very high power, very short pulse length, high-frequency klystrons suitable for this purpose may be developed.

ii. A special case of a source of short-wavelength electromagnetic radiation is the wakefield of a high energy beam passing through a cavity system. This idea is being pursued theoretically and shows considerable promise and a special simplicity since the wakefield source cavity can be combined with a beam-accelerating cavity within a single structure.

iii. In the two-beam accelerator concept, a high power, low energy electron beam travels parallel to the desired high energy particle beam. Using a principle such as that of the free electron laser, the high-power, low energy beam radiates its power to the high energy beam, thus providing the acceleration.

iv. A more radical approach is to use the very short wavelength obtainable from a laser. In this case one cannot consider accelerating structures of conventional design; the dimensions are far too small. It appears possible, however, to use a suitable optical grating in place of a conventional cavity. The most extreme case would be replacement of the periodic grating by a periodic plasma, possibly formed over a grating surface, by which gradients as high as 1 TeV per kilometer could theoretically be attained. Such high and obviously desirable gradients can exist only in or near a plasma and not in or near any solid conductor or dielectric.

v. A particularly interesting idea is to expose a plasma to two laser beams of suitably close frequency. The beat frequency between the two lasers can be matched to the natural plasma frequency, to induce a strong periodic and moving charge modulation, which generates large electrostatic fields that could be used to accelerate suitably injected beams. Accelerating fields as high as 2 TeV per kilometer have been discussed, but great uncertainty remains about the stability and energy efficiency of such a mechanism and its suitability for use in a high energy linear collider.

Many other ideas have been suggested. Some of them may not work. Others may work but not have application for high energy physics. It is clear, however, that without some such idea, no great further step in energy will be economically possible after the SSC. On the other hand, with gradients of the order of 1 TeV/km theoretically possible, an accelerator of 100 TeV is not unthinkable. It is thus very important to the future of the field that these ideas be followed up in coming years.

Continued advance in the development of instrumentation and detection systems is the experimental high energy physicist's challenge. These systems provide both the raison d'être for accelerator facilities and the means by which theory can be confirmed by experiment. Support for the development of instrumentation will grow during the next decade, and basic research on new detectors will be increasingly recognized and supported as being fundamental to progress.

E. Funding and the Transition to the Superconducting Super Collider

A funding scenario for the future High Energy Physics Program must be based on specific programmatic and budgetary assumptions. It is, of course, not a simple task to predict the future. To get a feeling for what might be possible for High Energy Physics in the future, we take the results of a recent HEPAP study, to be found in "Report of the 1985 High Energy Physics Advisory Panel Study of the U.S. High Energy Physics Program 1985-1995." In the following, in order to emphasize that the details of program and budget in this 10-year period are recommendations offered by HEPAP, the use of "we" is maintained as in the report and refers to the HEPAP study group.

High Energy Physics 1985-1995. We are looking forward to the start of SSC construction in 1988 and an expected completion in 1994 after a 6-year construction period. Here we describe a strong U.S. High Energy Physics Program and show how the transition to the SSC might proceed. We look at "snapshots" of the program at three distinct times: in 1987, before the start of SSC construction; in 1991, during the SSC construction; and finally, in 1995, after the SSC has been completed and its operation for experiments has begun. Obviously, there are uncertainties associatd with anticipating the program 10 years in the future in a rapidly moving field such as high energy physics. The following is our expectation from today's vantage point.

1. Evolution of the Research Program

The Program in 1987. By this time, the new facilities now under construction will be completed. This will be the start of the rich utilization of all of our facilities. At Fermilab, the Tevatron 1 $p\bar{p}$ collider will be in early operation with one detector. The Tevatron II fixed-target programs will be in full operation. At SLAC, the e^+e^- collider, SLC, will be in its initial operation and PEP will be in full operation. The lower energy e^+e^- collider, SPEAR, will be utilized about half-time for high energy physics. At Brookhaven, the AGS will be in full operation. The majority of AGS time will be devoted to high energy physics, with a small portion of the time used by the nuclear physics program studying heavy ion collisions. At Cornell, the newly upgraded e^+e^- collider, CESR, will be fully utilized. There will also be a number of U.S. groups working at non-U.S. facilities.

Detectors for the e^+e^- collider, LEP, and the $e^{\pm}p$ collider, HERA, will be under fabrication, and there should be experiments running at the e^+e^- collider, TRISTAN, and the $\bar{p}p$ collider, the $S\bar{p}pS$. A variety of non-accelerator experiments will be taking data. A vigorous SSC R and D program in advance of construction will be in progress, both on the accelerator systems and the associated detectors. There will also be activity in advanced accelerator R and D.

The Program in 1991. By this time, we expect the SSC will be in the middle of its construction period. The construction of detectors for the SSC will be in full swing. At Fermilab, the collider will operate with both major detectors, and the full complement of fixed-target facilities will be utilized, both for experiments and for providing test beams for detector development. At SLAC, the SLC will be in full operation. At Brookhaven, a high intensity AGS with a completed Booster will be utilized partially by the High Energy Physics Program and partially by the Nuclear Physics Heavy Ion Program. At Cornell, CESR will still be operating. There will also be U.S. groups taking data in experiments at LEP and TRISTAN. Experiments at HERA, with U.S. participation, will be starting up. A variety of non-accelerator experiments will be taking data. The activity in advanced accelerator R and D will be continuing.

The Program in 1995. By this time, we expect that the SSC will be in its early stages of operation with at least a partial complement of detectors. At Fermilab, we expect that the collider will be running with a mature program; the fixed-target facilities will be operating for an experimental program at a reduced level in addition to supplying test beams for new detector development. The SLC at SLAC will still be in operation. The AGS at Brookhaven is expected to be almost fully occupied as an injector for the heavy ion collider, RHIC, and as a fixed-target facility for the Nuclear Physics Program and is not expected to be utilized significantly by the High Energy Physics Program. U.S. groups will probably still be involved at LEP II and HERA, but the TRISTAN involvement is expected to be winding down. A variety of non-accelerator experiments will be taking data. We anticipate a vigorous program in advanced accelerator R and D at this time. It will be difficult to maintain a forefront U.S. program if the operation of the SSC is delayed beyond this time.

2. Funding Required for the High Energy Physics Program

An estimate has beem compiled for a minimum budget required to carry out the program described here. For the pre-SSC program in FY

1987, these estimates are $435 million for operating, $90 million for equipment, $70 million for construction (for a total of $595 million non-SSC related) and $65 million for SSC R and D, for a total High Energy Physics budget from the DOE of $660 million. An additional $55 million is required from the National Science Foundation (all in FY 1987 dollars). For comparison, the FY 1985 High Energy Physics budget (converted to FY 1987 dollars) was $400 million in operating, $66 million in equipment, and $129 million in construction (for a total of $595 million non-SSC related) and $23 million for SSC R and D, for a total DOE budget of $618 million, with an additional $46 million from the NSF. We note that while the total non-SSC-related funds needed for FY 1987 are essentially the same as in FY 1985, the operating funds are up and the construction funds are down relative to FY 1985. This is necessary for the following reasons. Two of the major accelerator facilities of the High Energy Physics Program will be coming into full operation by 1987. These unique state-of-the-art accelerators (SLC and Tevatron) are not only the mainstay of the High Energy Physics Program but also prototypes for future higher energy accelerators. It is essential that incremental operating funds be provided for these facilities.

With these assumptions, the projection of the DOE High Energy Physics budget to FY 1995 is given in Table XIII. Here the SSC construction, equipment, and operating costs from the SSC Reference Designs Study have been used. It can be seen that a substantial pulse of incremental funding over the FY 1987 level is required for the SSC construction for the FY 1988-1993 period. After this period, the funding required for the program, including the operation of the SSC, returns to within about 10% of the FY 1985 level. During the SSC construction period, the non-SSC part of the budget declines, reaching a reduced level of about $400 million by FY 1995. The SSC operating costs are expected to be $215 million with an additional $50 million in equipment funds and $15 million in accelerator improvement (AIP/GPP) funds, for a total High Energy Physics budget from the DOE in FY 1995 of about $680 million (in FY 1987 dollars).

3. Scientific Manpower

Current estimates indicate that the U.S. scientific manpower (Ph.D. physicists or equivalent) working in high energy physics at the present time consists of about 1150 experimentalists, 750 theorists, and 330 accelerator scientists. The number of experimentalists now occupied in carrying out the present program is not a bad match to the requirement of the program during the SSC construction period and the period after the start of SSC operation. We do, however, anticipate a scarcity of accelerator physicists during the SSC construction period. A need for a 20 to 30% increase in the number of accelerator scientists is anticipated. A vigorous program of training present high energy physics experimentalists and theorists in accelerator science is of high importance in the near future. In addition, efforts should be made to make a career in accelerator physics attractive to young physicists entering the field of high energy physics. We anticipate a flow of experimentalists and accelerator

Table XIII Projected High Energy Physics Budgets 1985-1995
(in units of millions of FY 1987 dollars)

	FY 85	FY 87	FY 91	FY 95
Non-SSC Program				
Operating*	400	435	410	350
Equipment	66	90	70	40
Construction	129	70	20	10
Total Non-SSC	595	595	500	400
SSC Program				
R and D	23	65	40	0
Construction			700	15
Equipment			130	50
Operating			30	215
Total SSC	23	65	900	280
Total DOE Program	618**	660	1400	680
NSF	46	55	48	48

* Includes University budgets, part of which will be spent on SSC experiments.
**546 in FY 85 $.

physicists from the present program to the program including the SSC. We expect that in addition to the U.S. physicists, there will be a substantial number of non-U.S. experimentalists working at the SSC. This analysis shows that the existing high energy physics manpower, with an anticipated 10% growth over the next decade, can accomplish the program described here.

In summary, the major points are the following:

i. A budget slightly above the FY 1985 level would be adequate to permit effective research with the new facilities at present being completed and would allow the needed R and D for the SSC in the period prior to the initiation of construction.

ii. During the SSC construction phase, there will be needed an increment, or pulse, in capital funds for construction and detectors.

iii. After the SSC is in operation, a productive program can be achieved with a funding level in constant-year dollars not much larger than that prior to SSC construction. This is possible because some of the present facilities will have reached maturity by then and there can be an orderly transition from these activities to research on the SSC.

These points are shown schematically in Figure 25, with budget figures given in constant 1984 dollars.

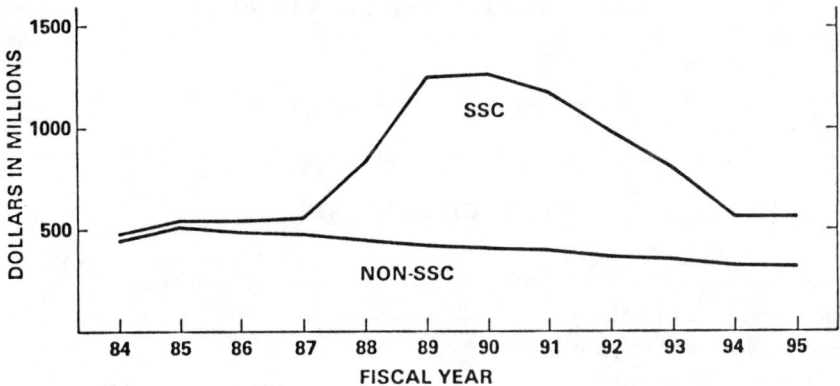

Figure 25. High energy physics funding scenario (FY 84 dollars) for the DOE-supported U.S. High Energy Physics Program. Indicated is the base program (including operation of major facilities in existence or under construction, advanced technology R and D, and the facility user groups at universities and laboratories) and the funding for the SSC including preconstruction R and D, R and D in support of construction, construction costs, pre-operating costs, and the design and fabrication of the initial complement of detectors.

CERN AND HIGH-ENERGY PHYSICS

H. Schopper
CERN, Geneva, Switzerland

TABLE OF CONTENTS

1 Introduction 75
2 Main Activities of CERN 76
3 The p$\bar{\text{p}}$ Program and Possible Improvements 79
   3.1   Possible p$\bar{\text{p}}$ Improvements 83
4 LEAR 85
5 The LEP Machine 87
6 LEP Experiments 94
7 Future Options for LEP 95

0094-243X/85/1340074-24$3.00 Copyright 1985 American Institute of Physics

CERN AND HIGH-ENERGY PHYSICS

H. Schopper
CERN, Geneva, Switzerland

1 INTRODUCTION

CERN was founded in 1954 when 12 European countries joined forces to allow European physicists to participate in research on elementary particle physics in a competitive way. The Member States today are Austria, Belgium, Denmark, France, Germany, Greece, Italy, the Netherlands, Norway, Sweden, Switzerland, and the United Kingdom; Spain will join in 1983. Although formally CERN is a European laboratory, with respect to its physics program it has become international, as may be seen from Fig. 1, a map of the world on which each dot indicates a university or national laboratory participating in the CERN program. Small dots indicate some individual scientists participating in the program; larger dots denote more formal collaborations. Besides the CERN Member States, many dots can be seen in the USA, representing about 200 physicists who work at CERN. There are also smaller numbers from the Soviet Union, Eastern Europe, China, Japan, India, and elsewhere.

Fig. 1. World map showing location (red dots) of universities and laboratories participating in CERN research program.

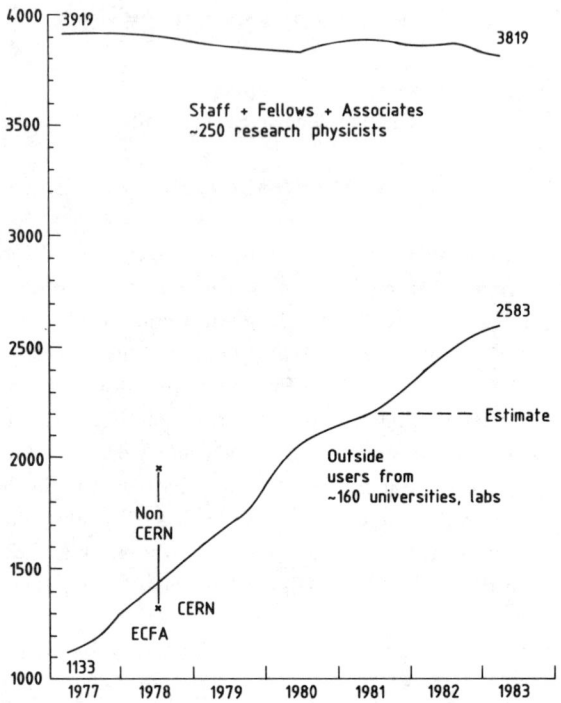

Fig. 2. CERN personnel numbers 1977–1983.

CERN has a total staff of about 3500 people, of whom only about 100 are staff research physicists, both experimental and theoretical. In addition there are about 250 Fellows and Associates paid by CERN. The great majority of the research work is carried out by some 2600 scientists (users) from universities and national institutes or laboratories. Figure 2 shows the evolution of these numbers over the past few years, and in particular the striking increase in the number of users. In 1981, at the beginning of the LEP (large electron–positron storage ring) construction period, it was thought that the number of users would level off; in fact it turns out that the attraction of CERN for physicists from non-Member States and also from non-high-energy physics parts of the community [in particular low-energy physicists interested in LEAR (the low-energy antiproton ring)] was considerably underestimated.

2 MAIN ACTIVITIES OF CERN

The main research activities of CERN are summarized in Table I, which is self-explanatory. Figure 3 shows the location of the different accelerators on the CERN site. One remark is relevant here. The PS (proton synchrotron) is more or less a copy of the Brookhaven AGS (alternating-gradient synchrotron), whilst the SPS (super proton synchrotron) is

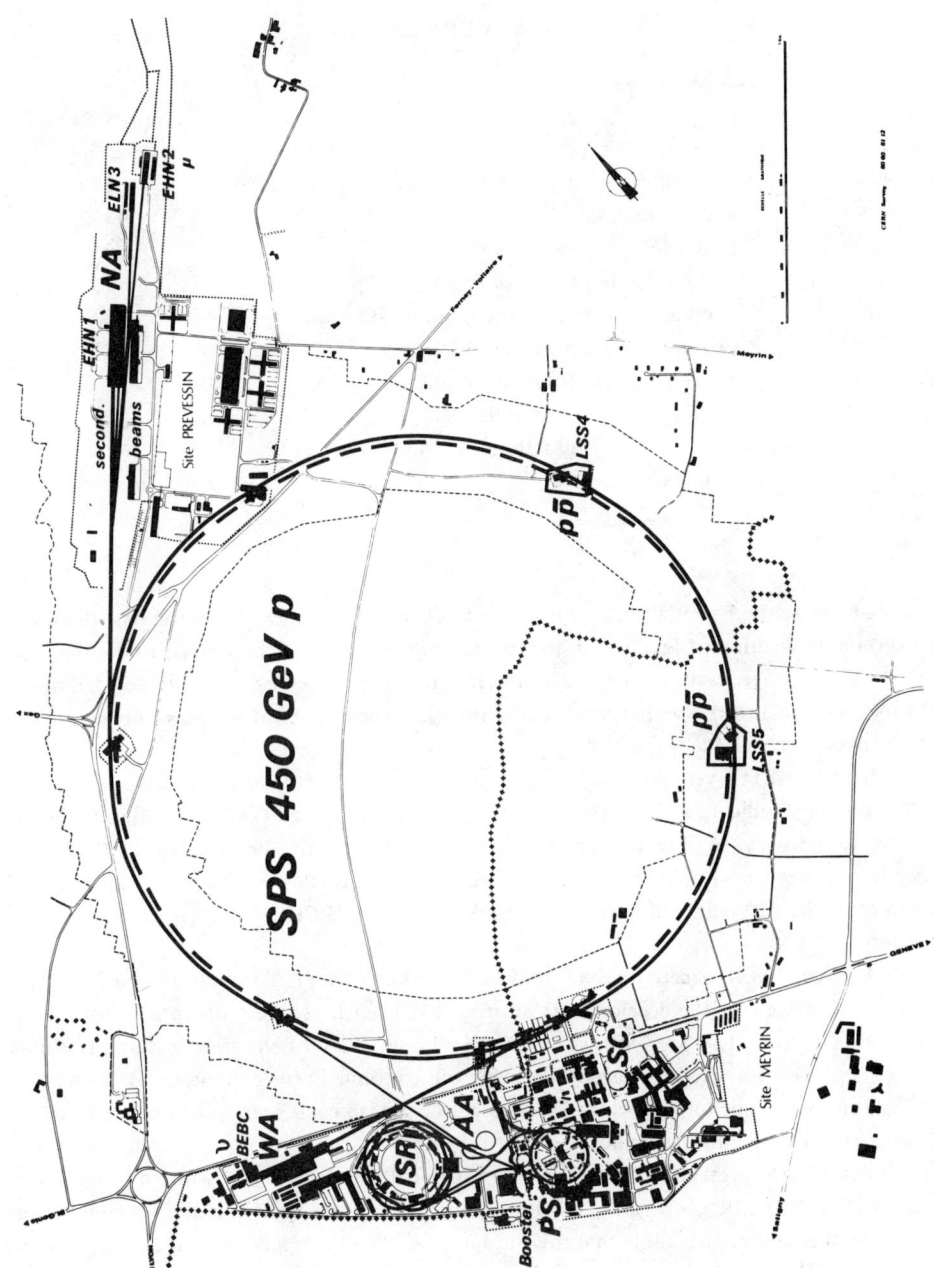

Fig. 3. Plan of present CERN site showing accelerators and experimental areas.

Table I CERN activities

		Approx. No. of physicists involved
SPS	p$\bar{\text{p}}$ collider 2 × 270 GeV (highest energy available)	250
SPS	Fixed-target 450 GeV	1100
LEAR	Low-energy antiprotons	260
SC/ISOLDE	(ISOLDE 3 in preparation) 600 MeV p	150
ISR	Collider, 2 × 31 GeV (high luminosity, p$\bar{\text{p}}$) (will be closed down end of 1983)	270
	Miscellaneous: PS ν oscillations proton decay gravitational waves	(few)
LEP	Four experiments	over 900

very similar to the Fermilab accelerator, which illustrates the fact that in the past there was a certain duplication of facilities in the two regions. Now we are in a phase of complementarity of the programs in the United States and in Europe, because Fermilab decided to upgrade their synchrotron with a superconducting ring (the Tevatron) whereas Europe decided to go ahead with LEP.

In the past few years the main development at CERN has been the conversion of the SPS into a p$\bar{\text{p}}$ collider, with the two major experiments UA1 and UA2 installed in LSS5 and LSS4, respectively, as shown in Fig. 3. The PS is the injector for the SPS, for the ISR (intersecting storage rings) operation, and for LEAR. This is a considerable worry as the PS is now more than 20 years old, and if it broke down the whole CERN program would come to a stop.

The year 1983 is exceptional not only because the $W^\pm$ and Z^0 were found but also from the point of view of the accelerators; all the CERN machines were operating more hours than ever before. The SPS will work for 5600 hours, 40% of the time in the p$\bar{\text{p}}$ collider mode and 60% for fixed-target physics. This will continue in the coming years as one long block of p$\bar{\text{p}}$ and one long block of fixed-target operation each year, which seems to be the most efficient way of functioning.

The PS is at present operating for more than 6500 hours per year with an efficiency of 91 to 95%. The ISR is operating for its last year and about 3500 hours are scheduled for physics. It is quite remarkable that the rings can be filled to collide protons against protons or protons against antiprotons and the beams can circulate for about a week with one filling, and that within that week an integrated luminosity of about 3×10^{34} cm^{-2} can be achieved. Although the ISR is the only high-luminosity pp colliding machine in the world, and is still producing good physics, it will have to be closed down at the end of the year in order to allow LEP to be built within a constant budget and with a constant (or even decreasing) number of staff. In fact it will even be dismantled as the space in the ISR tunnel is required as assembly space for LEP, particularly for the magnets. However, there will be a few months of one-ring ISR operation in 1984 with a gas jet target to do charm physics before the final shutdown around Easter 1984.

The low-energy antiproton ring LEAR will start its physics runs in August of this year; more details will be given in Section 4 of this report. The SC (synchro-cyclotron) will continue to run, mostly with protons for ISOLDE (isotope separator on line) but also with light ions.

The SPS fixed-target program, in which about 1100 physicists are involved, will continue to be an important element of the CERN program even after LEP comes into operation. The future of this program was extensively discussed at a Workshop held at CERN in December 1982, when it was concluded that interesting things would remain to be done also after the Fermilab Tevatron comes into operation. However, because of the constant budget situation there will be no major upgradings of the fixed-target facilities during the LEP construction period—which in particular means no civil engineering, no extraction of antiprotons from the SPS, and no polarized protons in the SPS. The East Area and half of the West Area have been closed to experiments, and just recently two other painful decisions have been taken, namely: to end the operation of BEBC (the big European bubble chamber) in one or two years time; and to close down the muon area. This is not only to save money and manpower but also to keep the remaining fixed-target program competitive; the idea is to concentrate it into fewer experiments and to have some money available for new activities and new experiments. It has been estimated that, when LEP comes into operation in 1988, about 300 of the 1100 physicists mentioned above will be involved in LEP experiments so that it is indeed reasonable to reduce the fixed-target program somewhat over the next few years. However, some improvements were foreseen. For example, the SPS can now be operated in a routine way at 450 GeV, and the circulating beam intensity has been increased. The West Area has been reconstructed to accept 450-GeV particles instead of 200-GeV as in the past. The Omega spectrometer has been improved, and a new spectrometer for beauty physics, using mainly existing equipment, has been installed. In addition, some new experiments have been proposed and are under consideration. Lastly, it may be of interest to mention that an experiment has recently been approved that will use oxygen ions accelerated in the PS; these ions could also eventually be transferred and accelerated in the SPS. The ion source and preaccelerator will be built by the Gesellschaft für Schwerionenforschung in Darmstadt in collaboration with Berkeley and some other US laboratories; it should be ready in two to three years time. It is too early to say whether or not this kind of physics will become part of the future fixed-target program.

3 THE p$\bar{\text{p}}$ PROGRAM AND POSSIBLE IMPROVEMENTS

The way in which antiprotons are produced and used at CERN is shown in Fig. 4. Protons from the PS strike a target to produce antiprotons, which are then accumulated, cooled, and stacked at 3.5 GeV/c in the AA (antiproton accumulator ring). From there they

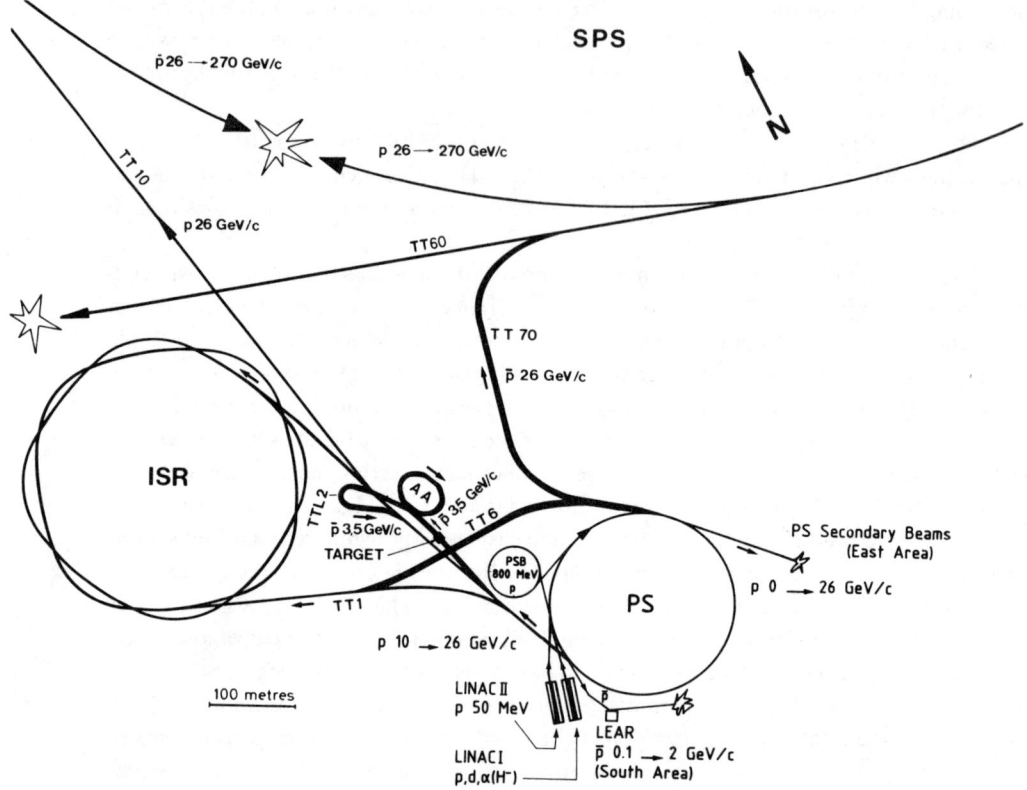

Fig. 4. Plan of the antiproton facility. Antiprotons produced by 26-GeV/c protons from the PS striking a target are 'cooled' and stored at 3.5 GeV/c in the AA. Stacks of antiprotons are then sent to the PS, where they are accelerated to 26 GeV/c travelling in the opposite direction to normal proton acceleration. They can then be transferred to the SPS or the ISR for colliding-beam experiments.

are brought back into the PS, whence they can be sent to the SPS or to the ISR or, more recently, after being decelerated, to LEAR. This very complicated system has had to be consolidated over the past two years because, when the project was initially approved, it was intended to be only a short exploratory experiment. In fact it has become a major program, and now the AA is supposed to operate for between 5000 and 6000 hours per year. After some test runs in 1981 the first physics runs took place in late 1982, followed by a long run from April to the end of June 1983. Figure 5 shows the steady improvement with time; good average luminosities have been achieved regularly, with peak luminosities close to 2×10^{25} $cm^{-2} s^{-1}$ at the beginning of a run. Figure 6 shows the integrated luminosity for the 1983 run; the aim was to achieve 100 nb^{-1} but the final figure was about 153 nb^{-1}. The amount of data collected by the two main experiments UA1 and UA2 is considerable, with about one magnetic tape being written every 15 minutes. The data analysis will keep the physicists

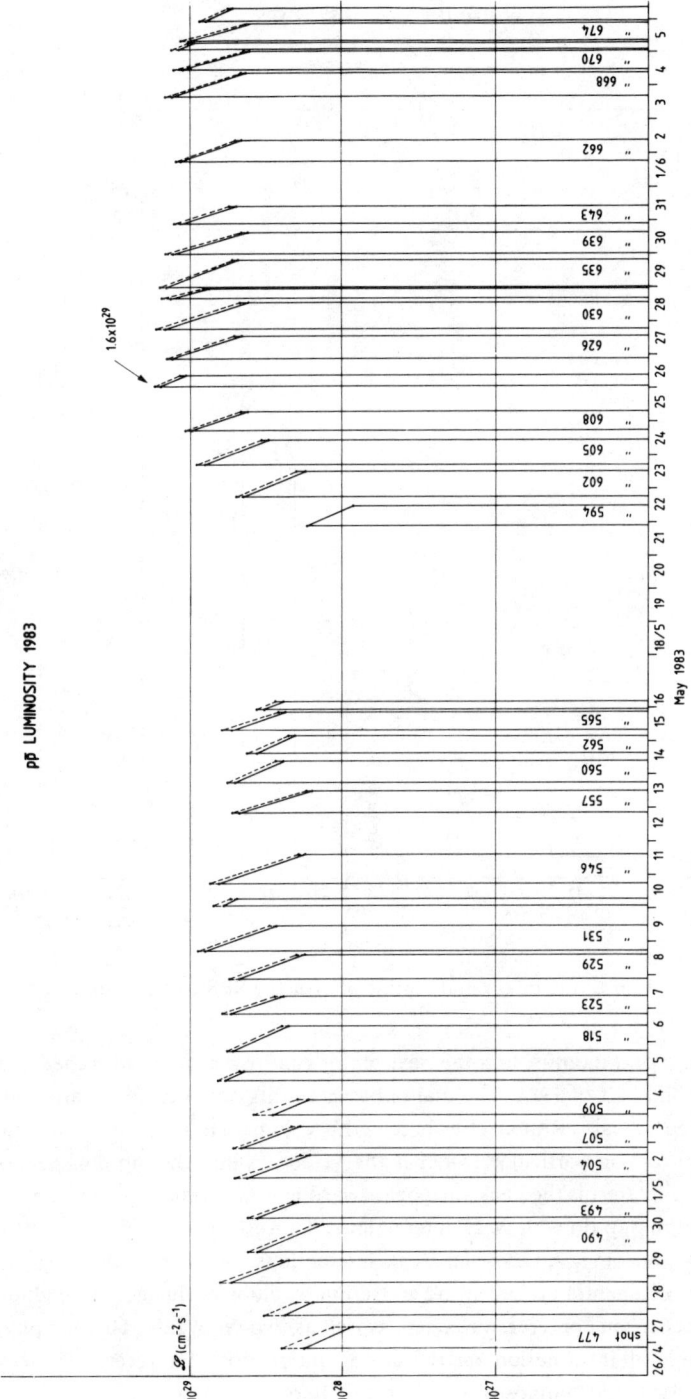

Fig. 5. Histograms of luminosity recorded during the SPS p̄p collider run of May–June 1983.

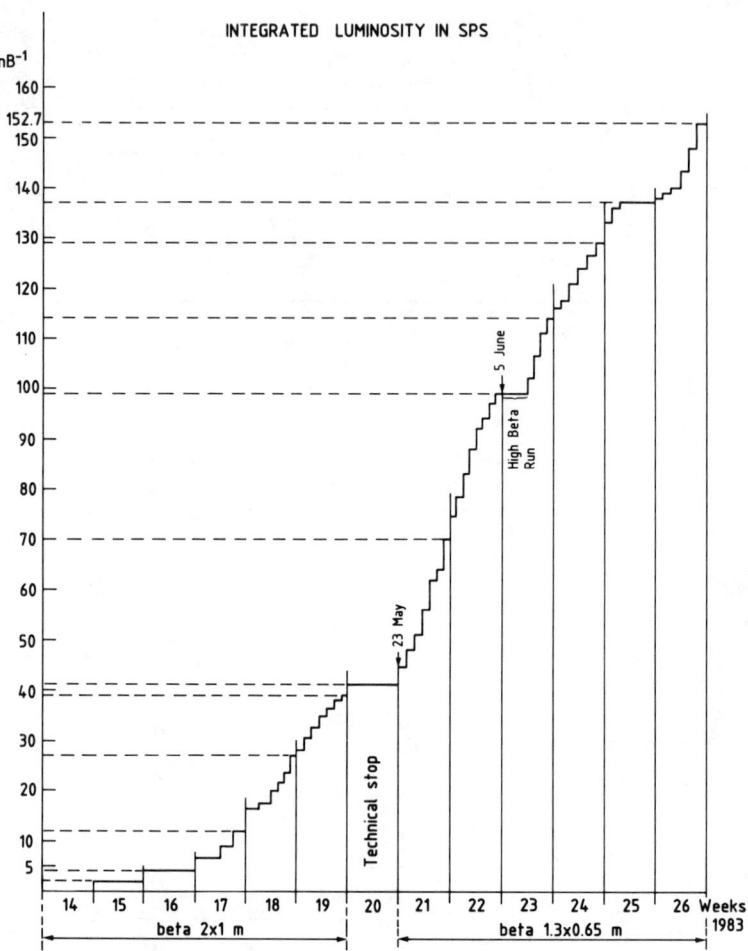

Fig. 6. Plot of the integrated luminosity for the SPS collider run in 1983.

very busy for the next months, and the next major run of the collider is scheduled to take place in the second half of 1984. The main physics results obtained so far are well known, and are described by Carlo Rubbia elsewhere in this volume. However, there are many more things to search for—in particular, whether the W decays into the top plus beauty quark. Another topic of interest is the backward or forward lepton asymmetry to prove that the W really has something to do with weak interactions. This effect has been seen but more statistics would be welcome. Apart from weak interactions, one can study the dynamics of QCD, i.e. the fundamental process of $q\bar{q}$ scattering by gluon exchange. Since gluon bremsstrahlung can occur, one observes three jets as well as two-jet events. This 'jet physics' will provide a wealth of information on the strong interaction. However, all these studies require an increase in the luminosity by a factor of 10 or even more.

Before considering ways and means of increasing the luminosity, one more experiment is of interest. This is the UA5 experiment which uses a streamer chamber to study multiparticle production in pp̄ collisions at 2 × 270 GeV. In order to go to higher energies so as to be closer to the cosmic-ray energies at which the so-called Centaur events have been observed, it is proposed to operate the SPS in a pulsed mode at 2 × 450 GeV. This will probably be done in due course.

3.1 Possible pp̄ Improvements

The present system has reached all its design values or even surpassed them except for the p̄-accumulation rate. However, with some additional effort, small increases in luminosity can be obtained in a number of ways. One way is to squeeze the beams evenly, i.e. to go to lower betas, which already has been partly done. Another possibility is to get longer beam lifetimes, which might increase the average luminosity by 20 to 50%. Another possibility for the continuous operation mode, which requires improvement to the power supplies and to the cooling systems, is to increase the energy from 270 to 310 GeV per beam. This would increase the luminosity because of the gamma dependence, and, even more, the yield of W and Z particles and of jets, since the cross-sections rise sharply with energy. Another variable is the accumulation rate of antiprotons per hour, where small improvements may be made possible by using a focusing target and a lithium lens. In total, what we might hope to obtain from all these factors is an increase in the average luminosity by a factor of 3 to 4. One further possibility is the cooling of the antiprotons in the SPS itself, which would prevent beam blow-up due to the beam–beam interaction. This is also being studied.

To make a big step forward—that is, to increase the luminosity by a factor of 10 or more—we need to improve the collection of antiprotons. At present, the antiprotons produced from the target are collected in the AA ring and cooled in 2.5 s every time the PS produces a pulse. These pulses are stacked in the AA ring at a different position in its vacuum chamber, which takes about 24 h. One limitation is that these two functions are performed simultaneously in the same ring; the system could be considerably improved if they were separated. One possibility under study is to construct a new accumulator ring called ACOL, which would have a large acceptance and a larger momentum bite, and would cool the antiprotons in a few seconds on a pulse-to-pulse basis and then transfer them to the AA ring with improved cooling systems to cope with the larger number of antiprotons to be cooled. In this new system, therefore, the collection and storage processes would be performed in separate machines.

Figure 7 is a plot of the behaviour of average luminosity during the coast as a function of run duration, showing that some improvement in the accumulation rate could be achieved without the introduction of a new machine. Increased energy and further squeezing of the low beta would give an antiproton accumulation rate of 0.5×10^{10} h^{-1}, but it is generally thought that a rate of 10^{10} antiprotons per hour is the limit with the present system.

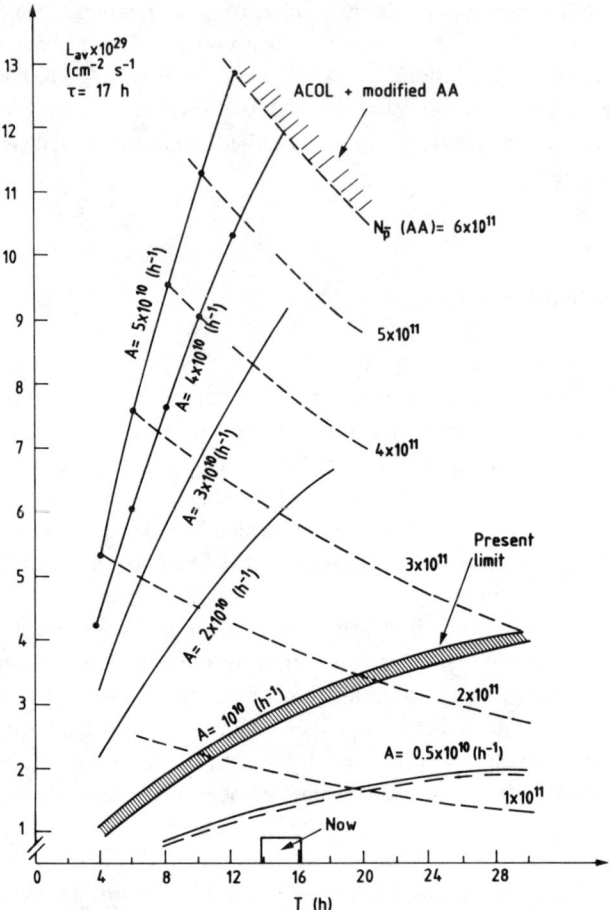

Fig. 7. The behavior of the average luminosity during the coasting period as a function of run duration.

By autumn 1983 it is hoped to complete an assessment of the technical feasibility of ACOL. An upper limit is given by the total number of antiprotons stored in the AA ring, and the figure 6×10^{11} (see Fig. 7), accumulated in bunches, is very close to the number of protons now available in the SPS.

The approach to adopt with respect to the different possibilities described above is now under discussion. We could make a number of small improvements reasonably soon and hope to achieve a factor of 3 increase in luminosity; we could start operation of ACOL, which would certainly give a factor of 10 by, say, 1986–1987. However, ACOL cannot be financed from within the present constant-budget framework with first priority being given to LEP, and whether or not the Member States are willing to give additional money for it remains to be seen.

4 LEAR

The low-energy antiproton ring LEAR is at the other extreme of the energy scale for antiprotons. Antiprotons coming back from the AA ring (see Fig. 4) can be slowed down in the PS to about 300 to 600 MeV/c momentum. They are then injected into LEAR, but by slowing them down the phase space is blown up so that they have to be cooled again. Thus LEAR is also a cooling ring but, at the same time, it is used as a beam stretcher ring. After cooling and stretching, the antiprotons can be ejected and distributed to a number of experiments, as shown in Fig. 8. There are 16 approved experiments on the floor, and the physics program will start in August 1983. About 260 physicists from 45 institutes are involved in the program, forming a new community at CERN. Table II shows the properties of an antiproton beam from LEAR compared with those of a standard antiproton beam as obtained from present machines. The duty cycle is close to one, and in machine tests it was possible to extract the beam over more than 30 minutes, so the beam is almost d.c. There is no pion contamination as they all decay during the cooling, and the momentum definition, owing to this cooling, is better than 10^{-3}. The flux of antiprotons per 2×10^{12} initial protons is more than a factor of 1000 better than present-day beams.

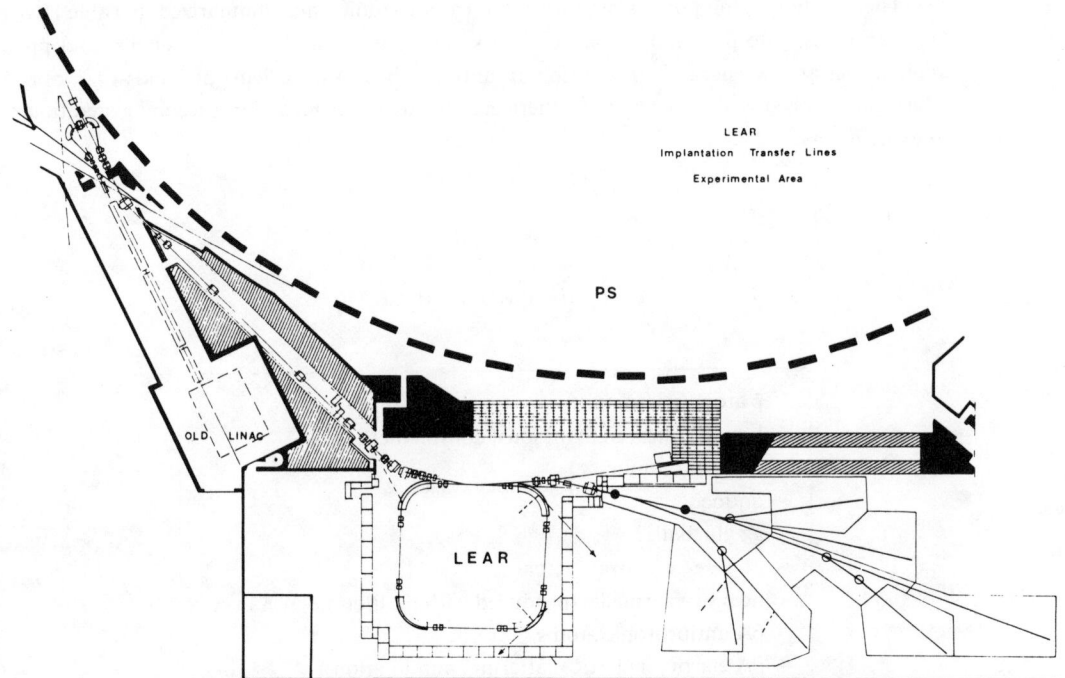

Fig. 8. A plan of the Low-Energy Antiproton Ring LEAR and its experimental area.

Table II Beam quality of antiprotons at 400 MeV/c

	Standard p̄ beam	LEAR	
Duty cycle	~ 0.1	~ 1	Extraction time > 1 h
π contamination	~ 10^{-2}	~ 0	
Δp/p	~ 1.5×10^{-2}	< 10^{-3}	
p̄ flux per 2×10^{12} p	< 10^3	> 10^6	
ε { divergence	~ 10 mrad	< 1 mrad	
ε { diameter	~ 1 cm	< 0.1 cm	

The widely ranging physics questions that can be studied are summarized in Table III. One can investigate protonium, various quark-antiquark bound states, glueballs, and pp̄ annihilation at low energies. By putting an antiproton into a nucleus, all kinds of exotic nuclei can be produced. Furthermore, there are already many ideas for a second generation of experiments.

Table III Physics with LEAR

- p̄ atom (protonium)
- bound states, light mesons, confinement
- qqq̄q̄ baryonium } do they exist?
- qqqq̄q̄q̄
- gg glueballs?
- pp̄ → e^+e^- form factor
- heavy hypernuclei (produced with p̄), lifetime of Λ in A
- p̄A antiprotonic atoms
- p̄A elastic, inelastic scattering, annihilation

Physics will start in August 1983.

5 THE LEP MACHINE

Let us now consider LEP. Some people have questioned the building of LEP now that the W and Z^0 intermediate bosons have been found, and the suggestion has even been made that LEP should be dropped in favour of a proton ring in the LEP tunnel. However, even with the most optimistic improvements in the $p\bar{p}$ Collider one could only hope to get some 20 to 30 Z^0 events per *year*. LEP Phase I will be able to produce 10 000 Z^0 events per *day*. The Z^0 is a source of all kinds of particles that are lighter than the Z^0 itself. The heaviest particle we knew of before 1983 was the b quark with 5 GeV, and a whole new mass range up to 90 GeV is available with the Z^0. There are many things to be looked for, such as supersymmetric particles, heavier leptons, Higgs particles, and so on. The LEP machine is really a Z^0 factory to produce all these kinds of particles, if they exist—and, of course, one hopes also for surprises. To go, after Phase I, to energies in the region of 100 GeV per beam will also be very interesting. The W pair production threshold is now known to be at about 80 GeV per beam. This process will allow the gauge field character of the theory to be checked, since it involves the three-boson vertex $Z^0 \rightarrow W^+ W^-$. There is a good chance that some of the new particles will be produced in association with a Z^0, requiring higher energies than in Phase I. Such particles could decay into others with relatively low p_T and hence might be very difficult to detect at pp and $p\bar{p}$ machines.

The definition of LEP is that it is an e^+e^- ring optimized for about 100 GeV per beam; 50 GeV per beam is LEP Phase I, but the dipole magnets can take beams up to 125 GeV. With 100 GeV as the optimum energy, a circumference of about 26.7 km is obtained. The ring will be 80 to 125 m underground. The present accelerators, both the PS and the SPS, will be used as injectors. In Phase I, four interaction regions will be equipped with experiments.

The cost of LEP is 910 million Swiss francs, or roughly 460 million dollars, at 1981 prices. The project was approved unanimously by the Member States in December 1981, with the condition of a constant budget and constant staff during the construction period. Figure 9 shows a map of LEP in relation to the present CERN site. It is quasi-tangential to the SPS so that the latter can serve as an injector with short transfer tunnels. There will be eight access shafts, as indicated by the numbers, and the first four experiments will be installed at the even-numbered places. The rf accelerating cavities will be installed at points 2 and 6. Later it would be easy to open up interaction regions at points 1 and 5, whereas this would be a little more difficult at points 3 and 7 because, for various reasons, short access tunnels have to be built.

Figure 10 shows a simplified geological section across a diameter of the ring from the Jura to the Geneva airport. Most of the LEP tunnel will be excavated in molasse sandstone, which is very good for tunnelling and in which the SPS was built. Unfortunately the roof of this rock slopes down towards the airport, and there is moraine, gravel, and sand above, which is bad for tunnelling. Hence, in order to stay in the molasse, the tunnel has to be rather deep underground. In the opposite direction the limestone of the Jura rises; this limestone could contain cracks filled with water, with much rock above. There could be

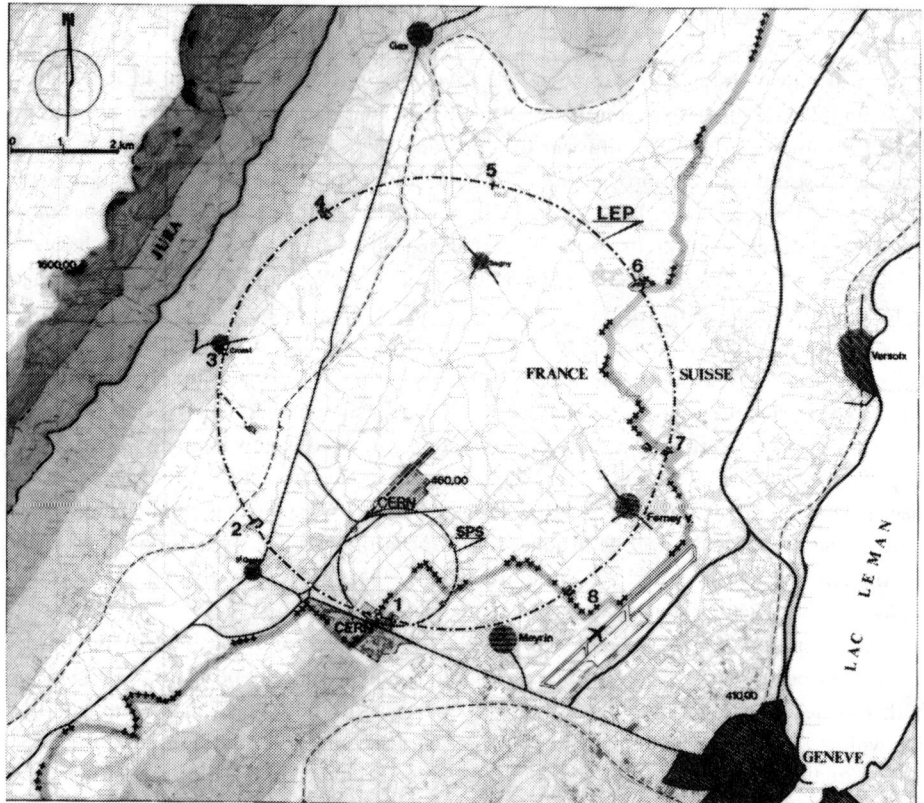

Fig. 9. Map of the Geneva region showing the position of LEP in relation to the present CERN site.

rather high water pressures, which would be a nuisance when tunnelling; therefore, the aim was to have not more than 150 m of rock above the tunnel. In order to achieve these different requirements LEP is designed to be built on an inclined plane; it is probably the first accelerator to be built in such a way. The slope is about 1.5%, which seems negligible but is not. For example, the water pressure difference from one side of the ring to the other would be 15 atm, which has to be taken into account, as well as the fact that some of the experiments will be on an inclined plane.

The tunnelling will be done in the following way (see Fig. 11). From point 8, two tunnelling machines will start in opposite directions. The molasse can be excavated with full-face boring machines. The part of the tunnel that is in hard limestone (lot C) will be excavated in the conventional way, with explosives.

Figure 12 shows details of an interaction region which is one of the more complicated ones, since it has a large experiment and one rf station. There is an underground hall for the experiments, and a shaft for lowering the apparatus for assembly in the 'garage' position

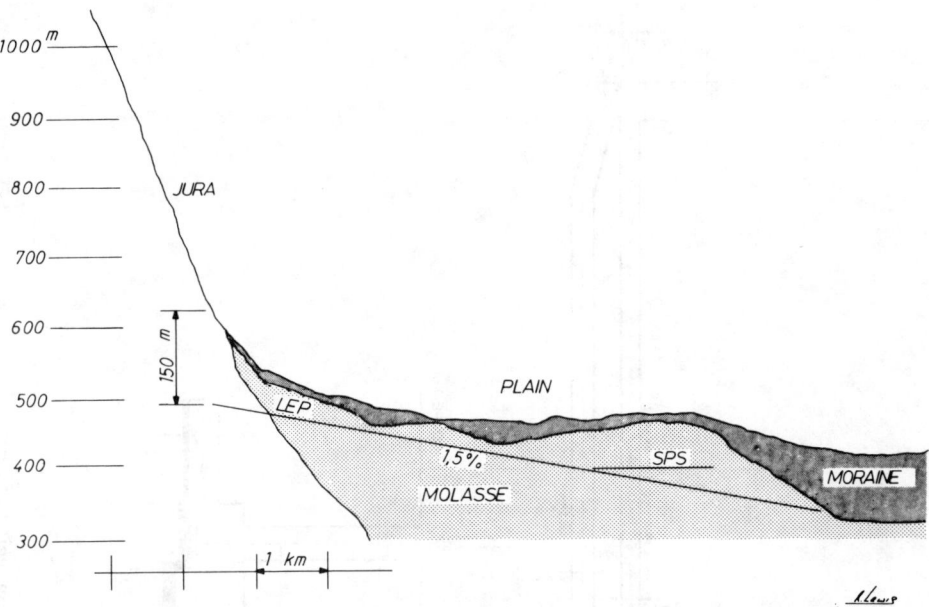

Fig. 10. Sectional view of the LEP site. Thanks to the inclination of the plane of the machine, the tunnel passes almost entirely through molasse rock favourable for drilling operations.

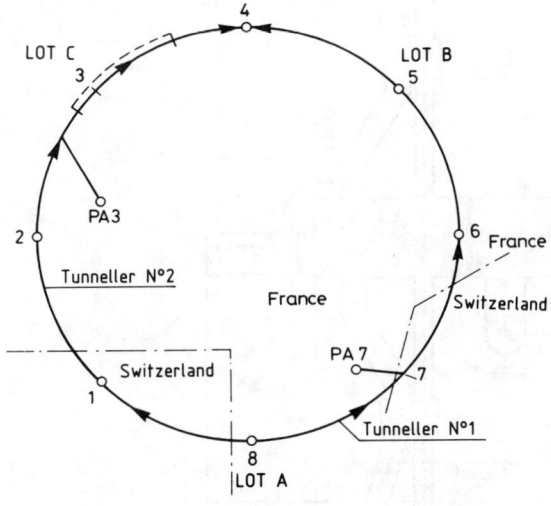

Fig. 11. Proposed plan of the LEP tunnelling operations.

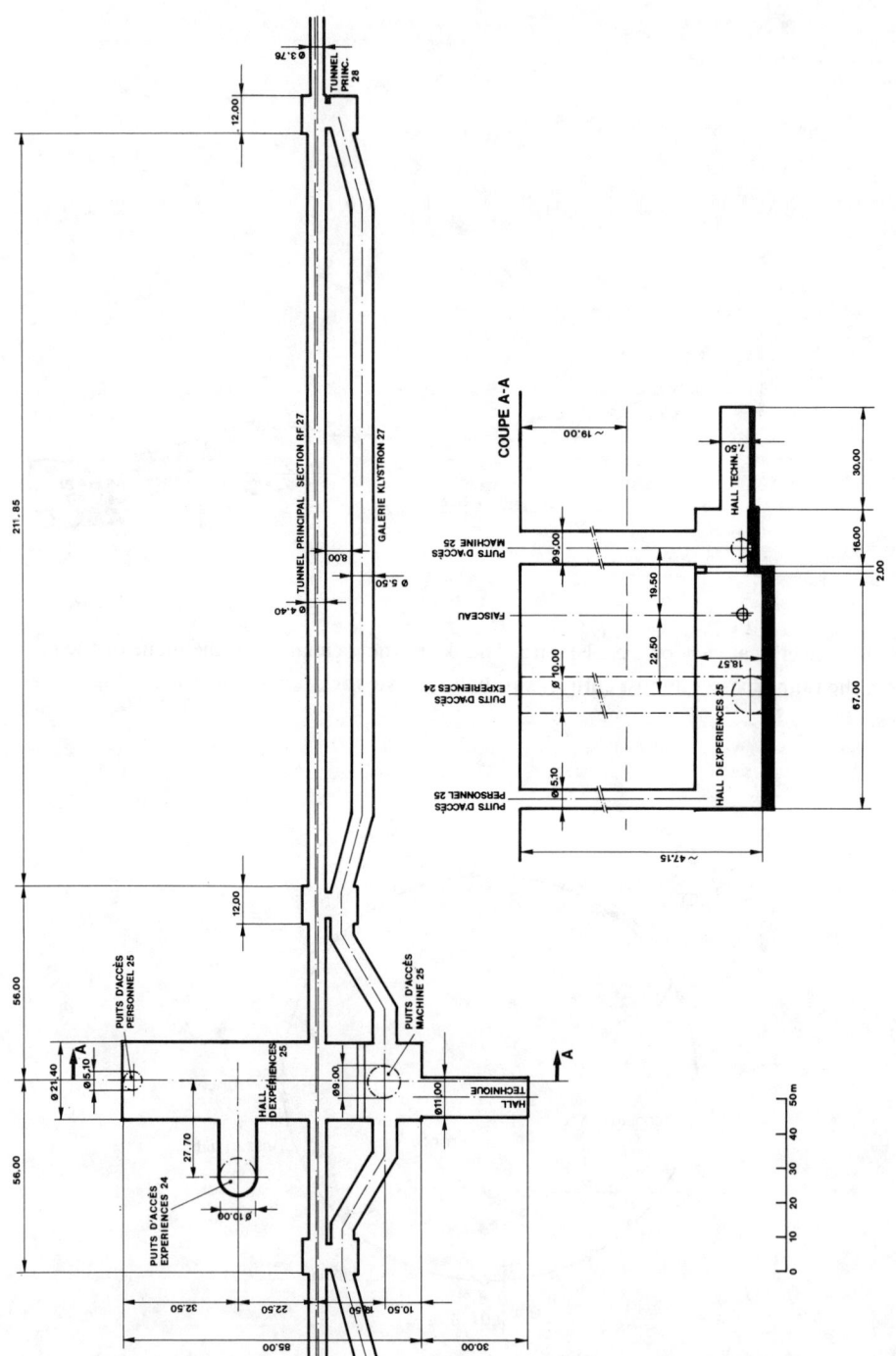

Fig. 12. Plan of an underground interaction region of LEP.

before moving it into the beam. There is a second shaft for the machine components and a small shaft for personnel. The galleries for the rf klystrons are also shown.

The injection system is shown in Fig. 13. There are two linacs; one is a 200-MeV high-current linac to produce electrons which will then produce positrons from a target; the second linac will accelerate both electrons and positrons to 600 MeV, and this is followed by an electron-positron accumulator (EPA). The electrons and positrons are then accelerated in the PS to 3.5 GeV, after which they are transferred to the SPS to be accelerated to 20 GeV before finally being injected into LEP. The two linacs are being built in collaboration with LAL, Orsay, France.

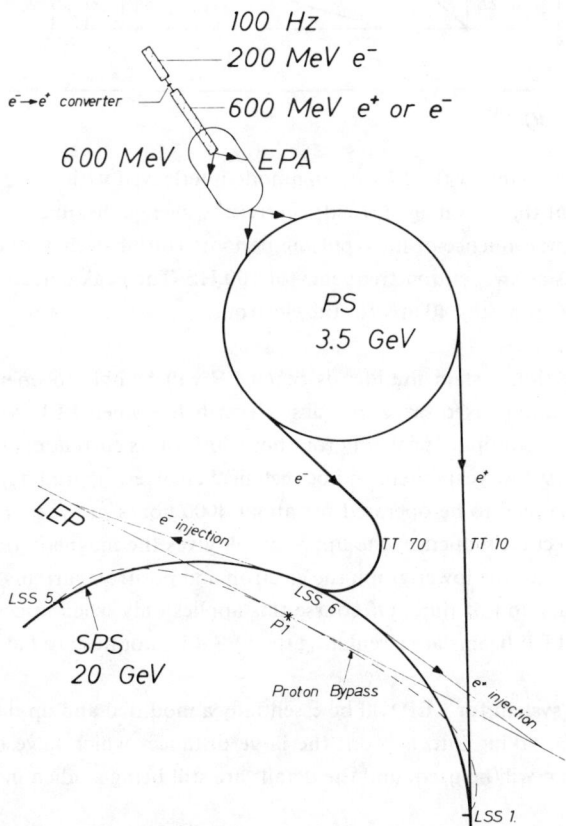

Fig. 13. Schematic diagram of the LEP injector chain. The low- and high-intensity linacs in tandem feed electrons and positrons into an Electron-Positron Accumulation Ring (EPA) whose circumference is one fifth that of the PS, the next stage of the pre-acceleration process. Before injection into LEP the particles are accelerated to 20 GeV in the SPS.

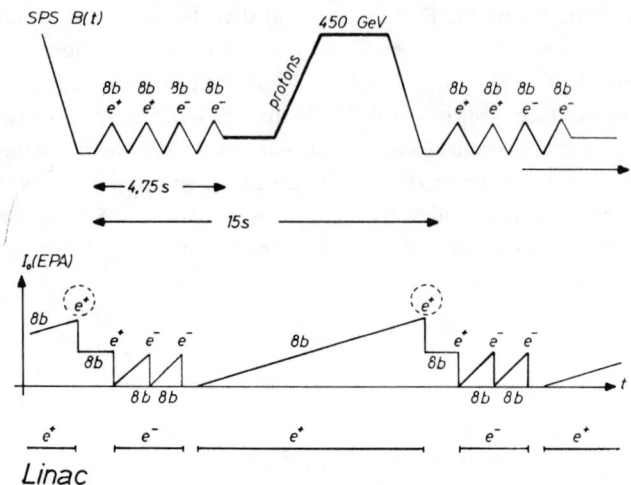

Fig. 14. Basic scheme of the LEP filling mode interleaved with proton operation. Top: a 15 s supercycle of the SPS magnet field; centre: the average beam current in EPA (8b = 8 bunches); bottom: sequence of linac pulsing periods. During each period the linac produces 12 ns long pulses at a repetition frequency of 100 Hz. The peak currents during a pulse are 8 mA for the positrons and 40 mA for the electrons.

For the injection system the idea is that LEP will be able to operate simultaneously while the SPS is doing fixed-target physics. Nevertheless when LEP comes into operation the p$\bar{\text{p}}$ Collider will continue, probably for about 40% of its current operating time, since in any case in the first few years there will be technical changes, upgrading, and completion of LEP, which is planned to be operated for about 4000 hours per year. Figure 14 shows the details of the injection sequence: the upper graph gives the magnetic field of the SPS as a function of time, and the lower graph the electron and positron currents in the accumulator ring, also as a function of time. Of course this applies only when injection is taking place into LEP; while LEP beams are circulating the SPS does not need to have such a complicated supercycle.

The control system for LEP will be essentially a modified and up-dated version of that used for the SPS, taking into account the large distances which have to be covered. The same control room will be used, and the details are still being studied in collaboration with industry.

One major difficulty will be the infrastructure; there are only eight access shafts with 3.5 km of tunnel between each of them. The solution adopted for the transport of both components and people is a monorail system suspended from the ceiling of the tunnel. In order to speed up the installation in the tunnel there will be various devices for transporting long items, which will be preassembled at the surface as much as possible; for example, one unit of 14 m length would be two bending dipoles with a common vacuum chamber. Figure 15 shows two such units.

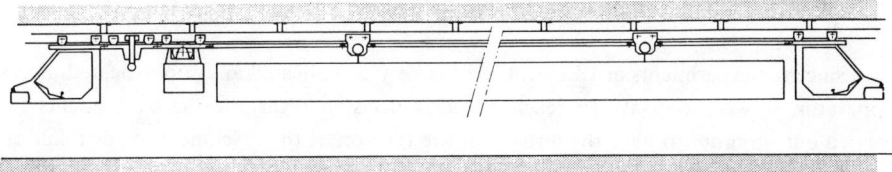

Une motrice: 3750 daN. à 12 km/h. ou 7500 daN. à 6 Km/h.

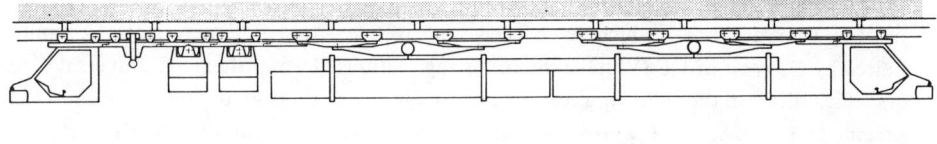

Deux motrices: 7500 daN. à 12 km/h. ou 15000 daN. à 6 km/h.

Fig. 15. Monorail system for the LEP showing transport of dipole magnets.

The installation schedule has been planned in considerable detail. The civil engineering contracts for the whole tunnel, both in the plain and in the Jura, have been signed. It will take about a year—or a little longer—to make the access shafts, which will be between 80 and 120 m deep. To complete the shafts and the tunnel should take 48 months as from summer 1983; during this time there are 21 fixed contractual dates to be met by the civil engineering firms. Tunnel installation (power, light, etc.) should start in spring 1986, and the installation of ring components early in 1987. Injection tests should be started in autumn 1987 with the aim of a first start-up of LEP at the end of 1988. The construction of the injector building for the linacs, adjacent to the PS, is already well advanced. Contracts for the main magnet and rf systems have been placed and, in fact, the total value of all the contracts placed so far, including those for the civil engineering, is some 450 million Swiss francs, i.e. about 50% of the total cost of LEP (1983 prices). All prices thus far are within the estimates made, and it is hoped that the same will be true for the remaining contracts.

Some of the unique features of LEP, such as the 'concrete' dipole magnets and the storage cavities for the rf accelerating cavities, have been reported on at recent accelerator conferences and will not be considered here. Work is being actively pursued on superconducting cavities, and a five-cell niobium cavity was tested at PETRA in the spring of 1983 and worked quite well. By the time of the turn-on of LEP it is hoped to have four or five rf superconducting cavities in place in LEP so that experience can be gained in their use. Any increase of the energy of LEP beyond 50 GeV per beam will probably be achieved by using such superconducting cavities, as the results of the present development program are rather encouraging.

6 LEP EXPERIMENTS

Since the experiments at LEP will require very complicated and sophisticated pieces of apparatus, it was necessary to reach an early decision regarding the experiments to be carried out, in order to have them ready at the turn-on of the machine. One difficult question that had to be discussed was whether to fill four intersection regions right from the beginning or to reserve some for unforeseen things. The decision was to fill all four areas because the four accepted experiments are thought to be sufficiently diverse both in physics aims and/or in detection techniques. Furthermore, if necessary, two more areas, 1 and 5, could be opened up.

The four detectors, which have been approved by a LEP Experiments Committee chaired by G. Wolf of DESY, have the following features. Two of them are universal detectors which differ in the following respects: OPAL is more or less an upgrading of the JADE detector at PETRA, has a warm coil, and uses well-proven techniques. ALEPH is also a universal detector but uses some novel techniques; in particular it has a very large time projection chamber (TPC), much bigger than the one built for PEP. It might, in the long run, be a more powerful detector but much more risky than OPAL. Then there are two specialized detectors. One is L3, which is specialized in calorimetry. It is planned to have a very powerful electromagnetic calorimeter using the new material BGO (bismuth–germanate), and a very fine grain hadron calorimeter; the momenta of muons will also be very precisely measured. DELPHI is specialized in the identification of hadrons and uses the new technology of so-called ring imaging Cherenkov (RICH) counters. As an example of the size of these detectors, Fig. 16 shows a sketch of ALEPH.

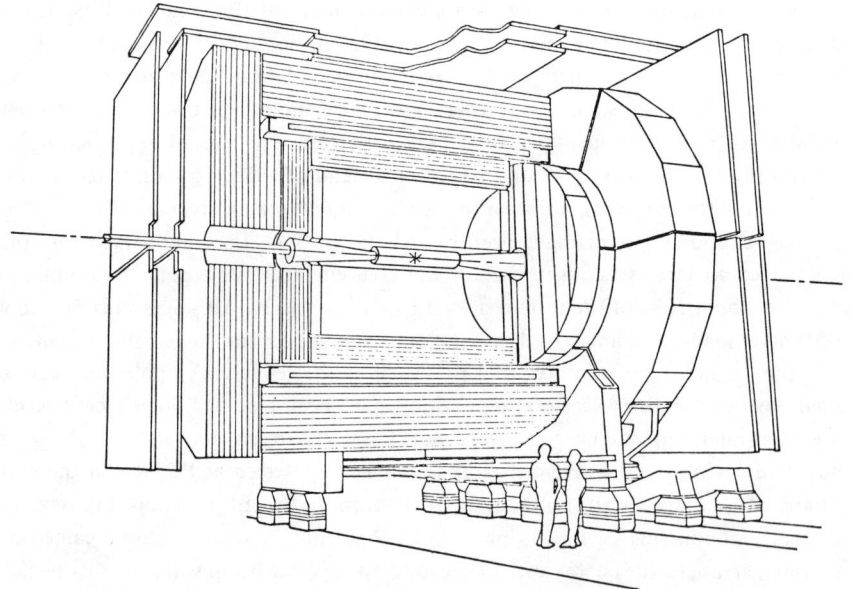

Fig. 16. Artist's impression of the ALEPH detector for LEP.

The total number of physicists and engineers involved in these four experiments is now close to a thousand. Table IV gives some details. The total cost of the experiments is estimated to be about 320 million Swiss francs; Member States will contribute about 140 million, CERN about 50 million (plus 10 million for the infrastructure), and about 130 million is expected from non-Member States such as the USA, the USSR, Japan, and smaller countries. The majority of these contributions will, of course, be in terms of equipment, not cash.

Table IV Physicists in LEP experiments

OPAL	142
ALEPH	260
L3	263
DELPHI	268
Total	933
Member States	664
CERN	81
Non-Member States	188
Total	933

7 FUTURE OPTIONS FOR LEP

Phase I of LEP will give 50 GeV per beam, and the addition of superconducting cavities could increase the beam energy to about 100 GeV, for which LEP was optimized. Eventually it could go to 125 GeV per beam, although the rf power required for this energy is very considerable.

Other possibilities are to put a proton ring in the tunnel at a later date, or two proton rings, either for p$\bar{\text{p}}$ or pp collisions. The great advantage is that the tunnel would already exist with its infrastructure, and part of the rf of LEP could be used for a proton machine. In addition, thanks to the ISR, CERN is the only laboratory to have experience with a high-luminosity pp and p$\bar{\text{p}}$ colliding beam machine.

There are two extreme possibilities. One approach would be to try to build a fast and 'cheap' machine of relatively low energy, using present technology and NbTi magnets of about 5 T which would give 4.5 to 5 TeV per beam. By doing some technological development work we might hope to reach 10 T using Nb_3Sn magnets which would give about 9 TeV per beam.

Earlier it had been foreseen that electrons from LEP would be collided with protons from the SPS. This will not be done, as HERA, at DESY, will be much better in this respect. However, if there were a proton ring in the LEP tunnel, one could foresee collisions between the electrons in LEP and the protons in that ring.

Some basic parameters for $p\bar{p}$ and pp machines in the LEP tunnel, based on preliminary calculations and assuming the same lattice as for LEP, are shown in Table V. For the $p\bar{p}$ case, for which 56 bunches have been assumed as a realistic number, it is necessary to provide some bunch separation. One can separate the two beams on one side of an experimental crossing (using electrostatic means) and arrange them in such a way that the number of bunches matches the betatron wavelength, with one bunch per betatron wavelength. Thus, where the bunches meet they are separated, whereas where the two beam trajectories cross, the bunches are absent, and in this way they always stay separated, as illustrated in Fig. 17.

Table V Hadron collider

Parameter Particles:	$p\bar{p}$	$p\bar{p}$	pp	pp
Dipole field (T)	5 to 5.5	10	5 to 5.5	10
BEAM ENERGY (TeV)	4.5 to 5	9	4.5 to 5	9
Number of bunches	56	56	400	400
Protons per bunch	1.5×10^{11}	1.5×10^{11}	5.1×10^{10}	5.1×10^{10}
Proton current (mA)	15.2	15.2	$(37.3) \times 2$	$(37.3) \times 2$
Antiprotons per bunch	1.5×10^{10}	1.5×10^{10}	-	-
Antiproton current (mA)	1.52	1.52	-	-
Norm. emittance ($\pi\mu$rad·m)	20	20	20	20
β^* at crossing (m)	1	1	1	1
LUMINOSITY (cm^{-2} s^{-1})	1.2×10^{31}	2.2×10^{31}	10^{32}	1.8×10^{32}
Luminosity per encounter (cm^{-2})	1.9×10^{25}	3.5×10^{25}	2.2×10^{25}	4×10^{25}
Number of intersections	8	8	8	8
Revolution time (μs)	89	89	89	89
Injection energy (GeV)	400	400	400	400
Synchrotron Radiation				
Energy loss per turn (kV)	1.6	16.8	1.6	16.8
Power loss (both beams) (W)	27	280	2×60	2×625
Longitudinal damping time (h)	77	13	77	13
Transverse damping time (h)	144	26	144	26

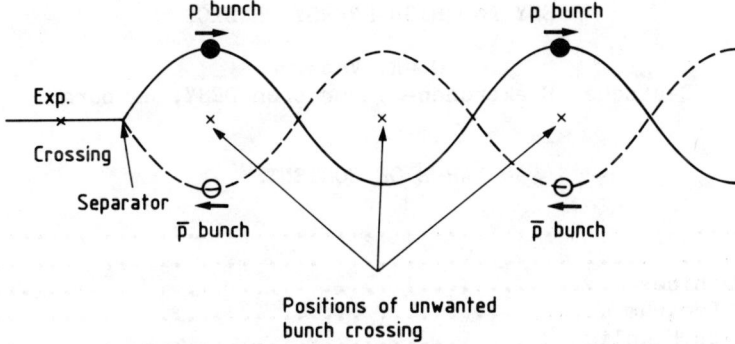

Fig. 17. Separation of proton and antiproton beams for a possible p$\bar{\text{p}}$ machine in the LEP tunnel.

All this is rather preliminary, but in principle it seems possible to separate the beams even at 10 TeV. Quantitatively for example, the separation of the two bunches has to be of the order of 1.5 mm, whereas at injection energy (400 MeV) a separation of about 5 mm is required. This is not very large, and in fact that kind of aperture is needed anyway for the closed-orbit distortion.

The second possibility is to have pp in two rings, where we could learn from the 'two-in-one' ideas developed at BNL. In this case there is no problem of bunch separation, and one could go to 400 bunches for example (see Table V). Thus luminosities of up to 10^{33} cm^{-2} s^{-1} seem possible. However, in Table V it was assumed that the experiments can take only one event per bunch crossing (L per encounter $\leq 4 \times 10^{25}$ cm^{-2}), which limits the luminosity to about 2×10^{32} cm^{-2} s^{-1}.

Whether all these ideas will ever be turned into reality is another matter, but it is clear that for quite some time CERN has interesting programs and interesting future options, and our main difficulty will be to choose between all these possibilities within the constraints of limited budgets and limited manpower.

DESY AND HIGH-ENERGY PHYSICS

G.-A. Voss
Deutsches Elektronen-Synchrotron DESY, Hamburg

TABLE OF CONTENTS

```
DORIS ............................................................. 99
PETRA ............................................................ 102
Other Machines ................................................... 105
Physics Program .................................................. 106
    Physics Highlights ........................................... 106
    Future Program ............................................... 109
New Projects ..................................................... 110
    HERA ......................................................... 110
    DESY-II ...................................................... 116
    RF Superconductivity ......................................... 117
    Wakefield Acceleration ....................................... 118
Discussion ....................................................... 120
```

DESY AND HIGH-ENERGY PHYSICS

G.-A. Voss
Deutsches Elektronen-Synchrotron DESY, Hamburg

I will first talk about the DESY laboratory, then about DORIS I and DORIS II, the present status of PETRA, and some auxiliary machines like the two injection linacs and the positron accumulator. After that, I would like to say a few words about the physics program of DORIS and PETRA; then about the new machines, HERA and DESY II (a second electron synchrotron which we are now building); and, finally, about far future projects: superconducting radio-frequency systems and wakefield acceleration.

DESY has a staff of 1000 people and a budget of $56 million per year including operating and investment costs but excluding HERA and new projects. It is financed 90% by the federal government and 10% by the state of Hamburg. DESY was founded in 1959 -- next year we will celebrate our 25th anniversary. The number of particle physicists doing high energy physics is about 400; about 180 of them come from abroad.

Figure 1 shows the original 7.5-GeV electron synchrotron which gave the whole laboratory its name. It shows the storage ring DORIS, which I am going to talk about, and PETRA with its four experimental halls, which has radio-frequency areas. Two injection linacs are seen: linac II produces 450-MeV positrons and is connected with the positron accumulator PIA, and there is a small electron linac. DESY serves as injector for DORIS and PETRA. Part of the HERA ring is shown, which we hope will be authorized so that we can start construction at the end of the year. The physics program at the synchrotron itself has ceased to exist; the synchrotron is used only for test beams in two experimental areas, and otherwise it serves only as an injector of positrons and electrons into DORIS and PETRA. There is not much space for expansion: around the site is private property with houses, factories, etc.

DORIS

DORIS originally was a two-ring arrangement, the name standing for <u>do</u>uble <u>r</u>ing <u>s</u>torage apparatus, and the arrangement was as follows: there were two rings, one on top of the other, and two interaction points: the beams would cross vertically. DORIS was built for 3.5 GeV in each beam, with the idea that it could be used to collide electrons with electrons, positrons with positrons, and electrons with positrons. After the upsilon resonance had been found at Fermilab, DORIS was rebuilt as a single-ring machine by keeping the beam in the upper ring halves. After going down at the interaction points, the beam is brought up again so that beams are circulating only in the upper ring halves. The two-ring arrangement had 480 bunches in each ring. The new single ring has only two, one electron and one positron bunch.

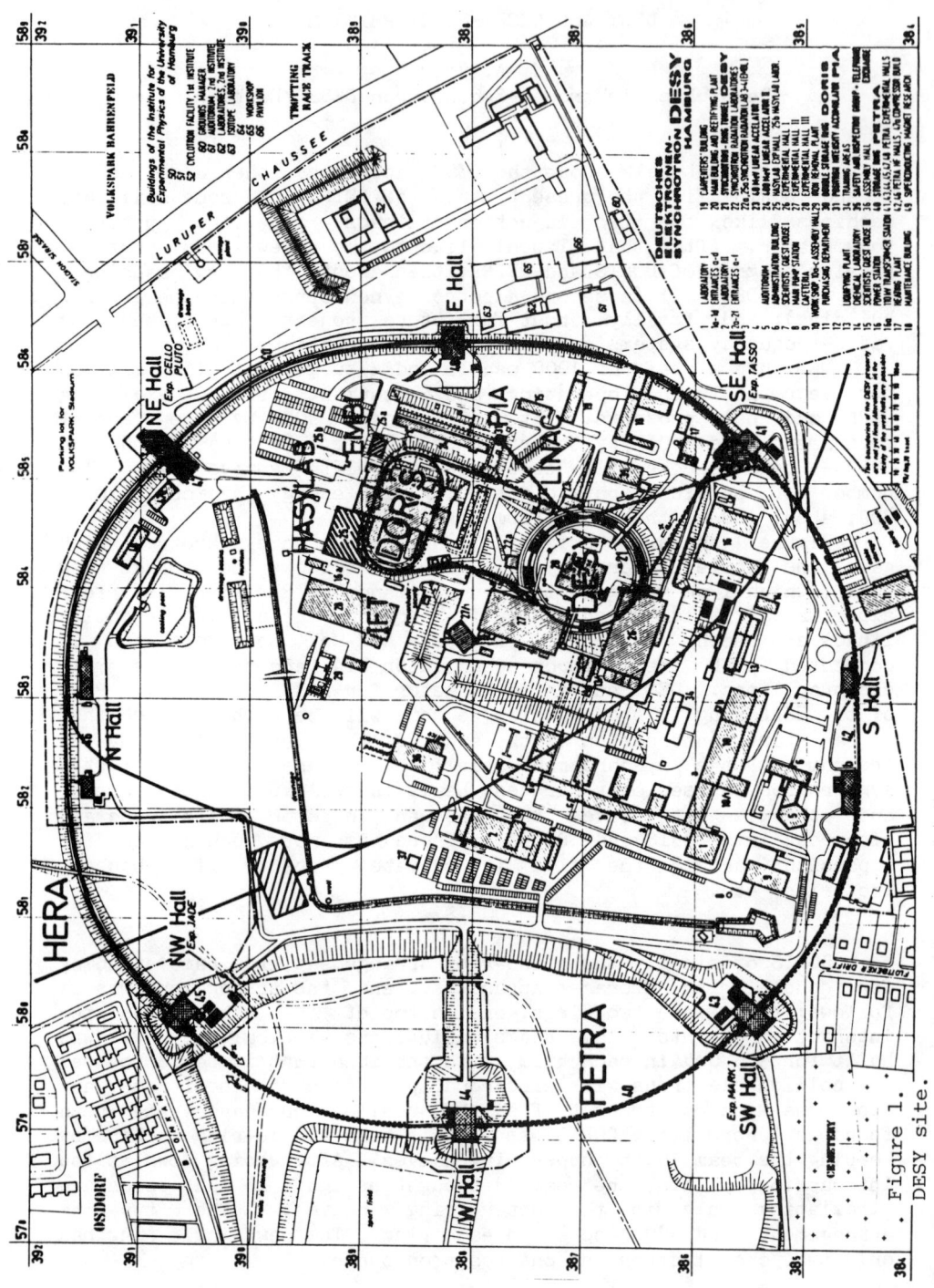

Figure 1. DESY site.

This made it possible to raise the energy of DORIS to 5.1 GeV for each bunch. But this machine turned out to be very power hungry -- after all, it was designed for 3.5 GeV. Also, the luminosity was not very high. Therefore we completely rebuilt this machine into what we call DORIS II, with a circumference of about 288 meters. The rebuilding had three purposes:
1) To increase the luminosity by at least a factor of 10 in the high energy region, to 2×10^{31}.
2) To increase the peak energy, which was limited to 5.1, up to 5.6 GeV so that we would cover the whole upsilon region.
3) To reduce power consumption by at least a factor of 2; it is actually down to 40%.

All rebuilding was done in less than 6 months. The whole ring was taken out and all magnets were rebuilt. Six months later, the machine was back in operation.

We took one magnet ring completely out, took the coils from its magnets, and added these coils to the other magnets. To do this, we put some spacer in the back leg, added the coil, and added pole pieces. The new pole pieces had a more appropriate shape and produced a better field. They also produced a narrower gap, and, by reducing the total magnetic flux, we could get the extra 10% in energy. By adding the coils we cut the power consumption down.

The quadrupoles also were shimmed, and more sextupoles were put in. Next we removed all the single cavities that had originally been used and installed, instead, 10 PETRA cavities, which have a much higher shunt impedance. Thus we could produce higher voltages with much less rf power. The vacuum system was completely rebuilt. We know now that the vacuum chambers have to be very smooth to avoid longitudinal and transverse field excitation by the circulating bunches. We also rebuilt the injection channels such that beams can now be injected at the operating energy, whatever it is, up to 5.6 GeV.

Two new detectors were installed in DORIS. In the Argus detector, we installed a mini-beta arrangement: quadrupoles very close to the interaction point so that the beta values at the interaction points can be greatly reduced without unduly increasing the chromaticity of the machine. Now, by moving quadrupoles very close, we are locating magnets almost in the detector field. We therefore have compensating coils to keep the quadrupoles free of the detector field, and this seems to work nicely. In the other interaction region, we have the crystal ball from SLAC.

The present parameters of DORIS are listed in Table I. The maximum luminosity we have reached is 1.8×10^{31}, but what is more important is the average luminosity per day. Since we can inject very fast -- in <30 sec we can fill the ring with electrons and positrons -- integrated luminosity has been as high as 1045 inverse nanobarns per day for each interaction region. The filling rate is about 3 mA/sec. We have had a current of 50 mA, but typically we now run with 30 mA/bunch. An interesting feature of this machine is that beams are strongly polarized and it is not so easy to destroy the polarization. This is the Sokolov-Ternov mechanism of polarization produced by synchrotron radiation, and we think it

Table I DORIS II Parameters

Circumference	288 m
Bending radius	12.2 m
Q_x, Q_z	7.21, 5.16
Max. energy	2x5.6 GeV
Injection energy	= operating energy
Emittance at 1 GeV	2.5×10^{-8} π m
β_x/β_z at iP (mini-beta quadrupoles!)	0.6 m/0.04 m
Max. luminosity (at 5.1 GeV)	1.8×10^{31} cm^{-2} s^{-1}
Max. accumulated luminosity/day	$2 \times 10 45$ nb^{-1}
Filling rate	3 mA/s
Max. current: single bunch	50 mA
multibunch	120 mA
Lifetime: 60-mA multibunch	>12 h
2x35-mA single bunch	>3 h
Polarization	P = 82 ±10%
Exact energy calibration of Y and Y' energies through artificial depolarization	

could be close to 92%, the theoretical maximum. We have managed to depolarize electrons in this machine by exciting depolarizing resonances, and we plan to report, at the Photon Conference, on the exact energy value of the T' as determined by the depolarization process. Two-thirds of the DORIS running time is given to high energy physics and one-third to dedicated synchrotron radiation use.

The new Hamburg Synchrotron Radiation Laboratory (HASYLAB) has 36 stations where experiments can be done. DORIS is a very powerful x-ray machine because the radius of curvature is only 12 m. At 5.6 GeV, we have 30 kW of synchrotron radiation with a characteristic wavelength of half an ångstrom. Two other synchrotron radiation laboratories are attached to DORIS. One is run by the European Molecular Biology Organization (EMBL), and the other by the "Fraunhofergesellschaft," a group doing industrial research, mostly applied physics.

PETRA

PETRA has been in operation for 5 years. It was turned on July 17, 1978, after a construction time of two years and 8 months and built at a cost of $40 million (not including staff salaries). Table II is taken from our proposal, with a column added to show corrected numbers as they are today. The planned energy was 5 to 19 GeV, but with our energy upgrading program we have actually reached (and taken data at) 22.5 GeV. We have not reached the hoped-for maximum luminosity of 10^{32}. We have reached only 1.8×10^{31} because beam-beam interactions and space-charge blow-up have been much stronger than anticipated. We have not reached the ΔQ = 0.06 which we (and everyone else) had assumed, but only about 0.03 or even less. The circumference of 2304 m has not changed. The beta

Table II PETRA Proposal

	Original	Corrected
Energy	5-19 GeV	5-22.5 GeV
Max. luminosity (10 m interaction region)	10^{32} cm^{-2} s^{-1}	1.8×10^{31}
Circumference	2304 m	
Bending radius	192 m	
Free length for experiments	15 m	
Focusing structure	FODO	
Q_x/Q_z at 23 GeV	27.14/23.11	
β-values at interaction points	3.00/0.15 m	1.3/0.08
rf frequency	500 MHz	
rf power	4.8 MW	10
Length of rf structure	96 m (64 cavities)	157 157
Single-bunch current (max.)	20 mA ($\sim 10^{12}$ particles/bunch)	6
No. of bunches	2×(1 to 4)	
Injection energy	7 GeV	
Authorization	Oct. 20, 1975	
First stored beams	Sept. 1978	July 1978
Begin experiments	April 1979	Nov. 1978
Costs	<100 MDM	

values are now 1.3 and 0.08 m. With our mini-beta arrangement we tried to regain some of the lost luminosity (by moving quadrupoles very close to the interaction points). In order to get up in energy we increased the radio-frequency power to 10 MW CW (continuous power), and we increased the total length of the rf structure from 96 m to 157 m.

There are now 96 five-cell and 61 seven-cell cavities installed. Before we put in all these cavities, we could reach single-bunch currents of 20 mA, which is 10^{12} particles/bunch, but with these many cavities we have difficulty getting up to 6 mA because the increased higher-order mode excitation of the cavities sets instability limits. We have to inject from the synchrotron at an energy of 7 GeV. This is one of the particular problems of PETRA, that you have to inject a stable beam at a relatively low energy, and then you have to accelerate the beam up to 21.5 GeV without losses. The other problem is excitation of satellite resonances, i.e., coupling resonances between betatron and synchrotron oscillations. There are up to 5 satellite sidebands to the main resonances and they are so strong that once we touch them, we lose the beam. Therefore, when we accelerate from 7 to 21 GeV, we require extremely tight Q-control of betatron and of synchrotron frequencies so as to stay exactly at the right place in the Q-diagram. But this has all been perfected now and we have conditions we can live with.

Table III presents the PETRA performance data. The maximum luminosity was 1.7×10^{31}, maximum luminosity per day was 790 inverse nanobarns. The average is about 54% of this best number. At 11 GeV, luminosities are 8×10^{30}; at 7 GeV, 1.8×10^{30}. We also have polarization in PETRA, but, since PETRA is a much larger storage ring, the polarization does not come automatically, one has to work on it. To get polarization we require an extremely well-corrected vertical orbit and we must particularly tune out those depolarizing resonances which would depolarize the beam. But with this technique we can reach a polarization of single beams of 82%, and with two beams we have actually reached polarizations of

Table III Present PETRA Performance
[after installation of mini-beta (1.3/0.08 m)]

E_{max} 21.5 GeV

E_{max} for colliding beam physics   2×21.5 GeV

L_{max} (at 17 GeV, current limited) 1.7×10^{31} cm^{-2} s^{-1}

∫Ldt / day   790 nb^{-1} ($\bar{L} = 9.1 \times 10^{30}$ cm^{-2} s^{-1}, i.e. 54%)

∫Ldt / 19 days   8908 nb^{-1} ($\bar{L} = 5.4 \times 10^{30}$ cm^{-2} s^{-1}, i.e. 32%)

L_{max} (at 11 GeV, ΔQ limited to 0.025) 8×10^{30} cm^{-2} s^{-1}

∫Ldt / day   306 nb^{-1} ($\bar{L} = 3.5 \times 10^{30}$ cm^{-2} s^{-1}, i.e. 44%)

L_{max} (at 7 GeV, ΔQ limited to 0.02) 1.8×10^{30} cm^{-2} s^{-1}

∫Ldt / day   128 nb^{-1} ($\bar{L} = 148 \times 10^{30}$ cm^{-2} s^{-1}, i.e. 82%)

∫Ldt / 25 days   1942 nb^{-1} ($\bar{L} = 9 \times 10^{29}$ cm^{-2} s^{-1}, i.e. 50%)

($L \sim E^{2.5}$, very roughly)

Polarization:
     Single beam $0 < P < 82 \pm 10\%$
     Two beams $0 < P < 60\%$ with $L \leq 3 \times 10^{30}$ cm^{-2} s^{-1}

Exact energy calibration through artificial depolarization

60% in the colliding beam mode at a luminosity of about 3×10^{30}. Also, we have used the depolarizing method to get a very accurate energy calibration of PETRA. To determine polarization, one measures the vertical asymmetry of Compton back-scattered photons from a laser. It is amazing that the accuracy with which the energy can be determined is of the order of 10^{-5} -- this is only 1% of the natural energy width of the stored beam! Polarization decreases as the current of an opposing beam increases. Thus, we have beam-beam depolarization which seems to coincide with vertical blow-up of the beams. We think that blow-up and depolarization may be more directly connected.

OTHER MACHINES

DESY has two linacs. The 50-MeV electron linac is being upgraded by the addition of 2 five-meter sections in order to raise the injection energy to 220 MeV. The positron linac II consists of two parts: one makes 250-MeV electrons which bombard a tungsten target, and the other accelerates the positrons coming from the tungsten target to 450 MeV. It has a total of 13 five-meter sections, and all these klystrons are now being equipped with SLED cavities, an invention made at SLAC which doubles the rf power by pulse compression. By the end of this year we will have all the klystrons equipped with SLED, and that will allow us to take two waveguides out, put them into the electron injector, and still have some as spares. Also, fewer klystrons will be necessary.

One thing that has been very successful at DESY was the installation of a positron accumulator ring, PIA. A problem with a linac is that it provides a bunch train that may be 30 m long and must be compressed to fit into one rf bucket in the synchrotron. Another problem is that injection into a storage ring like PETRA can be done at most 8 times per second because at the low energy of 7 GeV radiation damping is not very strong. This accumulator ring will actually solve both problems. It will first take the bunch trains from the linac and, with a first-harmonic cavity, collect all the positrons into single bunches about one meter long. After collection of some 9 injection pulses, a cavity which runs on the 12th harmonic is turned on and further compresses the positrons so that they fit into a single rf bucket in the synchrotron.

Another thing recently done at DESY was to put all machines -- both linacs, PIA, the DESY synchrotron, DORIS, and PETRA -- on the same computer control system. This solved a big technical problem, and also a phychological and sociological problem. We have a common control room, and operators can now see each other and argue with each other directly. We have a smaller crew and we can be sure that all the machines are up to the best technical standard. Much of this control we learned from CERN. We have the same computer system, North 10 computers, and everything is done on color TV displays with touch panels and trackerballs. The computers can also display the history of machines. We are now doing an energy scan and have reached a beam energy of 20.810 GeV. Since we are still hoping to find the toponium, we scan the energy in 15-MeV steps and accumulate some 60 or 80 inverse nanobarns at each step.

PHYSICS PROGRAM

Physics is done by collaborations, and we have strong participation by groups from German universities and from abroad (see Table IV). Of the total of about 400 particle scientists, 180 are from abroad. PETRA now has four detectors; it used to have five. When we started PETRA, we moved the Pluto detector from DORIS to PETRA. We later moved Pluto out and in its place installed the CELLO detector, which also has superconducting coils, drift chambers, and argon calorimeters. We have just approved an upgrading of CELLO to include better tagging for two-photon physics. All our detectors are pretty open. We have no shielding walls. We think that detectors shield themselves pretty well, and we use only local shielding, so that all the detectors are more or less accessible even during the runs.

In another area we have the Mark J detector, on which Professor Ting is the expert. This detector has magnetized iron and was built mainly for μ-pair detection to show interference effects of electromagnetic with weak interaction forces, but it has also been used for much physics in other areas.

The special feature of the JADE detector is the very high pressure drift chamber called the jet chamber which, besides having good spatial resolution, allows accurate measurements of dE/dx. It has 3000 lead glass counters for shower measurements.

The special feature of the TASSO detector is that the magnet is open to the sides and it has Aerogel and Crenkov counters so that it provides good particle identification over an angle of something like 2π steradian.

Physics Highlights

After the J had been discovered at Brookhaven and the ψ at Stanford, the ψ' was discovered at SLAC. It was then that DORIS came into operation, and DORIS was lucky enough to show the first cascade decay of the ψ'. Pearl at SLAC found the heavy electron τ, but very nice measurements were done on DORIS to determine the exact mass of the heavy electron. After Lederman at Fermilab had found the T, we rebuilt DORIS (as mentioned above), and we have determined those resonances and studied them.

Figure 2 shows the original Fermilab points of the γ-resonance. It also shows the first data from DORIS, which enabled us to make the determination of the charge of the quark, which is most likely one third.

Another finding at DORIS was a first indication of the gluon decay of the γ: If the probability of the decay is plotted against the angle between the two highest energy jets, then a three-gluon decay is expected to give the solid curve shown in Figure 3 (left).

Table IV Elementary Particle Physics Groups at DESY

FR Germany	Technische Hochschule Aachen
	Universität Bonn
	Universität Dortmund
	Universität Erlangen-Nürnberg
	Universität Freiburg
	Universität Hamburg
	Universität Heidelberg
	Kernforschungszentrum (KfK) Karlsruhe
	Universität Karlsruhe
	Max-Planck-Institut (MPI) München
	Gesamthochschule-Universität Seigen
	Universität Wurzburg
	Gesamthochschule-Universität Wuppertal
USA	Carnegie-Mellon University Pittsburgh
	MIT Cambridge
	University of Maryland College Park
	University of Michigan Ann Arbor
	University of Kansas Lawrence
	University of Wisconsin Madison
	University of South Carolina Columbia
	Harvard University Cambridge
	CalTech Pasadena
	Princeton University Pennsylvania
	SLAC Stanford
	Stanford University Stanford
Canada	York University Toronto
P.R. China	Institute for High Energy Physics Peking
France	LAL Orsay
	University VI Paris
	CEN Saclay Gif-sur-Yvette
Great Britain	University of Glasgow
	University of Lancaster
	Imperial College London
	University of Manchester
	Oxford University
	Rutherford Appleton Laboratories Chilton
Netherlands	NIKHEF Amsterdam
	University of Nijmegen
Israel	Weizmann Institute Rehovot
	Tel Aviv University Tel Aviv
Italy	University of Florence
	University of Rome
Japan	University of Tokyo
Norway	University of Bergen
Poland	Institute for Particle Physics Krakau
Spain	JEN Madrid

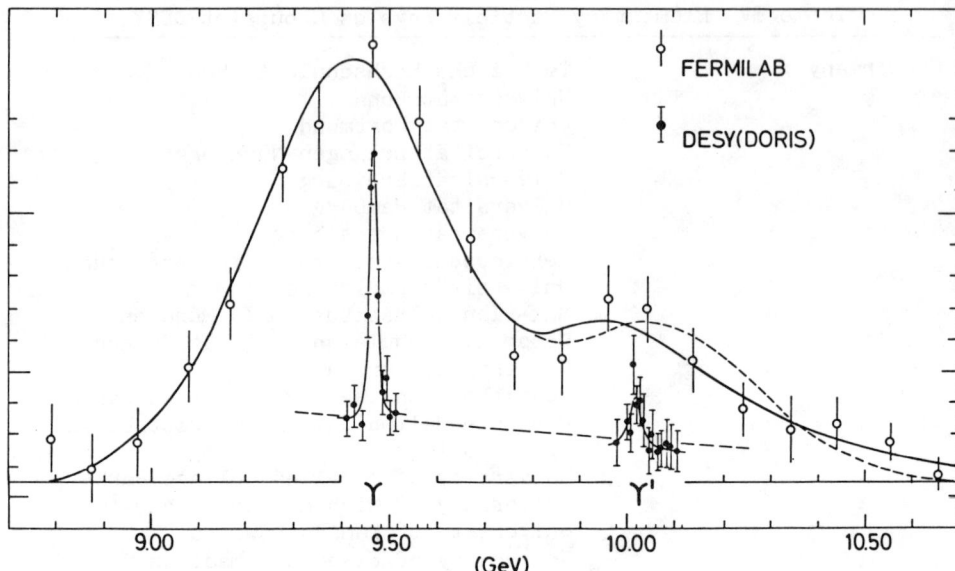

Figure 2. Comparison of the data on the ϒ resonances from the Fermilab experiment, where the particles were discovered, and from the DORIS storage ring. Note how much cleaner the signals from electron-positron annihilation appear.

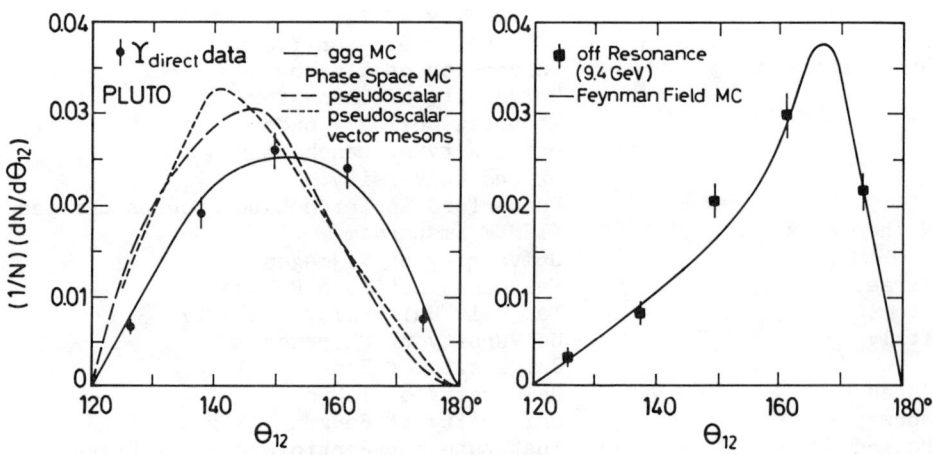

Figure 3. Gluon decay.

The experimental data verify that. The dashed and dotted curves are for other models. The distribution of the jets off resonance is shown in the right-hand panel. These observations are accepted as providing a fairly good description of three-gluon decays.

Perhaps the most important discovery at PETRA was three-jet decay, which has been interpreted as gluon beamstrahlung in the production of quark pairs. The angular relationship between the three jets shows that the relationship actually is very consistent with the gluon being a vector particle with spin 1, as it should be.

Another important field is study of the weak interference effects in μ-pair and τ-pair production, and in quark production, to test the electroweak theory. The results show that μ-pair production becomes asymmetrical because of the weak interaction effects, and similar results are seen for τ production. Similar observations have been made for charmed quark production.

Quantum electrodynamics (q.e.d.) and its limits has been a basic topic. Plotting the cross sections as a function of center-of-mass energy shows that indeed electrons, μ mesons, and τ mesons all behave according to q.e.d. They must all be point-like. Q.e.d. seems to work. Limits can be obtained, described by breakdown parameters of the order of 100 to 300 GeV. Maybe even more important is the measurement of R, the ratio of hadron production to μ-pair production. HERA has extended the energy range of R to 42.7 GeV, and R can now be determined with a very small error.

The results indicate that quarks, like leptons, also must be point-like (smaller than 10^{-16} cm). The absolute cross section seems to agree with the theory of quantum chromodynamics (q.c.d.). R has a small energy dependence because of strong interaction effects predicted by q.c.d., and this correction can now be determined. This is the first correction term, and it gives directly the strong interaction coupling constant, which has been found to be 0.24 (with an error of about 25%).

Another wide area of study at PETRA is two-photon physics.

A newly developing field is the analysis of quark and gluon jets. A first surprising result seems to be that these jets contain many heavy particles. At higher storage ring energies there are relatively more heavy particles like p's and $\bar{p}$'s than at lower energies.

Future Program

The future physics program at DORIS will include more T spectroscopy. At PETRA we will finish our energy scan. We'll be able to go to 21.5 GeV, and we will have this by the end of the month. Then we will add more cavities, and that will carry us to about 22.5 GeV or 45 GeV in center of mass. We hope to reach that by the end of the year. That is as high as PETRA can go, and this is the end of the search for the top. If we don't find toponium, we will concentrate on weak coupling constants of charm and bottom quarks, and on searching for new particles.

NEW PROJECTS

HERA

Our main new project is HERA (hadron electron ring accelerator), a machine in which 820-GeV protons collide with 30-GeV electrons. With its circumference of 6.3 km it is similar to the Fermilab Tevatron in size and in energy. The physics case for HERA has been discussed in ECFA and many workshops and meetings and has generated many reports. HERA basically allows studies of the interaction of the neutral and the charged currents at a mass that is large compared with the characteristic mass of the weak interaction, and studies of the interaction of these currents with the constituents of the protons. It is hoped that HERA will have a polarized electron beam and we will be able to turn the spin parallel or antiparallel to the direction of the electrons. This should provide a good way of studying weak interaction effects. We might be able to look for right-handed currents, and we will be able to search for new particles up to masses of about 200 GeV. For instance, if the toponium were at 50 GeV, we would still have more than 100 events per day. HERA will be a unique machine where one can study charged current interactions at high Q^2. Therefore, people are confident that they will find something new with this kind of collision.

Figure 4 shows HERA and its surroundings: the Hamburg airport, a public park, the PETRA ring, a horse racetrack, and a soccer stadium. The HERA ring will be under private property in one corner, but most of it will occupy public land or land owned by DESY.

A machine of this size requires a tunnel fairly far below ground. No radiation problems -- or any other problems -- are expected. Tunneling methods will be used for the tunnel, and cut-and-fill for the experimental areas, which are also all under ground and have only an access shaft. The small surface building conforms to the ideas of the environmentalists. We have to be very careful not to antagonize people, and we are in close contact with all our neighbors. The earth is all sand. Half the tunnel is below the water table, but drillers in Hamburg are used to this and have a standard procedure that does not require pressurizing the tunnel even though the water pressure may be as high as 2 atmospheres. Compressed helium comes from one central compressor and is pumped around the ring.* The expanded helium is pumped back. Each of the four experimental areas will have a refrigerator, cold boxes, turbines, and pipes for water and other things.

The parameters of HERA appear in Table V. The energy will be 820 GeV on 30 GeV. The luminosity at those energies is supposed to be 0.5×10^{32} per cm^2-sec. Both machines will

---

*The concept of refrigeration has changed (1984) to one central liquifier with compressors and three cold-boxes and a He-distribution system around the ring.

Figure 4. HERA and its surroundings.

Table V HERA Parameters

	Proton Ring	Electron Ring
Energy, GeV	820	30
Momentum transfer (Q^2), (GeV/c)^2	98400	
Luminosity, cm^{-2} sec^{-1}	0.5×10^{32}	
Polarization time, min		19.5
Total number of particles	6.3×10^{13}	7.4×10^{12}
Number of bunches	210	
Number of interaction points	4	
Free space for experiment, m	15	
Circumference, m	6336	
Lattice	FODO	
Injection energy, GeV	40	14
Filling time, min	8.5	5
Peak bending field, T	4.53	0.1832
Magnet aperture, mm	60x60	80x40
RF frequency, MHz	208.189	499.667
Energy loss per turn, MeV	0.14×10^{-9}	140.2
Critical energy, KeV	10^{-6}	111.0
Straight section, M	4x360	
Geometric radius of arcs, m	779.2	
Diameter of tunnel, m	3.2	
Depth of tunnel below surface, m	10...20	
Floorspace for experiments per interaction region, m^2	875	
Height of beam in halls above floor, m	5.5	

have a large number of bunches, 210 bunches in each ring, and that means that both machines must have a crossing geometry such that each bunch does not see the other beam more than four times per turn. There are four interaction regions. The circumference is 6.3 km. The injection energy into the proton ring is 40 GeV. We do not have such a large proton accelerator, but we are convinced that we should be able to accelerate protons in PETRA, if we do some modifications, and that the PETRA bending magnets are good for 40 GeV. The electron injection energy will be 14 GeV and the energy loss for the electron ring, 140 MeV per turn. We now have ample rf, 10 MW, and almost 150 m of cavities in PETRA. This was attained at great expense, but we already had HERA in mind. As soon as we start to build HERA we will move 7/8 of all the rf out, and move it just over to HERA. That will give us already 30 GeV in HERA and leave enough rf in PETRA to provide 14 GeV for injection into HERA. There are 700 bending magnets. The field is 4.5 tesla, very close to the FNAL numbers.

Some of the interaction region numbers (Table VI) are interesting: We have low betas at the interaction point, 3 m and 15 cm in one ring, 3 m and 30 cm in the other. The crossing angle is 20 mrad, as in the ISR at CERN. The beam-beam Q-shifts are very low.

Table VI Interaction Parameters

	Proton Ring	Electron Ring
Horizontal emittance, m	0.47×10^{-8}	3.2×10^{-8}
Vertical emittance, m	0.24×10^{-8}	0.32×10^{-8}
Minimum β (horizontal), m	3.0	3.0
Minimum β (vertical), m	0.3	0.15
Maximum β (horizontal), m	400	318
Maximum β (vertical), m	436	445
Dispersion (H & V), m	0/0	0/0
Beam size at waist (Σ_x), mm	0.12	0.31
Beam size at waist (Σ_z), mm	0.027	0.022
Bunch length (Σ), m	0.095	0.011
Crossing angle (horizontal), total, mrad	20	
Total number of particles	6.3×10^{13}	7.4×10^{12}
Particles per bunch	3.0×10^{11}	3.5×10^{10}
Number of bunches	210	
Horizontal beam-beam Q-shift	0.0003	0.007
Vertical beam-beam Q-shift	0.0004	0.013
Luminosity, $cm^{-2} sec^{-1}$	0.5×10^{32}	
Depolarization factor caused by the spin rotation		0.63%

For the proton beam they are 3×10^{-4} and 4×10^{-4} -- at present the SPS runs with Q-shifts of 0.003, so this design number is about an order of magnitude lower. The beam-beam Q-shifts for the electron beam are also lower by a factor of 2 or 3 than in other electron rings. With the crossing beam geometry, however, the Q-shift limit could be considerably less, so this is still an uncertain number. If we have problems, we are prepared to go to head-on collisions.

In the straight section layout, the beams come from the tunnel, the protons on top of the electrons. The protons come down to the plane of the electrons and make a horizontal crossing. The electrons go straight through. This geometry has the advantage that there are no common magnets between two rings. The energy of each ring is independently adjustable. Since there are no bending magnets near the electron interaction points, synchrotron radiation at 30 GeV does not upset experiments.

Close to the 200-m point is what we call a spin rotator, where the electron spin, which has been transverse, is rotated in a series of horizontal and vertical bending magnets, so as to be longitudinal at the interaction point. If you want to have the reverse spin, then you have to make an appropriate modification in the bending configuration. On the other side, the spin direction is restored. There is no dispersion, either vertical or horizontal, in the straight section. In the very low beta configurations, there is some vertical dispersion in the spin rotator, but it is zero everywhere else. The spin rotator is very complicated, and a new one is designed about every 2 months. The latest idea is a series of horizontal and vertical bending magnets with a total length of about 30 m, and there are no quadrupoles in the whole range. This is an important aspect, and the vertical deflection is only about 16.8 cm. With this newest version of spin rotator, we hope to be able to build magnets that need only be changed in polarity to reverse the spin direction.

The superconducting magnets have been of great concern, but we have already learned a great deal, and we are still pursuing two lines. We built in-house a version that looks very much like the Fermilab magnet. It has Rutherford cables and a collar, which in our case is not stainless steel but aluminum. We have a cold beam pipe, and we have correction coils on the beam pipe. Then we have, as in the Fermilab magnet, one-phase and two-phase helium flow.

The second line of superconducting magnet development is that of a magnet with "cold iron" similar to the BNL magnet for the CBA. This magnet uses the same kind of Rutherford cable as the "warm iron" magnet but looks mechanically somewhat simpler than the latter. Because of its proximity, the iron contributes more to the field at the beam location so that some superconducting cable can be saved. Finally, this magnet allows for a simpler quench protection system, which here consists of simple diodes right on the magnet (at low temperature). The "warm iron" magnet requires an active quench protection system that involves a considerable amount of electronics. Against these advantages for the "cold iron" magnet must be put the disadvantage of the iron saturation, which requires considerably more sophisticated correction schemes at high fields. An interesting aspect of the "cold iron" magnet development is that it is being done entirely by industry, in close contact with DESY.*

We are not yet ready to decide between these two lines.

Injection into HERA has already been under consideration to some extent. Electrons will be accelerated in PETRA to an energy of 14 GeV and then transferred to HERA. Considerably more complicated is the injection scheme for protons. Protons are something new at DESY. We would first need a 50-MeV proton linac very similar to that at CERN. We may, however, take advantage of

---

* The most attractive features of the "cold iron" and the "warm iron" magnet have been combined in the so-called "hybrid magnet" (1984), which now seems to be the most favored candidate for the proton magnet system of HERA. In the hybrid magnet, as in the "warm iron" magnet, the coil is clamped by aluminum collars, but these collars are then directly surrounded by a "cold iron" yoke. In this way, the iron contribution to the field is made much larger than in the "warm iron" magnet, without getting into the saturation problems of the "cold iron" magnet. The quench protection can be as easily done with diodes as in the "cold iron" magnet, while the integrity of the coil clamping equals that of the "warm iron" magnet.

some modern developments and use, for instance, a radio-frequency quadrupole for preacceleration to 750 keV instead of the bulky Cockcroft-Walton generator. 50-MeV protons will then be injected into the DESY synchrotron and accelerated to an energy of about 8 GeV. The present synchrotron would require a large number of modifications for proton acceleration, and it would be very difficult to maintain the electron capability.

For this reason, we decided to build a new electron synchrotron (DESY II, discussed below) which, after its commissioning, would free the present synchrotron to be rebuilt as a dedicated proton accelerator (DESY III). Protons from DESY III will be transferred to PETRA and accelerated to an energy of 40 GeV. For this, PETRA will require some modifications. Fortunately, the bending magnets are good for the necessary high field strength. Only the quadrupoles are somewhat weak, so that the number of betatron oscillations per turn will be somewhat smaller. As a consequence, dispersion in the cells will be much larger, so that transition occurs below an injection energy of 8 GeV. This, in turn, is highly desirable if one wants to avoid phase-space dilution of the very dense proton bunches at transition.

We will need a new proton acceleration system for PETRA, but the required voltages are very modest. A bigger problem is the electron acceleration system, which has to compensate for synchrotron radiation. I mentioned above that 7/8 of the rf system of PETRA will be transferred to HERA and the remaining 1/8 will occupy only half of one of the four long straight sections. But the shunt impedance of that rf section is still far too large to be tolerated by the high current proton bunches. For this reason, we are going to build a bypass for the protons at the location of the electron rf cavities.

Proton injection in HERA at an energy of 40 GeV still seems to be difficult. But tracking calculations, which assume realistic field errors in the superconducting magnets, indicate that, even at the low injection energy of 40 GeV, acceptances should be large enough.

The electron ring of HERA involves standard technology, and a good deal of expertise is available at DESY. At HERA, we want to reduce the excitation coil for the magnetic dipole field to a single bar of aluminum. And, in order to speed up installation and greatly simplify the survey and adjustment of the magnets, we want to combine dipole, quadrupole, and sextupole into magnet modules that will be internally prealigned and will have only one support per unit, right under the quadrupole.

HERA is a very large project, requiring more than six times as much money and effort as PETRA. It would strain the resources not only of DESY but also of our sponsors, the Federal Government of Germany and the City of Hamburg. We are therefore anxious to get outside help. Interest in doing physics with HERA seems to be very high, and many institutes outside Germany seem to be willing to contribute to the HERA construction. Their contributions would be in the form of components built and paid for by them and then brought to DESY. In many talks with our foreign friends, we were convinced that it would be realistic to assume that one-third of all the neccessary effort could come from outside Germany. Particularly in areas of new technology (e.g. superconducting magnets), interest in taking over a major portion seems to be quite large (one institute offered to have all superconducting dipoles built in their country).

We estimate that HERA will cost 654 million Deutschmark, based on prices as of December 31, 1980. This does not include salaries of the staff already at DESY (which would have to be paid even if HERA were not built). We assume a total effort of 3000 man-years, of which two-thirds would come from DESY. If the construction period goes beyond 6 years, the equivalent of half the DESY staff would be needed to build HERA.

We estimate the total construction time to be 6 years. This means that, if we don't have to wait too long for the authorization, we will have e-p physics before the end of the decade!

DESY-II

The new synchrotron DESY II has been approved and is well under way. Half the money has already been spent. We are putting a second synchrotron in the same tunnel. This is possible because the DESY design was very generous in terms of space -- we have space for more synchrotrons if necessary. DESY II needs to have only a repetition rate of 12.5 Hz because, with our positron accumulator, it doesn't matter how fast we cycle the synchrotron. We will install PETRA cavities and PETRA rf systems in DESY II which should get the energy up to about 9.2 GeV. DESY II is a separated-function machine. The present combined-function synchrotron is radiation antidamped, and at high energy (7 GeV), the beam quickly spreads out. This is one of the problems now with our injection into PETRA. We cannot go up to very high energy because the emittance will grow very large. But the new synchrotron, as a separated-function machine, is radiation-damped in all three modes of oscillation. Thus, we could go up to 9.2 GeV and actually even store the beams there for a while before they are injected. This machine with a lower repetition rate will also allow an all metal vacuum chamber.

Table VII lists the main parameters of DESY II. The present plan is to have this machine installed during the winter shutdown in 1984-85 (about 1-1/2 years from now) and then to take a whole year to check out the new machine and get it going. In the winter of 1985/86, we will connect this machine to all the injection channels of PETRA and DORIS and start rebuilding DESY I. DESY I will then become DESY III and be a proton accelerator.

Two projects not in the immediate future but still of interest are superconducting radio-frequency systems and wakefield acceleration.

Table VII   DESY II Parameters

$\bar{R}$	46.6 m
Rep. rate	12.5 Hz
h	488
$\hat{P}_{rf}$	1.2 MW
r_{rf}	336 MΩ (=$\hat{u}^2$/2p)
Q_x, Q_z	6.575, 5.623
α	0.0242
E_{max}	9.2 GeV
ϵ_H (9 GeV)	7.7x10^{-8}m
Separated function lattice	
FODO	(24 cells)
Costs	∫15 MDM

RF Superconductivity

In the field of rf superconductivity we started off with a collaboration with CERN and Karlsruhe. A single-cell cavity at 500 MHz that had been built at Karlsruhe was installed in PETRA. It operated at a field level of about 2.5 MV/m, maybe even 2.8. With only this cavity in operation we accelerated and stored electrons at 5 GeV. Since this cavity has been in the PETRA storage ring for 2 or 3 months without any sign of deterioration, we are confident that superconducting rf is feasible. A field of 2.5 MV is about 2.5 times as much as that in conventional cavities. So, why is it important to have superconducting rf? To me, the most important aspect is that you need only one-third the number of cavities in the ring, if you have three times the gradient. This means that all the instabilities normally produced by the rf cavities are down by a factor of 3; i.e., you can have three times as much current before you hit the instabilities, and thus you can have 10 times the luminosity if you work in an instability-limited region. That is the strongest argument for going to superconductivity. The other argument is that, to get high voltage, you need not dissipate so much power in the cavity. In PETRA, at present, we dissipate 10 MW of rf power in the cavities; that means 16 or 17 MW of main-line power, which is quite undesirable. Together with CERN we are also in the process of testing 5-cell cavities built at CERN. The 500-MHz cavity also has operated at 2.3 to 2.8 MW/m in PETRA. Because it had a small helium leak from the tank into the vacuum, we had to take it out after 4 or 6 weeks, but it had worked quite well. It stored 7-GeV electrons with no other systems being on.

At DESY, we are building cavities at 1000 MHz. Besides the 500-MHz system, PETRA also has a 1000-MHz system which is used for bunch lengthening. In our superconducting effort, we decided to try 1000-MHz cavities with elliptical shape. A single cell has reached 5.5 MV/m and a multi-cell cavity has reached 3.5 MV/m. We are now testing a 9-cell cavity in a cryostat.

Wakefield Acceleration

Everybody concerned about the far future of high energy physics is thinking about how to get super-high energy accelerators with very large accelerating fields. A line that we want to pursue is to test the wakefield principle. At this Summer School, Perry Wilson discussed wakefield accelerators (see AIP Conf. Proc. No. 127). Imagine a structure with a slot in the shape of the ring. Suppose you have an electron beam that would just fit through the slot; this beam would radiate electromagnetic fields which are reflected at the wall and travel back to the center. As they travel to the center, the field strength is greatly increased, by factors of 10. A second beam which goes through the center sees very high strength. (See Fig. 5.)

Tom Weiland has done computer calculations with his BCI and TBCI codes and has nice pictures that have been shown at various conferences and workshops. If the outer ring has a charge of about 1 mC, you could reach in the center a field strength of about 200 MV/m. If you wait for the field to travel again to the wall and be deflected a second time, then you get a reverse polarity and you can also accelerate positrons. This looks very good on a computer. Table VIII lists the parameters for a wakefield accelerator. There are about 6×10^{12} particles in a ring with a diameter of 8.4 cm. The ring is very short, only about 4 mm long. This ring will lose about 17 MV per meter -- not per centimeter -- through its own wakefields. The secondary beam has about 10^{11} particles per bunch, and in this example it has about 170 MV/m acceleration.

Table VIII Wakefield Accelerator

Driving beam	6×10^{12} particles/ring bunch
Ring diameter	8.4 cm
Ring length	0.4 cm
Maximum energy loss	16.8 MeV/cm
Secondary beam	10^{11} particles/bunch
Maximum acceleration	170 MeV/m
Maximum energy loss	15 MeV/m

If the wakefield accelerator principle should turn out to be practical, a large linear collider could look like the sketch in Figure 6. Two regular linear accelerators accelerate the beams up to about 30 GeV. Then, the beams go to a wakefield accelerator where they transfer their energy into new beams and accelerate these beams up to about 300 GeV. Thus, you would have a linear collider.

In order to find out where the real problems are, we are now engaged in a test of this principle. We take accelerating cavities from the synchrotron that are not needed there and build a linear accelerator having a hole so big that we can accelerate rings up to an energy of about 10 MV. These rings come from a laser cathode. Then, we will compress these rings through longitudinal magnetic fields and use them to accelerate a second beam in a short section.

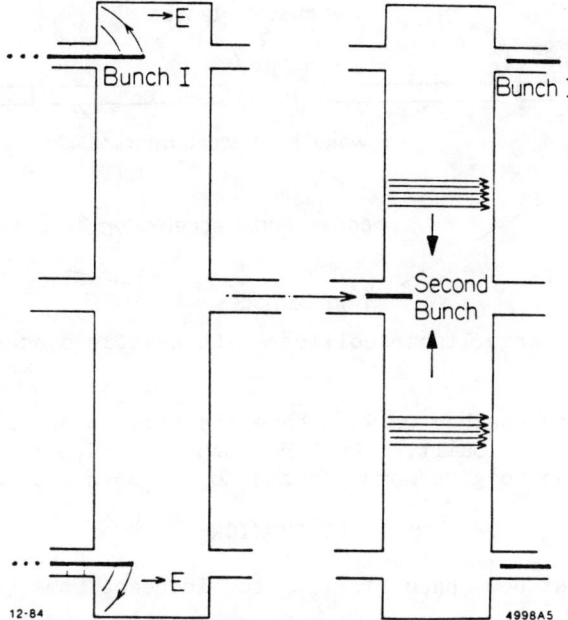

Fig. 5. Qualitative field inside a pillbox transformer showing the collapsing field packet.

Figure 5 shows an example in which a driving hollow beam passes through a slot in the outer perimeter of a pillbox cavity, followed at an appropriate distance by a secondary bunch of particles moving along the axis. As the driving beam enters the cavity a wave packet is generated which subsequently travels towards the axis. During this time the volume occupied by the wave decreases linearly with the distance from the axis. The electric and magnetic field strengths must increase approximately as $r^{-1/2}$. A secondary beam can then be accelerated along the axis of the transformer by these fields.

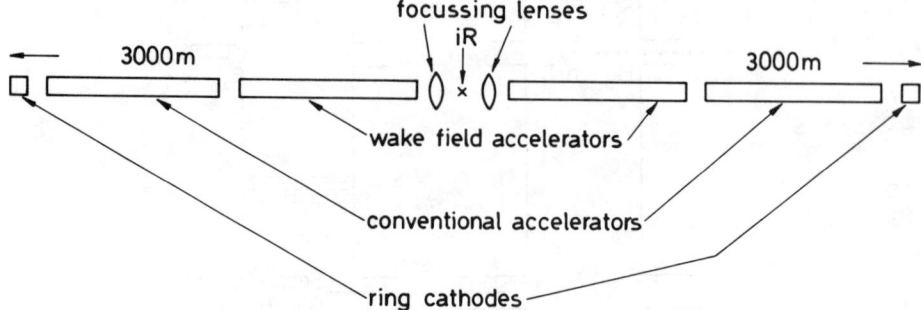

Figure 6. Large linear collider with wakefield acceleration.

I don't know whether we will have the time to do all these things if HERA is to be built. On the other hand, it seems to be important also to give some thought to the more distant future.

DISCUSSION

Question: How hard is it to increase the energy beyond 800 GeV?
Answer: 820 GeV corresponds to 4.5 tesla and, as I said, the magnets go up to 5.2. But I don't know whether this is an operational possibility. I think that 820 seems like a safe number.

Q: Could you indicate the cost differential between the two types of magnets?
A: Regarding the cost differential between the "cold iron" and the Fermilab magnet, I don't want to quote a number. The "cold iron" magnet does not have the extra collar, and that should save money, maybe 20%. But one magnet is built by industry and the other one in-house, and that may more than make the differential go the other way.

Q: Which country offered to build superconducting dipoles?
A: Italy.

Q: You use low currents. Is that because of single-beam effects? The single beam seems to be very low.
A: There are various limitations. At the high end of 30 GeV we are surely limited by rf power in the electron ring. At lower energy, the beam-beam effects limit the current. As far as the proton current is concerned, this is just an assumption that seems to be on the safe side as far as ΔQ for electrons go.

Q: On your HERA time-scale, what is the time left for PETRA physics operation?
A: There has been no policy decision. Time will certainly be needed to put in the bypass for the protons, and to test proton acceleration in PETRA. We also think some of the experiments will want to be moved from PETRA and modified and then perhaps installed in HERA. All things considered it may make sense to stop e^+e^- operation sometime in 1987, but this is still an open question.

Q: Can you say a word about detectors for HERA?

A: I don't have any pictures with me. There have been discussions about detectors. At a workshop at Amsterdam sponsored by ECFA, coinciding with the Woods Hole meeting, various ideas were presented. These detectors first of all will be asymmetric, they have a central detector but most of the momentum goes in the proton direction. By the way, the interaction point is also asymmetric in the hall. We will have a central track detector. Important in all detectors is calorimetry, which seems to be more effective than magnetic field analysis in determining energy.

Q: Why is the spin rotator necessary in the electron ring?

A: If you are interested in longitudinal polarization for electrons, it is needed to separate weak interaction effects due to charged currents. Depending on the spin direction, you may produce a left-handed W and make a weak interaction, or you can turn off the weak interaction with a left-handed W by having the spin in the opposite direction. Thus, by controlling the longitudinal spin, you can turn the weak interaction part on or off, and by putting it in the wrong direction, you might look for new effects, e.g., whether there is a right-handed W particle.

Q: What are civil engineering costs?

A: The civil engineering for HERA costs about 210 million marks, which is $84 million. The underground halls may account for something like 40% of the total. Most of the rest goes into the tunnel, i.e. around 100 million marks or $40 million. That means the cost is about $7000 per meter.

PHYSICS OF e^+-e^- COLLIDERS: PRESENT, FUTURE, AND FAR FUTURE[*]

MICHAEL E. PESKIN
Stanford Linear Accelerator Center
Stanford University, Stanford, CA 94305

TABLE OF CONTENTS

1. Introduction 123
2. Features of e^+-e^- Collisions, according to the Standard Model .. 125
3. Particle Search Experiments 127
4. A Digression: 3 Kinds of 1 TeV Physics 146
5. Technics for 1 TeV e^+-e^- Collisions 151
6. Physics of 1 TeV e^+-e^- Collisions 154

   References 165

---

[*] Work supported by the Department of Energy, contract DE-AC03-76SF00515.

PHYSICS OF e^+-e^- COLLIDERS: PRESENT, FUTURE, AND FAR FUTURE

MICHAEL E. PESKIN
Stanford Linear Accelerator Center
Stanford University, Stanford, CA 94305

1. Introduction

Electron-positron colliders have emerged in the last ten years as tools of major significance in the study of fundamental physics. They have allowed the detailed study of the ψ and Υ resonances which has given striking confirmation to the quark model of strong interactions, and they have provided the setting for the discovery of the D and B mesons and the τ lepton. One could easily fill the time I have been allotted in reviewing this glorious history. The history of e^+-e^- colliders, however, has been recounted on many occasions and one may read of it in a number of excellent review articles;[1-3] I have little to add to this story. Indeed, I doubt that this is what one should present to a summer school devoted to the physics of accelerators. You are, I am sure, much more interested in the future of e^+-e^- colliders. You want to know what is the importance of the continued development of e^+-e^- colliders and, especially, what physics can be done with the machines you will design. I have therefore taken as my topic the glorious future of e^+-e^- colliders, reserving just enough time for review to find some principles to use in extrapolating to higher energies.

Before beginning this discussion, I should offer two warnings to the reader. The first is that I am, by trade, a theorist, and worse, a rather speculative one. In some sense, I stand at the opposite end of the profession from designers of accelerators. But I feel that this is an example of the kinship of extremes: Both wide-eyed theorists and creative accelerator designers find themselves drawn to thinking about the highest achievable energies and link their aspirations to the exploration of these elevated realms. The second warning is that I have been asked by the organizers of this school to begin at the very beginning and proceed in one lecture to a high level of sophistication. I have tried to err in the direction of clarity, with the result that most of the material to be presented here will be familiar to most workers in the field of e^+-e^- collisions. Still, though the physics of e^+-e^- colliders in the regime of present and nearby future energies has been rather thoroughly studied, the physics of the realm of very high energies has been given serious attention only relatively recently.[4-6] Whatever here is new is the result of a collaboration with John Ellis, whom I thank for many discussions.

The presentation of this lecture will proceed as follows: In Section 2, I will review the features of e^+-e^- collisions according to the standard gauge theory of strong, weak, and electromagnetic interactions. This discussion will review a few of the most important features of e^+-e^- collisions at currently accessible energies and the expectations for e^+-e^- reactions which produce the intermediate vector bosons Z^0 and $W^\pm$. In Section 3, I will review some of the experimental work done at the current generation of e^+-e^- colliders; this discussion will emphasize the search for new types of elementary particles. Section 4 will be a theoretical digression, introducing a number of ideas about physics at the energy scale of 1 TeV. Section 5 will discuss (rather superficially) a number of technical aspects of electron-positron colliders designed to reach the TeV energies. Finally, in Section 6, I will discuss various possible effects which could appear in e^+-e^- collisions as the result of new physics appearing at 1 TeV or above.

As an introduction to this discussion, however, I should quickly review the present status of e^+-e^- colliders, if only in the compact form of Fig. 1. The

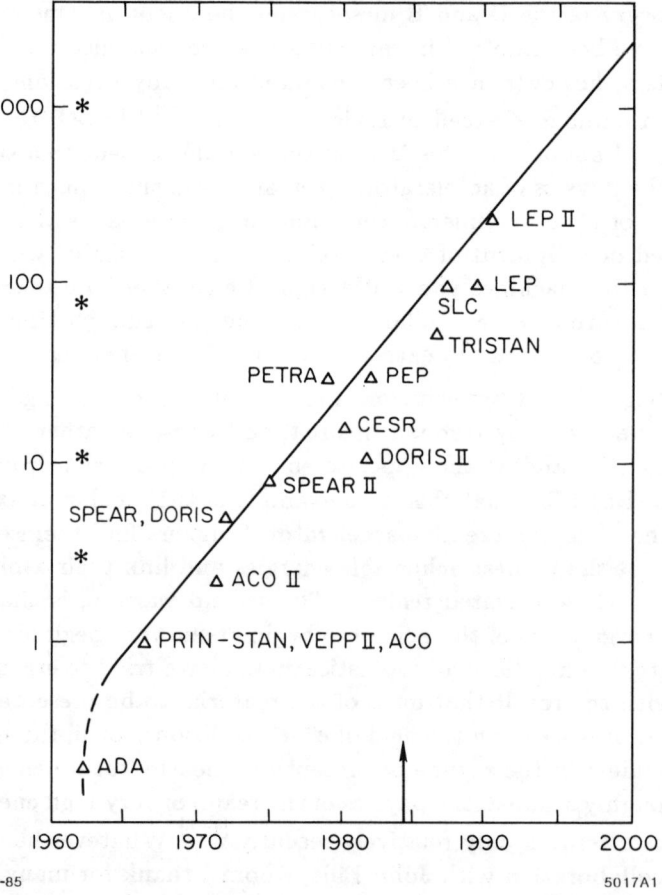

Fig. 1. Past, present, and proposed e^+-e^- colliding-beam accelerators, after Ref. 7.

arrow on the horizontal axis separates operating machines from those under construction. The stars on the vertical axis denote the places where spectacular new physics has been found or is to be expected; I will explain the interest of these energies as we proceed.

2. Features of e^+-e^- Collisions, according to the Standard Model

I will begin this discussion by reviewing, along the most basic lines, the physics which we observe in e^+-e^- collisions at currently accessible energies. From this base, it will be possible to extrapolate forward a certain distance in energy by studying the predictions of the gauge theories which we now believe describe the structure of the strong, weak, and electromagnetic interactions. Earlier in this school, Chris Llewellyn-Smith has described the successes of the theory which models the strong interactions as a gauge theory based on the group SU(3) and the weak and electromagnetic interactions as a gauge theory based on the group SU(2)×U(1). In this section, I will treat this conglomerate gauge theory as established and refer to it as **the standard model**. Please allow me to postpone discussion of how far in energy this model remains an adequate description of Nature. I will discuss that question at some length in § 4.

We might start by considering the very most basic process which occurs at e^+-e^- colliders, the QED process $e^+e^- \to \mu^+\mu^-$. The cross-section for this reaction, computable from the Feynman diagram shown in Fig. 2, is given by

$$\sigma_{\mu\mu} = \frac{4\pi}{3}\frac{\alpha^2}{E_{\text{cm}}^2} = \frac{86.8\text{nb}}{(E_{\text{cm}})^2} \quad (E_{\text{cm}} \text{ in GeV}) \quad = 1 \text{ R unit}. \quad (2.1)$$

This equation defines the quantity I will refer to as 1 unit of R. The quantity (2.1) is the basic pointlike annihilation cross-section, and so the R unit sets the scale for the rates of all processes, leptonic or hadronic, which take place at e^+-e^- colliders.

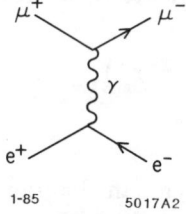

Fig. 2. Feynman diagram contributing to the process $e^+e^- \to \mu^+\mu^-$.

Throughout this lecture, whenever I quote a cross-section, I will quote it in units of R. Let me warn you from the outset that this is cheating. One more conventionally imagines building accelerators to obtain a fixed luminosity, and, indeed, the e^+-e^- colliders constructed since 1970 have shown a constant or slowly decreasing luminosity as the collision energy has increased. However, this is not an appropriate course for the future. The R unit falls rapidly with energy, as one can see by converting this unit to events/year for a fixed luminosity of 10^{31} cm^{-2} sec^{-1} characteristic of the current generation of colliders. Defining a practical year to be 10^7 sec, one finds:

E_{cm} (GeV)	events/yr
4	500,000
10	90,000
30	10,000
100	900
250	140
1000	9

Thus one must somehow devise a way to keep the number of R units per year fixed as the collider energy is increased. For the most part, I will simply assume that this can be done. I will, however, comment briefly on the problem of working at constant R/yr in § 5.

The first reason that the R unit is of interest is that the cross-section for the production of hadrons from e^+e^- is observed to be a number of R units of order 1. The standard model predicts, via the Feynman diagrams of Fig. 3, the relation[8]

$$\frac{\sigma(e^+e^- \to hadrons)}{\sigma(e^+e^- \to \mu^+\mu^-)} = \sum_f Q_f^2 \cdot 3 \cdot (1 + \frac{\alpha_s}{\pi} + ...), \qquad (2.2)$$

where α_s is the strong interaction gauge coupling constant, 3 is the number of colors, f runs over quark flavors, and Q_f is the quark electric charge. This relation is quite well satisfied experimentally.[9,10] Eq. (2.2) predicts a relatively small rate for the production of hadrons, but one which is democratic between familiar and exotic flavors:

$$\sigma(\to charm) = \sigma(\to up) = 4\sigma(\to bottom) = 4\sigma(\to down)$$
$$= \frac{2}{3}\sigma(\to \tau^+\tau^-). \qquad (2.3)$$

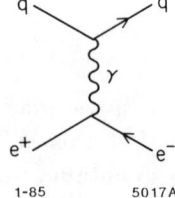

Fig. 3. Feynman diagram contributing to the process $e^+e^- \to hadrons$.

At present energies, the production of charmed quark pairs makes up almost half of the total hadronic cross-section, a situation rather different from that found in hadron-hadron collisions. The τ lepton is produced as copiously as the muon. This democracy of production provides a great advantage in studying the properties of new and exotic particles.

I might illustrate the experimental correctness of Eq. (2.2) by presenting in Fig. 4 the experimental data on the total cross-section for $e^+e^- \to hadrons$.

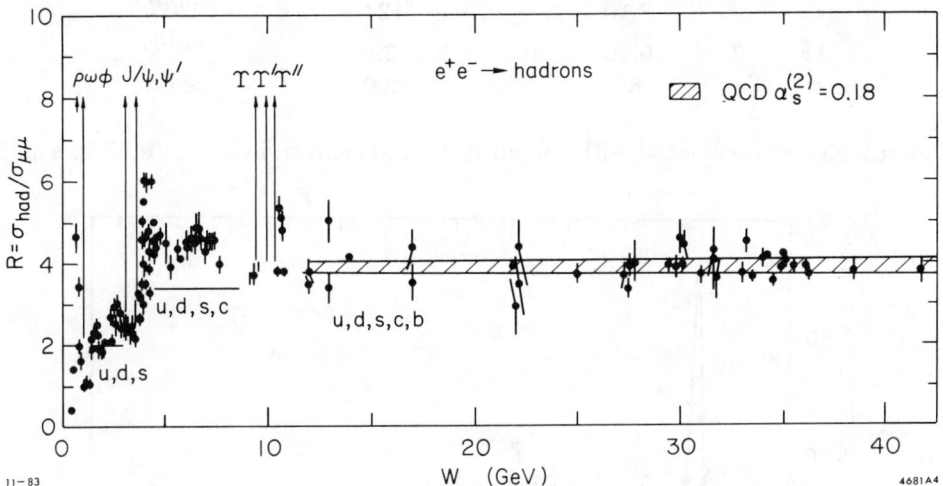

Fig. 4. Experimentally determined ratio of cross-sections for production of hadrons and of μ pairs in e^+e^- annihilation, from Ref. 11.

The ratio indicated in (2.3) is exceedingly constant over most of the energy range; however, at certain points this ratio increases markedly to show prominent resonances. The resonances indicated in Fig. 4 have properties which agree in detail with those predicted for quark-antiquark bound states of spin 1 and odd parity.[12] In general, any resonance with the right quantum numbers to couple to 1 photon (or directly to e^+e^-) should be directly visible as an increment of R. The size of this increment is given by

$$\int dE_{\text{cm}} \, \Delta\sigma_T(E_{\text{cm}}) = \frac{6\pi^2}{m_R^2}\Gamma(\Phi \to e^+e^-), \qquad (2.4)$$

where m_Φ is the resonance mass and Γ is the partial decay width of the resonance into e^+e^-. As an illustration of this result, let me display the size of the resonance associated with the lowest spin 1 ($^3S_0$) q-$\bar{\text{q}}$ bound state built, respectively, with charmed quarks, bottom quarks, top quarks. For the purpose of making estimates in the course of this lecture, I will assume that the top quark mass is 40 GeV.[13] The last column of the table below gives the observed (or to-be-observed) peak value of the increment of the hadronic cross-section, in R units. Since q-q̄ resonances are characteristically very narrow, with total widths of order 100 keV, it is the energy spread of the synchrotron which determines the observed width and height of these resonances.

state		mass (GeV)	$\int \Delta\sigma_T (\text{Runits}) \cdot 10\text{MeV}$	ΔR_{obs}
$c\bar{c}$	ψ	3.10	124	220[14]
$b\bar{b}$	Υ	9.46	29	15[15]
$t\bar{t}$	ς	80	400	≈ 60

The resonances associated with the $b\bar{b}$ system are shown in more detail in Fig. 5.

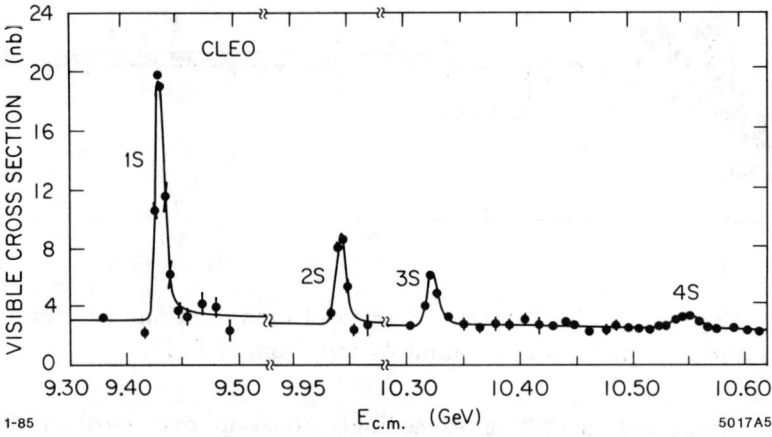

Fig. 5. The spectrum of $b\bar{b}$ resonances, as observed by the CLEO experiment, from Ref. 16.

In the energy region above these resonances, the process $e^+e^- \to$ *hadrons* resembles the process shown in Fig. 3 not only in the value of the total cross-section but also in the form of the hadronic states produced. These hadrons characteristically lie in narrow cones, roughly 15° in radius at PEP and PETRA energies, oriented about some axis. The orientation of this axis varies from event to event; it must be located anew in each event. A convenient systematic algorithm for finding this axis is that of maximizing a quantity called the *thrust*, defined,[17] for a given axis, by

$$T = \sum_i |p_\parallel^{(i)}| \qquad (2.5)$$

where the sum runs over all particles and $|p_\parallel^{(i)}|$ is the component of the momentum of the ith particle relative to the chosen axis. The distribution of particle momenta relative to the thrust axis obtained by the JADE experiment at PETRA is shown in Fig. 6. Another way to display this collimation of final-state particles, is to label the hadrons clustering about a particular axis as a jet and then to display the distribution of the invariant masses of these jets. The MAC experiment at PEP has done such an analysis, somewhat crudely defining a jet to be all the particles in a given hemisphere with respect to the thrust axis. The resulting distribution of jet masses is shown in Fig. 7. This distribution is sufficiently narrow that bottom quarks, of mass 5 GeV, can be recognized, at least on a statistical basis, by the fact that they produce jets of high invariant mass.

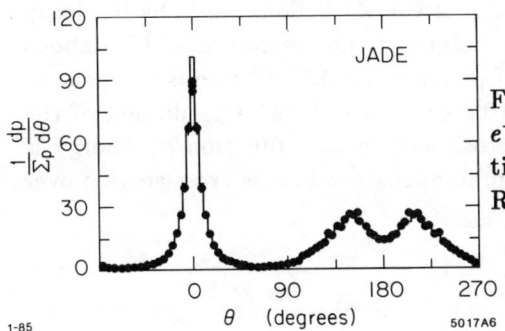

Fig. 6. Distribution of momentum in e^+e^- annihilations to hadrons, as a function of angle from the thrust axis, from Ref. 18.

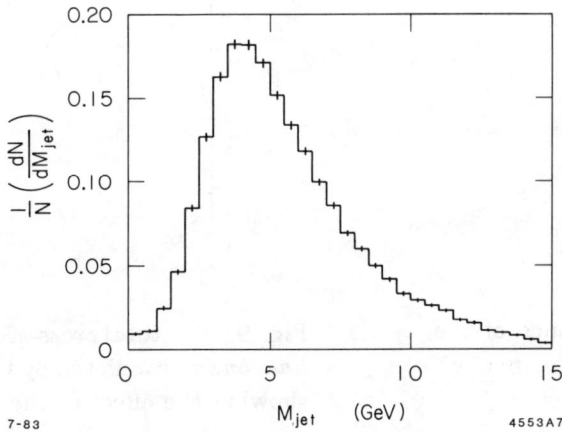

Fig. 7. Jet invariant mass distribution at $E_{\rm cm} = 29$ GeV, from Ref. 19. In this analysis, a jet was defined to be the collection of particles within 90° of the thrust axis.

Having now given a very brief review of the character of e^+e^- annihilation events at currently available energies, let me turn to a review of the features of e^+e^- annihilation at energies soon to be accessed, as these features are predicted by the standard model. The most striking features involve the production in e^+e^- reactions of the weak-interaction intermediate vector bosons $W^\pm$ and Z^0. These bosons couple as strongly to e^+e^- as the photon; thus, their influence becomes comparable to that of the photon when the center-of-mass reaction energy reaches the level of the masses of these bosons, respectively 83 and 94 GeV.[20,21]

The most important effect of the weak bosons arises from the fact that the Z^0 has the right quantum numbers to be produced singly as a resonance in the e^+e^- annihilation cross-section; this process is indicated in Fig. 8. The resonance is an enormous one. If the width of the Z^0 is dominated by its decay into the known quarks and leptons, the height of the resonance will be about 3000 units of R. Such a resonance would produce 3×10^6 Z^0 events per year at an e^+-e^- collider with a luminosity of 10^{31} cm^{-2} sec^{-1}. The prediction of the standard model for the total hadronic cross-section as a function of energy is shown in Fig. 9; the Z^0 is clearly the dominant feature in this cross-section over the energy range shown.

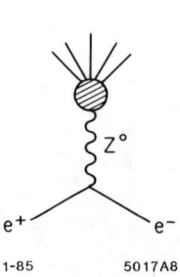
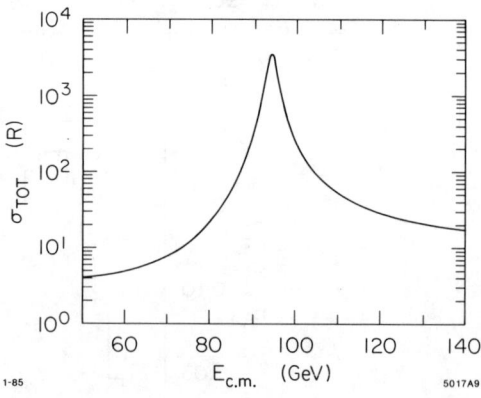

Fig. 8. Appearance of the Z^0 as a resonance in e^+e^- annihilation.

Fig. 9. The total cross-section for $e^+e^- \to$ *hadrons* as predicted by the standard model, showing the effect of the Z^0 resonance.

Like the photon, the Z^0 couples to quarks and leptons democratically between heavy and light flavors. Because the Z^0 is a component of the weak interactions, however, its coupling is chiral and depends on the fermion helicity. The Z^0 charges of the known fermions, in terms of their charges, weak isospins,

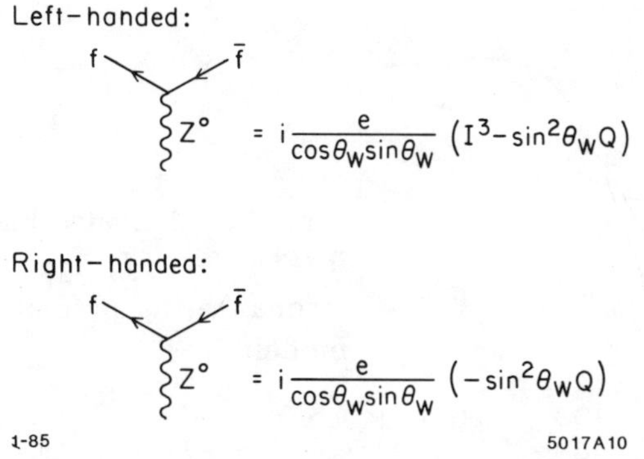

Left-handed:

$$Z^0 \underset{f}{\overset{\bar{f}}{\diagup\!\!\!\diagdown}} = i\frac{e}{\cos\theta_W \sin\theta_W}(I^3 - \sin^2\theta_W Q)$$

Right-handed:

$$Z^0 \underset{f}{\overset{\bar{f}}{\diagup\!\!\!\diagdown}} = i\frac{e}{\cos\theta_W \sin\theta_W}(-\sin^2\theta_W Q)$$

Fig. 10. Charges defining the coupling of the Z^0 to fermions. Q and I^3 denote, respectively, electric charge (in units of the electron charge) and weak isospin. e is the electron charge, and θ_w is the Weinberg angle which parametrizes neutral-current weak interactions.

and helicities, are indicated in Fig. 10. This helicity dependence of the e^+e^- annihilation cross-section produces a number of remarkable effects. The first of these is an energy-dependent forward-backward asymmetry, different for different types of fermions. This effect arises because the processes which produce left- and right-handed fermions from the electron-positron annihilation depend differently on the direction of the new fermion:

$$\text{for}\quad e^-(L) + e^+ \to f(L) + \bar{f}, \quad e^-(R) + e^+ \to f(R) + \bar{f}:$$

$$\frac{d\sigma}{d\Omega} \sim (1 + \cos\theta)^2;$$

$$\text{for}\quad e^-(L) + e^+ \to f(R) + \bar{f}, \quad e^-(R) + e^+ \to f(L) + \bar{f}:$$

$$\frac{d\sigma}{d\Omega} \sim (1 - \cos\theta)^2,$$

(2.6)

where θ is the angle between the original electron and the f. The various helicity cross-sections are all equal at low energies, where annihilation through a single photon dominates, but they receive very different weights as one passes through the Z^0 resonance, producing the forward-backward asymmetry shown in Fig. 11. The small asymmetry in lepton pair production at the relatively low energies now accessible has already been observed at PEP and PETRA.[22] A second effect of the helicity-dependent cross-sections is that the fermions produced in

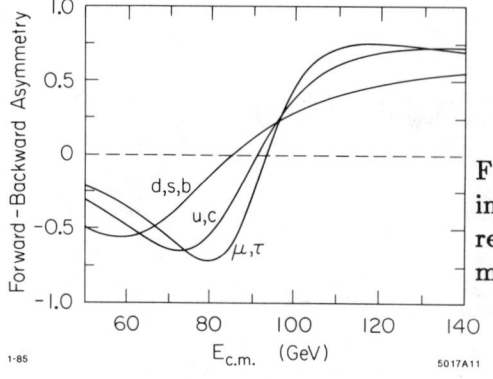

Fig. 11. Forward-backward asymmetry in $e^+e^- \to f\bar{f}$, in the vicinity of the Z^0 resonance, as predicted by the standard model.

Z^0 decays will be longitudinally polarized. Figure 12 shows the standard model prediction for the polarization of leptons pair-produced in e^+e^- annihilation, as a function of energy. To show the correlation between helicity and angle, I have plotted also the polarization of those leptons emitted into the forward hemisphere. The same dramatic energy dependence appears in the cross-section for producing lepton pairs from polarized electrons.

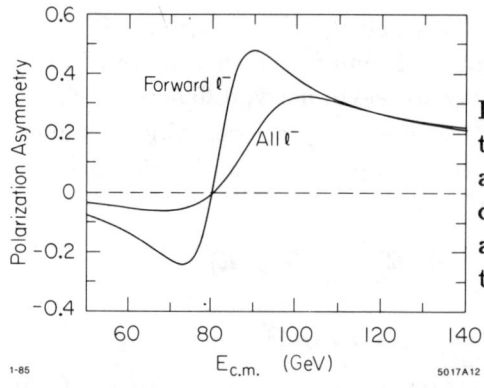

Fig. 12. Longitudinal polarization of leptons pair-produced in e^+e^- annihilation, according to the standard model. The two curves give the polarization integrated over all leptons and also integrated only over those in the forward hemisphere.

In addition to the striking qualitative features of the weak interactions near the Z^0 resonance, e^+e^- annihilation in the vicinity of this resonance offers the possibility of a dramatic improvement in the quantitative comparison of the standard model of the weak interactions with experiment. Two measurements, in particular, can be performed with high accuracy: It should be possible to use the well-defined beam energy available in e^+-e^- reactions to measure the position of the peak of the Z^0 resonance to better than 100 MeV. In addition, the dramatic effect of the lepton polarization shown in Fig. 12 depends on the quantity $(\frac{1}{4} - sin^2\theta_w)$, so that a precise measurement of this effect—which might be obtained from the polarization of τ leptons defined by their decay distributions, or from the measurement of the dependence of the cross-section for $e^+e^- \to \mu^+\mu^-$ near the Z^0 on the initial electron polarization—determines

$sin^2\theta_w$ to an accuracy an order of magnitude better. These two quantities should provide two independent measurements of $sin^2\theta_w$ good to an accuracy of 0.04%. The rapport between these measurements will provide a stringent test of the complete structure of the standard weak-interaction theory, including its radiative corrections.[23]

The Z^0 resonance decays dominantly into quarks and leptons, in proportions determined by their quantum numbers through the charges shown in Fig. 10. The rate of the Z^0 decay to the fermions of the standard model, ignoring their masses for the moment, is given by

$$\Gamma_Z = 90\,\text{MeV} \cdot [1.0\,N_\ell + 3.5\,N_u + 4.5\,N_d + 2N_\nu], \qquad (2.7)$$

where N_ℓ is the total number of charged leptons light enough to be pair-produced at the Z^0, and N_u, N_d, and N_ν are the corresponding numbers of up quarks, down quarks, and neutrinos. Assuming that one finds only the three generations already observed, and a top quark of mass 40 GeV, and taking into account the suppression due to the top quark mass, one would expect to see 4% of the observed Z^0's decay to each charged lepton, 14% to each up quark (9% for top), and 17% to each down quark. Note that a measurement of the width of the Z^0 resonance to 50 MeV allows one to count directly the number of light neutrinos. Since neutrinos account for such a large fraction of the Z^0 decays, this counting might also be done by studying the radiative process:

$$e^+e^- \to \gamma + \text{unobserved neutrals}$$

in the energy region just above the Z^0.

In addition to the known quarks and leptons, one can well expect that new, heavier particles will be pair-produced in Z^0 decays. In principle, the Z^0 can produce any particle that couples to the weak interactions which has a mass less than half the Z^0 mass, that is, any mass up to about 45 GeV. The phase space suppression for heavy states is offset by the large number of Z^0 events expected. Let us label the velocity of such a new particle by

$$\beta = \left(1 - \frac{4m^2}{E_{cm}^2}\right)^{\frac{1}{2}}. \qquad (2.8)$$

Then any new fermion is produced in Z^0 decays with the partial width

$$\Delta\Gamma_Z = 90\,\text{MeV} \cdot N_c \cdot 8 \cdot [(I_3^L - Q\sin^2\theta_w)^2 + (I_3^R - Q\sin^2\theta_w)^2] \cdot \beta\left(\frac{3-\beta^2}{2}\right), \quad (2.9)$$

where I_3^L, I_3^R are the weak isospin quantum numbers of the left- and right-handed components, respectively, and N_c is the number of color states. The corresponding formula for a new boson is

$$\Delta\Gamma_Z = 90\text{MeV} \cdot N_c \cdot [(I_3 - 2Q\sin^2\theta_w)^2] \cdot \beta^3. \tag{2.10}$$

A charged Higgs boson, for example, would appear in 1% $\cdot\beta^3$ of all Z^0 decays. One should keep in mind also that, because the number of expected Z^0 events is so large, particles which can only be produced by more indirect processes might still be visible experimentally down to branching ratios as small as 10^{-5}.

The evident interest in Z^0 physics has spurred the construction of two new accelerators designed to study e^+-e^- collisions at a center-of-mass energy of 100 GeV. This is not the place for a detailed discussion of these machines (such discussions may be found in Refs. 24 and 25), but I would like to briefly review their features. These two machines are of completely different design; they signal, in a certain sense, the end of one era and the beginning of another. The first of these machines to be proposed is LEP, a large synchrotron of 30 km circumference being built around the CERN site and under the Jura, for a cost of roughly 1 billion Swiss francs. The luminosity is projected to be a 10^{31} cm^{-2} sec^{-1}; the energy spread of the beam should be roughly 100 MeV. It is scheduled to be completed in 1988. It is not unlikely that LEP will be the last and largest electron synchrotron to be built. Its competitor is the first of a new type of colliding-beam accelerator: the linear accelerator collider. This device, the SLC, is being constructed at SLAC and is designed to collide directly beams extracted from the SLAC linac. A diagram of the apparatus is shown in Fig. 13. The SLC is designed to achieve a luminosity of roughly 6×10^{30} cm^{-2} sec^{-1}, not by the usual method of establishing circulating beams which interact repeatedly, but rather by extracting high-density bunches of electrons and positrons from the linac at a sufficiently high rate. The desired luminosity results from extracting 5×10^{10} particles/bunch at the SLAC linac rep rate of 180 Hz and focusing these particles to a bunch size of $1.2\mu \times 1.2\mu$ at the collision point. The estimate includes an extra factor of about 6 from the mutual attraction (and disruption) of the electron and positron bunches in the collision process. Since the electrons are extracted directly from the SLAC linac, they can be prepared as a polarized beam.

The design of LEP allows it, in principle, to reach even higher energies by the replacement of conventional with superconducting accelerating cavities. Center-of-mass energies up to 250 GeV are achievable in principle. At these elevated energies, one finds another set of new reactions predicted by the standard model—the pair production of weak-interaction bosons. The largest of these processes is the reaction

$$e^+e^- \to W^+W^-. \tag{2.11}$$

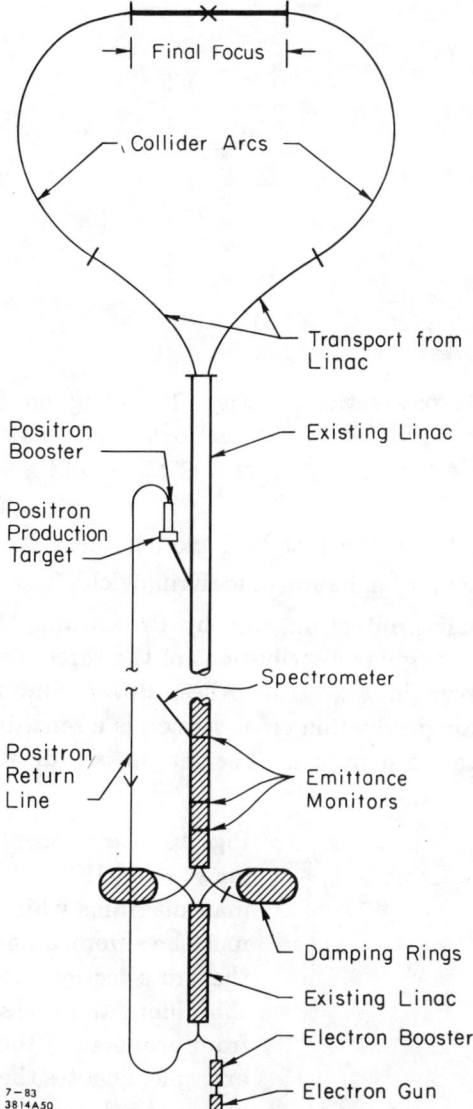

Fig. 13. Schematic plan of the SLC.

This process has its threshold at about 160 GeV but, as Fig. 14 shows, it quickly rises to become the major component of the e^+e^- annihilation cross-section. The dependence on energy and angle of this cross-section, and those of the related reactions

$$e^+e^- \to Z^0\gamma \text{ and } e^+e^- \to Z^0 Z^0, \tag{2.12}$$

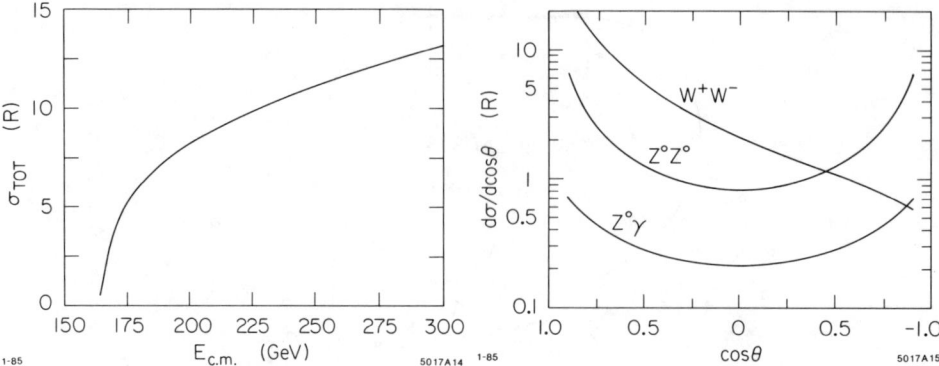

Fig. 14. The total cross-section for the reaction $e^+e^- \to W^+W^-$, in R units.

Fig. 15. Angular distributions for the reactions $e^+e^- \to W^+W^-$, $e^+e^- \to Z^0\gamma$, and $e^+e^- \to Z^0Z^0$, for $E_{cm} = 250$ GeV.

are predicted precisely by the standard model; these cross-sections were first computed by Sushkov, Flambaum, and Kriplovich[26] and Alles, Boyer, and Buras[27] for W pair production, and by Brown and Mikaelian[28] for the reactions (2.12). The angular distributions of the three reactions at the energy of 250 GeV are shown in Fig. 15. Alles, Boyer, and Buras stress in their paper that the W pair production cross-section is a sensitive test of the detailed structure of the standard model. The reason is indicated in Fig. 16. The

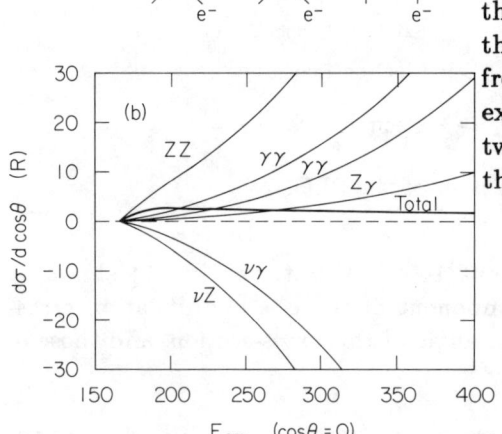

Fig. 16. Components of the cross-section for $e^+e^- \to W^+W^-$: (a) The three Feynman diagrams which (in unitarity gauge) must be summed and squared to produce the cross-section; (b). Contributions to the differential cross-section at $\cos\theta = 0$ from products of these diagrams. νZ, for example, denotes the interference term between the diagram with ν exchange and that with Z^0 exchange.

various diagrams which contribute to W pair production do not simply add; rather, they have an intricate cancellation of which the cross-section shown in Fig. 14 is the small residue.

Far above the Z^0 resonance, the cross-sections for production of quarks and leptons in e^+e^- annihilation returns to the simple form of Eq. (2.2), in which these cross-sections are essentially constant in units of R. One must, of course, include the modification that the electron and positron can annihilate through either of the light vector mesons γ or Z^0. Let me conclude my discussion of the predictions of the standard model, then, by tabulating the various contributions to the total cross-section for e^+e^- annihilation, at present energies, far below the Z^0, and at energies asymptotically far above:

Components of the e^+e^- annihilation cross-section
(in units of R)

Present Energies		$E_{\text{cm}} \gg m_Z$	
μ, τ	1	W^+W^-	≈ 15
d, s, b	$\frac{1}{3}$	$Z^0\gamma$	≈ 10
u, c	$\frac{4}{3}$	$Z^0 Z^0$	≈ 2
new fermions	$Q^2 \cdot N_c \cdot \beta\left(\frac{3-\beta^2}{2}\right)$	μ, τ	1.19
new bosons	$Q^2 \cdot N_c \cdot \frac{1}{4}\beta^3$	d, s, b	1.17
		u, c	2.04

In this table, Q represents the electric charge, N_c the number of color states, and β the velocity (2.8). The production of new heavy particles at high energies can usually be estimated from the 1-photon contribution, the Z^0 making a nontrivial but small correction. The reader should be aware that the cross-sections for (2.11) and (2.12) (expressed in R units) rise with energy. The other pair production cross-sections become constant in units of R at a level set only by the electromagnetic and weak charges of the produced particles, reflecting a perfect democracy between heavy and light species.

3. Particle Search Experiments

Now that we have reviewed the features of e^+e^- annihilation according to the standard model, let us begin to look beyond this model, toward the discovery of new elements of fundamental physics. In this section, I will discuss the search for new physics as it has been carried on in the current experiments at PEP and PETRA. The participants in these experiments have put a great deal of effort into the search for new types of elementary particles. This search for new particles is an important pursuit for deep reasons associated with the deficiencies of the standard model and the possibility that this model will be superceded at very high energies. For my discussion in this section, however, I will discuss this search on its own terms and for its own intrinsic interest. The main question which I would like to explore is the following: We have seen in the previous section that new particles with electromagnetic or weak interactions are produced in e^+e^- annihilation with cross-sections of order 1 unit of R. But if such particles are actually being produced, can their presence be observed?

You are well aware that no new particles have yet been discovered at PEP or PETRA. What I will be discussing in this section, then, is a record of the failure to observe new states. This record is, of course, disappointing for the present, but I feel that it is very encouraging for the future. In the course of the search for new particles at present energies, the experimenters at PEP and PETRA have been exploring the backgrounds which might hide the new particles accessible at future machines. They have shown that these background are, in fact, exceedingly small. In many cases, one can find cuts which would retain a substantial fraction of the signal for the production of a new particle while eliminating virtually all possible background events. I would like to discuss several particle searches in some detail, to show the remarkable extent to which new particles should make themselves visible. Please note that this will not be a systematic survey of all particle searches at PEP and PETRA; however, you find such a systematic review in papers of Wu,[3] Lau,[29] and Yamada.[30]

I will focus my discussion on three particular unusual states which have been sought at PEP and PETRA: a scalar particle with muon number $\tilde{\mu}$, a new heavy lepton L^+, and a charged Higgs particle H^+. The first of these is predicted by the notion of supersymmetry; the last appears in almost any extension of the standard weak interaction theory. All three are remarkable states which would be, in any event, of great intrinsic interest. Let us ask, then, how carefully experimenters at the present e^+-e^- colliders have been able to search for them.

Let us begin with the $\tilde{\mu}$. This is a charged boson, and so it can be pair-produced in e^+e^- annihilation with a cross-section corresponding to $\frac{1}{4}\beta^3$ R units. In the supersymmetric models which require its existence, it is expected to decay via

$$\tilde{\mu} \to \mu + \tilde{\gamma}, \tag{3.1}$$

in which the $\tilde{\gamma}$ is a fermion with the quantum numbers of the photon, called a photino. In practice, the $\tilde{\gamma}$ should be light and very weakly interacting. If both $\tilde{\mu}$'s of the pair decay by (3.1), $\tilde{\mu}$ production results in the observed signature:

$$e^+e^- \to \mu^+\mu^- + (\text{missing } E, \ p_\perp). \tag{3.2}$$

There is a background from μ pair production via the 2-photon process, in which each electron bremmstrahlungs a photon, and these photons interact to produce a μ pair, but this process only rarely gives substantial transverse momentum to the 2-μ system.

The CELLO collaboration[31] has reported a search for the signature (3.2); this search made use of 11 pb^{-1} of data at $E_{\rm cm}$ between 33 and 36.7 GeV. This corresponds to about 800 (R units)$^{-1}$, that is, a quantity of data in which a reaction with a cross-section of 1 R unit produces 800 events. They selected for events of the type (3.2) by placing the following relatively loose cuts on the data: (1) less than 8 charged tracks, of which 2 had $p_\perp > 800$ MeV, $35° < \theta < 145°$; (2) two tracks are identified as muons and have $p > 0.2\,E_{\rm beam}$; (3) these two tracks are acoplanar by more than 30°; and (4) no additional neutral shower is observed. No event in the CELLO data sample passed these cuts. Figure 17(a) shows the number of events to be expected if a $\tilde{\mu}$ indeed existed in the PETRA energy region, and the 95% confidence limits placed by this experiment. Along the bottom of the graph I have recorded the total number of $\tilde{\mu}$-pair events which would be present in the whole data sample for a $\tilde{\mu}$ of that mass. Over a wide range of masses, more that 10% of these events pass the cuts which exclude all background. Figure 17(b) shows the analogous bounds which this experiment places on the existence of a scalar electron.

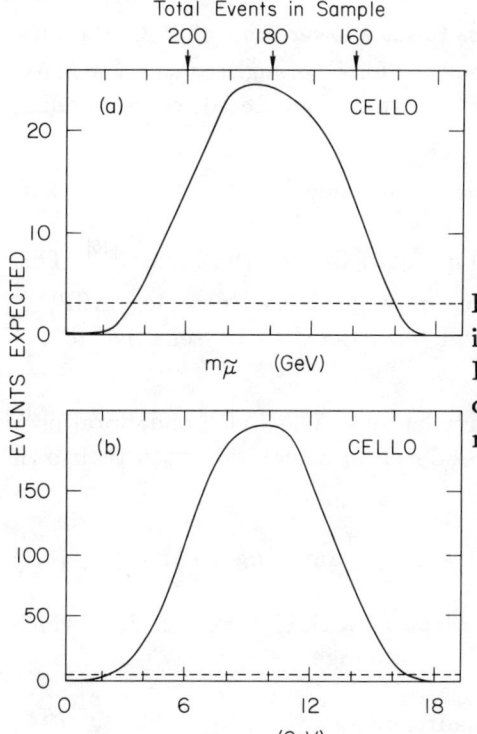

Fig. 17. Expected number of events passing the cuts of the CELLO experiment, Ref. 31, if there existed (a) a scalar muon or (b) a scalar electron of the indicated mass.

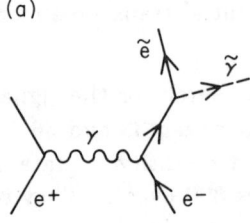
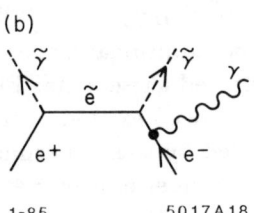

Fig. 18. Two processes sensitive to the presence of scalar electrons too heavy to be pair-produced in e^+e^- annihilation. The second process has no threshold in the mass of the $\tilde{e}$, though it is suppressed as the $\tilde{e}$ becomes heavy.

Let me digress to note that, with larger data samples, it is possible to search for supersymmetry even if the $\tilde{\mu}$ and $\tilde{e}$ are too heavy to be pair-produced. One way to do this is through the reaction shown in Fig. 18(a), a process introduced by Gaillard, Hinchliffe, and Hall.[32] The Mark II, MAC, and JADE collaborations[33–35] have used this process to set a lower limit of 25 GeV on the mass of the $\tilde{e}$ (assuming a very light $\tilde{\gamma}$). To probe for $\tilde{e}$'s of higher mass, the MAC collaboration has searched for the process shown in Fig. 18(b), corresponding to the signature

$$e^+e^- \to \text{wide angle } \gamma + \text{nothing} \qquad (3.3)$$

and has reported a (corresponding) lower limit of 37 GeV on the $\tilde{e}$ mass.[36] This process is now being sought at PEP with a new specialized detector, ASP;[37] the ASP collaborators expect, with 100 pb^{-1} of data, to be sensitive to the presence of $\tilde{e}$'s with mass up to 60 GeV.

Let us now turn to a search for new heavy leptons. The JADE collaboration[38] has reported a search for heavy leptons produced in association with their own neutrinos, leading to the process

$$e^+e^- \to L^+ (\to \text{hadrons} + \bar{\nu}) + L^- (\to \text{anything} + \nu). \qquad (3.4)$$

They have also searched for neutral heavy leptons with electron number, which might be produced from the electron by W^+ exchange:

$$e^+e^- \to E^0(\to \text{hadrons}) + \bar{\nu}. \qquad (3.5)$$

In either case, the emission of unobserved neutrinos leads to acoplanarity and missing energy. The JADE analysis (actually, I report only on the 'high-mass' analysis of Ref. 38) made use of 37 pb^{-1} of data at E_{cm} between 27.2 and 36.7 GeV, corresponding to 2800 (R units)$^{-1}$. Events were selected according to the following criteria: (1) a total energy > 3.0 GeV is deposited in the shower counters, or an energy > 0.4 GeV is deposited in each endcap shower counter; (2) the event contains at least 4 tracks, excluding the case of 1 isolated track and 3 opposite it; (3) $0.33 < E_{visible}/E_{cm} < 1.0$; and (4) the thrust axis points at an angle such that $|cos\theta_{thrust}| < 0.7$. For the sample of events which pass these loose cuts, one can divide the event into hemispheres, compute the thrust axis appropriate to each hemisphere, and plot the distribution of events as a function of the acoplanarity angle between these two axes. Figure 19 shows this distribution, together with the distributions which one would have expected from the processes (3.4) and (3.5). Placing a cut at an acoplanarity angle of 50° yields only 2 candidate events; this excludes the presence of new heavy leptons at the 95% confidence level for 6 GeV < m_L < 18 GeV and for 6 GeV < m_E < 24.5 GeV. A similar analysis carried out by the MARK J collaboration[39] excluded new heavy leptons for m_L < 16 GeV using a data sample of only 520 R^{-1}.

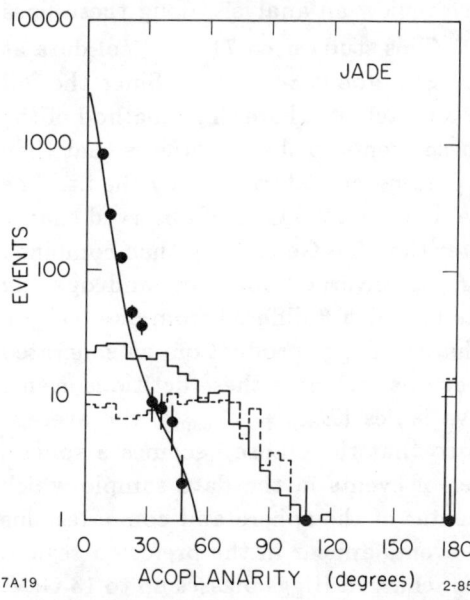

Fig. 19. Distribution of heavy lepton candidate events in acoplanarity angle, comparing data to Monte Carlo distributions for the processes (3.4) (dashed histogram) and (3.5) (solid histogram), from Ref. 38.

Finally, let us consider the search for a charged Higgs scalar H^+. The dominant decay modes of the Higgs should be those to heavy leptons ($\tau^+\nu_\tau$) and to heavy quarks ($c\bar{s}$ or $c\bar{b}$). The relative branching ratio into leptons versus hadrons is highly model-dependent, however, and it is necessary to search for the H^+ by different techniques depending on the assumed value of this ratio. If the branching ratios into hadrons and leptons are both substantial, one can search for the reaction

$$e^+e^- \to H^-(\to \text{hadrons})$$
$$+ H^+(\to \nu + \tau^+(\to \mu^+\nu\bar{\nu})).$$

(3.6)

The final signature is a hadronic jet opposite an isolated acollinear muon. The MARK J collaboration[40] has reported a search for events of this class, based on 40 pb^{-1}, or about 3000 R^{-1}, of data. They required (1) a muon opposite a hadronic shower; (2) more than 2 tracks in the inner vertex chamber; (3) thrust < 0.93; (4) $0.3 < E_{\text{visible}}/E_{\text{cm}} < 0.75$; (5) $|\sum p_\perp| > 0.4 E_{\text{visible}}$; and (6) $|\sum p(\perp \text{ to plane of } \mu \text{ and beam})| > 0.15\ E_{\text{cm}}$. No candidate event passed these cuts. An H^+ with a mass of 13 GeV and a branching ratio of 25% to $\tau^+\tau^-$ would have produced 3 events in this sample.

The case in which the Higgs decays dominantly into leptons is also quite straightforward to explore. The case in which the Higgs decays dominantly into hadrons, however, is rather tricky by the standards of e^+e^- annihilation, since the decay products of each Higgs superficially resemble hadronic jets. One can, however, make use of the narrowness and low invariant mass of typical e^+e^- jets to distinguish the Higgs. Let me review an analysis along these lines performed by the TASSO collaboration.[41] This study used 71.5 pb^{-1} of data at E_{cm} between 33 and 37 GeV, corresponding to about 5400 R^{-1}. Since the full analysis is rather complex, I will give only a sketch of it here. The method of the study was to select hadronic events, fit these events to the hypothesis that they contain 4 jets, and then place cuts on the parameters determined by the fit. The authors insisted that each jet should contain at least 2.6 GeV of observed energy and should reconstruct to an energy greater than 3.6 GeV. They then combined jets in pairs and computed the energy E_{pair}, invariant mass m, and opening angle θ of each pair. They selected events in which θ differed from one side to the other by less than 9°. Events with charged Higgs production, as generated by a Monte Carlo program, generally meet this criterion; these fictitious events also cluster into an ellipse in the three variables $(E_{pair} - E_{\text{beam}})$, the average m, and the average θ. Change variables so that the ellipse becomes a sphere; then Fig. 20 shows the expected number of events in the data sample which should fall at given distances from the center of the sphere and compares this distribution to the data. Clearly, no real events appear in the preferred region. The analysis gives a stringent constraint for charged Higgs masses up to 13 GeV.

Figure 21 summarizes the results of the various experiments which have searched for charged Higgs mesons at PEP and PETRA. It is remarkable that, despite the elusive nature of this particle and the vagueness of theoretical predictions for its decays, e^+e^- annihilation experiments are sensitive to the presence of the charged Higgs throughout almost the whole of the mass range which is accessible kinematically.

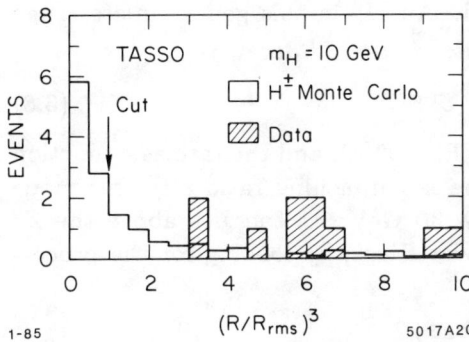

Fig. 20. Search for events containing charged Higgs pairs, Ref. 41: The figure shows the number of expected vs. actual events at various distances from the center of the sphere in parameter space defined by the analysis, for $m_{\text{Higgs}} = 10$ GeV.

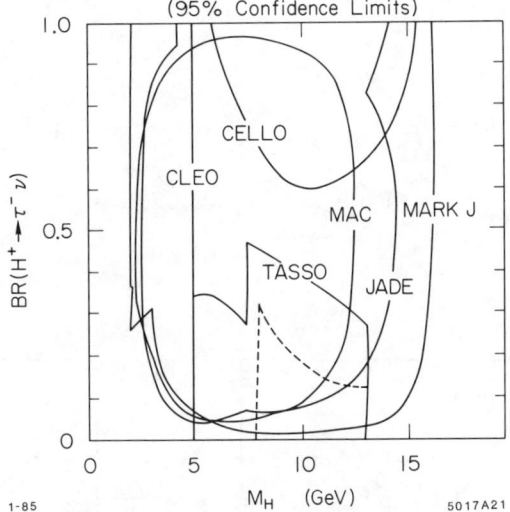

Fig. 21. Summary of the results of PEP and PETRA experiments which have searched for charged Higgs mesons, from Ref. 30.

I should properly note that the neutral Higgs meson H^0 is not easily produced in e^+e^- annihilation; this unfortunate property is, however, shared by all other types of high-energy accelerators. If the mass of the Higgs lives in specific regions, however, it does stand out if it is sought in certain elegant reactions. If the mass of the neutral Higgs is less than the mass of the lowest $t\bar{t}$ resonance, the Higgs should appear in the Wilczek process:[42]

$$\varsigma(t\bar{t}) \to \gamma + H^0. \tag{3.7}$$

If the mass of the ς is about 80 GeV, the branching ratio for this reaction in the simplest form of the standard model is 3% of all ς decays. The monoenergetic

γ is easily distinguished from background. An H^0 in this general mass range might also be found in the Bjorken process,[43] the Z^0 decay

$$Z^0 \rightarrow H^0 + \ell^+ \ell^-. \qquad (3.8)$$

The mechanism for this decay is shown in Fig. 22(a), and the rate as a function of Higgs mass is shown in Fig. 22(b); the Z^0 branching ratio into this mode is greater than 10^{-5} for H^0 masses below 30 GeV. At energies above the Z^0 resonance, one may search for a still heavier Higgs, by looking for the process shown in Fig. 22(c):[44]

$$e^+ e^- \rightarrow H^0 + Z^0. \qquad (3.9)$$

The rate of this process at $E_{\text{cm}} = 200$ GeV as a function of the mass of the H^0 is shown in Fig. 22(d).

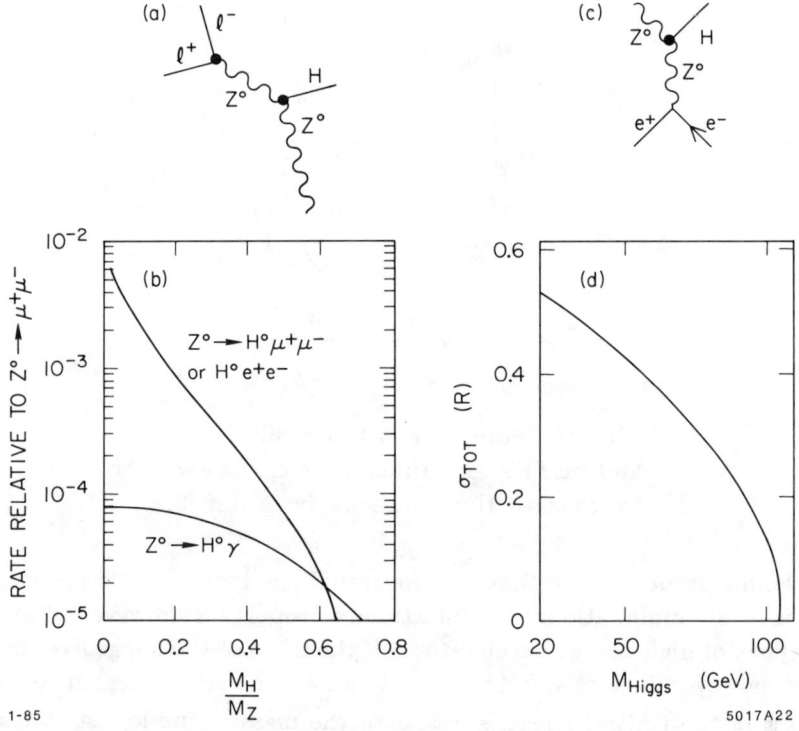

Fig. 22. Processes for producing a neutral Higgs meson H^0 in $e^+ e^-$ annihilation: (a) the Bjorken process; (b) the rate of the Bjorken process as a function of the H^0 mass; (c) production of the H^0 accompanied by a Z^0; (d) the rate of this process, in R units, at $E_{\text{cm}} = 200$ GeV.

A discussion of the status of the search for new particles would not be complete without a comment on a closely related endeavor, the search for substructure within the quarks and leptons. Like the search for new particles which would force us beyond the standard model, this study has so far been unsuccessful. No form factor effects have been seen in the hard scattering of quarks or leptons. For those two fermions for which the magnetic moment has been accurately measured—e and μ—the value of $(g-2)$ agrees quite precisely with the predictions of QED. Thus, if quarks and leptons are composite, their intrinsic size is much smaller than the distances we have currently probed; equivalently, the mass scale Λ which characterizes the internal momentum of quarks and leptons is much larger than the momentum transfers of 30-40 GeV which we may now consider well-explored. It is, then, interesting to ask, first, what experiments now provide the best lower bounds on Λ, and, secondly, how sensitively e^+e^- annihilation experiments can probe the existence of substructure.

Actually, two experiments on substructure stand out from the others in placing a lower bound on Λ which is close to 1 TeV. The first of these is the measurement of the muon $(g-2)$ value: The fact that this measurement agrees with the QED prediction to an accuracy

$$\frac{1}{2}|g-2| - \text{QED} < 1.5 \times 10^{-8} \tag{3.10}$$

requires, roughly, that $\Lambda > 800$ GeV.[45] The second measurement is one conducted in e^+e^- annihilation, the accurate test of the standard model prediction for Bhabha scattering at high energies. If electrons are composite objects, one would expect that electrons and positrons could scatter through a process not included in the standard model in which these particles overlap and exchange their constituents. This process would lead to an additional contribution to the amplitude for Bhabha scattering, of the form of a 4-fermion contact interaction, as shown in Fig. 23. Several experiments have searched for deviations from the standard model prediction. Figure 24 shows a comparison of the angular distribution in Bhabha scattering, as determined by the MAC collaboration,[46] to the standard model and to models with additional contact interactions. The bound which this experiment places on Λ is sensitive to the Lorentz structure assumed for the contact interactions, but in any case it is quite strong; the MAC group quotes, as 95% confidence limits, $\Lambda > 1.2$ TeV for purely left- or right-handed contact interactions and $\Lambda > 2.5$ TeV for vector-like contact interactions. Other analyses, giving bounds of similar quality, have been reported by HRS and JADE.[30,47]

We have now reviewed several different aspects of the search for physics beyond the standard model as it has been carried out at PEP and PETRA, the highest-energy e^+-e^- colliders now in operation. The e^+e^- annihilation experiments of the current generation have proved their ability to probe for the

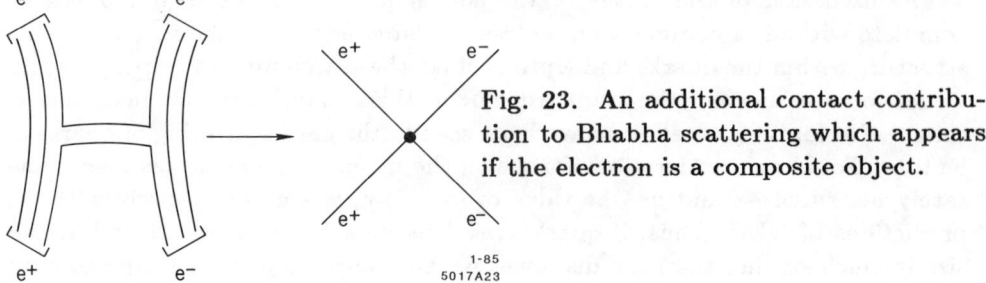

Fig. 23. An additional contact contribution to Bhabha scattering which appears if the electron is a composite object.

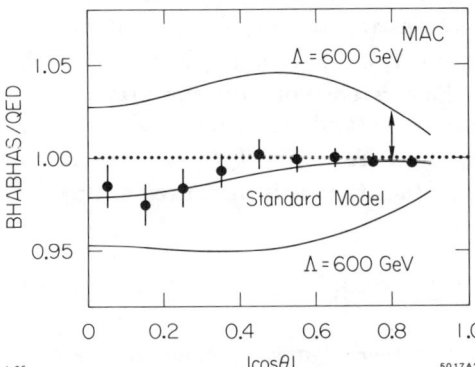

Fig. 24. Comparison of the angular distribution for Bhabha scattering obtained by the MAC collaboration (Ref. 46) with the prediction of the standard model and of a model with purely right-handed contact interactions with $\Lambda = 600$ GeV. The overall normalization is arbitrary, within the limits indicated by the arrow. The two predictions of the contact interactions correspond to the two possible signs of the interference.

presence of new particles definitively and unambiguously. That they have found none presumably reflects only the relatively low energies at which these experiments have been done. Let us now, therefore, turn to the question of how far in energy one must go in order to find new physics, the question of why one might expect to see new particles and new interactions and where, given this expectation, these new states must occur.

4. A Digression: 3 Kinds of 1 TeV Physics

To what extent does our present knowledge of physics call for the extension or improvement of the standard model? I know of no experiments which definitively contradict the standard model or require its extension; indeed, I have shown you in the previous section that the standard model has passed many stringent tests. Yet I do believe that a close examination of the details of this model reveals inadequacies whose correction would alter the model profoundly. In this section, I would like to discuss this issue. The question is essentially one of theoretical physics, in one of its more speculative manifestations, yet the answer to the question is obviously of great importance in choosing and planning for future accelerators. Allow me, then, this rather theoretical digression.

The standard model is essentially a theory of gauge symmetries and the interactions of gauge bosons. Its prediction for the couplings of these gauge bosons to fermions is reflected directly in the structure of the weak interactions, and, as such, is brilliantly confirmed by experiment. However, there is another aspect to the fundamental interactions which the standard model touches upon but does not at all illuminate. This is the production of masses for the quarks and leptons and for the weak bosons. It is a slight oversimplification, but, I think, a fair one, to say that the standard model offers as an explanation of the masses of the $W^\pm$ and Z^0 bosons and the quarks and leptons only a set of formulae of the form:

$$m_W = \frac{1}{2}g\langle\phi\rangle \qquad m_Z = \frac{1}{2}(g^2 + g'^2)^{\frac{1}{2}}\langle\phi\rangle$$

$$m_t = \lambda_t\langle\phi\rangle \qquad \ldots \qquad (4.1)$$

$$\ldots \qquad m_e = \lambda_e\langle\phi\rangle$$

Here g and g' are gauge couplings and therefore connected to other aspects of the theory; the λ_f, however, are new dimensionless numbers whose values range from 10^{-1} to 10^{-5}, and $\langle\phi\rangle$ is a new dimensionful parameter whose value

$$\langle\phi\rangle = 250 \text{ GeV} \qquad (4.2)$$

may be determined from the masses of the weak bosons. In the Weinberg-Salam theory, this is the vacuum expectation value of the fundamental Higgs scalar field which that model invokes. In the standard model all of these new parameters must be adjusted to their phenomenologically correct values by hand; more technically, they are all renormalized parameters, like the value of the fine structure constant in QED, whose values must be specified in order to define the theory. Grand Unified Theories, which introduce new structure at extremely large mass scales of order 10^{15} GeV, can explain the size of g and g' but say little about the other parameters. The values of the λ_f are a mystery from almost every perspective. If we wish to understand more deeply the sizes of the various quantities in Eq. (4.1), it seems that we find as our main clue the value of the dimensionful parameter $\langle\phi\rangle$.

It is tempting to suggest that the value of $\langle\phi\rangle$ is caused by something, that $\langle\phi\rangle$ appears as the calculable result of some physical process. If this is true, that process must add physics to the standard model. This new physics must somehow be associated with the mass scale set by $\langle\phi\rangle$, or, to speak roughly, the scale of 1 TeV. Let me sort the possibilities for what might appear into three broad classes, by saying that new physics at 1 TeV might be absent, weak, or strong.

Absent? I have already noted that the standard model, with the inclusion of a Higgs scalar meson, is an internally consistent theory which has so far been verified by experiments. It could be exactly correct. In that case, the parameters of Eq. (4.1) will remain a mystery, and, worse, we will see no new physics, except perhaps the Higgs boson itself, at any foreseeable new accelerator beyond the current generation. I do not consider this scenario likely, but one must still remember that it is possible.

Weak? I will use this term to refer to models in which the new scale at 250 GeV arises from new interactions which are essentially weak, in the sense that they can be described by Feynman diagram perturbation theory. To understand the structure of such a model in a general way, let us examine the Feynman diagrams which contribute to $\langle\phi\rangle$. Figure 25(a) shows the leading contributions to $\langle\phi\rangle$ in the standard model. The loop diagram is quadratically ultraviolet divergent; this is the same divergence which appears in the renormalization of the Higgs boson mass. If $\langle\phi\rangle$ and the Higgs mass are to be predicted by the theory, rather than being free parameters, we must add some other diagram which contributes to $\langle\phi\rangle$ and can cancel the divergent part of this loop. This new diagram would necessarily involve either new particles with masses in the range of 100 GeV - 1 TeV or new interactions which become active for particle momenta of this range.

Examples of models of this class are provided by unifying theories based on the idea of supersymmetry. Supersymmetry is a symmetry which transforms bosons into fermions and *vice versa*. Thus, if Nature is assumed to be supersym-

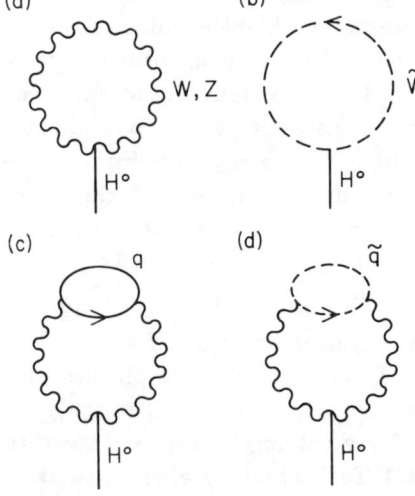

Fig. 25. Perturbative contributions to the Higgs field vacuum expectation value $\langle\phi\rangle$: (a) the leading contributions in the standard model; (b) an additional contribution which arises in supersymmetric models; (c) a higher order contribution involving virtual quarks, and (d) a related contribution containing the supersymmetric partners of quarks.

metric, even if this symmetry is broken spontaneously, there must be, for every boson that we see, a corresponding fermion with the same quantum numbers. In particular, the $W^\pm$ and Z^0 must have fermionic partners, and these can contribute to $\langle\phi\rangle$ through the diagram shown in Fig. 25(b). In still higher orders of perturbation theory, the quarks and leptons enter the calculation of $\langle\phi\rangle$, as shown in the graph of Fig. 25(c). The cancellation of the quadratic ultraviolet divergences of this diagram requires that the quark and leptons have bosonic partners which contribute, for example, the diagram of Fig. 25(d). If the theory is precisely supersymmetric, the sum of the contributions to $\langle\phi\rangle$ contains no quadratic ultraviolet divergences but only the more controllable divergences associated with coupling constant renormalizations and the wavefunction renormalizations of the various fields.

Strong? I will use this term to refer to models in which the new scale at 250 GeV is produced by new interactions by means of bound state formation or other effects characteristic of strong interactions. What is needed is that these effects break spontaneously the weak interaction gauge symmetry $SU(2) \times U(1)$. Studies of the phenomenology of QCD and of model field theories have offered us several mechanisms by which this symmetry-breaking might occur. Any one of these possibilities requires, again, the presence of new particles of mass roughly 1 TeV. At the most basic level, none of the interactions in the standard model can be strongly coupled at this new scale; thus, we must invoke either some change in the particle content of the standard model or, more likely, some completely new interaction. We might hope for new physics of considerable complexity and interest: There is no reason why these new forces should not build as rich a spectrum of particles as one finds in the conventional strong interactions.

Two distinct realizations of this idea have been discussed prominently in the literature. The first of these introduces so-called technicolor interactions and makes use of the analogue in these new interactions of the strong-interaction symmetry-breaking effects which are found in QCD. Some time ago, Weinberg[49] and Susskind[50] noted that the physics which generates large effective masses for the light quarks, and, at the same time, breaks the chiral $SU(3)$ symmetry of the strong interactions, breaks, at the same time, the weak-interaction symmetry $SU(2) \times U(1)$. This effect is rather small compared to the symmetry-breaking effects induced by $\langle\phi\rangle$, but the weak gauge bosons do receive contributions to their masses in the correct pattern:

$$m_W = \frac{1}{2}gf_\pi \qquad m_Z = \frac{1}{2}(g^2 + g'^2)^{\frac{1}{2}}f_\pi \qquad m_\gamma = 0. \qquad (4.3)$$

In (4.3), f_π is the pion decay constant, equal to 93 MeV. To produce a reasonable theory of W and Z masses, we need only postulate a higher-mass copy of the strong interactions, with new fermions and new bound states, such that the analogue of the pion decay constant in this new theory is equal to the value of $\langle\phi\rangle$. I should note that no completely satisfactory method is know for coupling this symmetry-breaking to the ordinary quarks and leptons to produce the parameters λ_f which appear in (4.1). The most straightforward such method[51,52] leads to a complex system of fermion-fermion couplings which produces, in particular, too large an amplitude for $K^0 - \bar{K}^0$ mixing.[53,54] A second possibility for this new dynamics is that it might be based on the formation of the quarks and leptons as bound states of more elementary constituents, with the weak-interaction symmetry-breaking being driven by the fermion bound-state formation.

Whether the new physics at 1 TeV is weak or strong, new particles and other novel phenomena should be visible to experiments which can provide such large momentum transfers. For the case of 'weak' 1 TeV physics, these new particles will be the primary manifestation of the physics which produces the new mass scale. For the case of 'strong' 1 TeV physics, these new particles provide at least one indication of the existence of new phenomena. Let me emphasize this point by showing you two conjectured mass spectra taken from the theoretical literature: Fig. 26 shows the result of a compilation done by John Ellis[6] of the masses of novel particles predicted in a number of supersymmetric unified

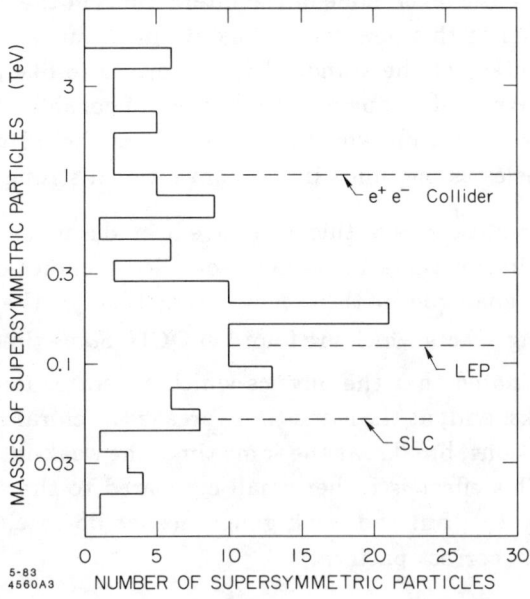

Fig. 26. Compilation of the predicted masses of novel particles in a number of supersymmetric grand unified theories, from Ref. 6

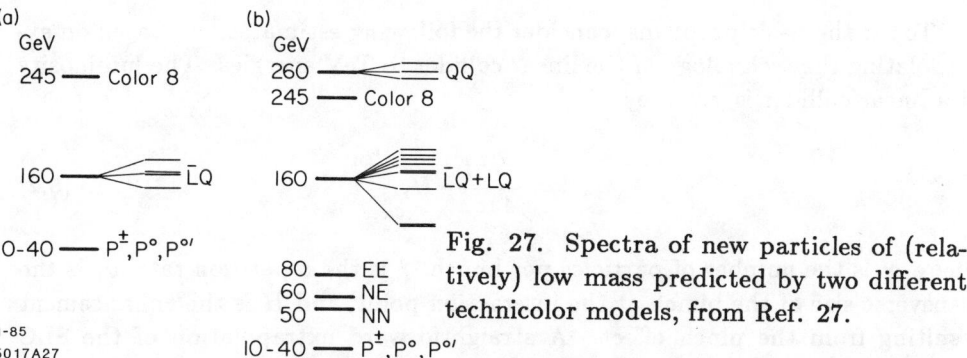

Fig. 27. Spectra of new particles of (relatively) low mass predicted by two different technicolor models, from Ref. 27.

models. Figure 27 shows the spectrum of the lowest-mass particles bound by the new forces in two classes of technicolor models.[27] In each case, there is a rich variety of new states to be expected in the mass region up to roughly 1 TeV.

In e^+e^- annihilation, all of these states which possess electromagnetic or weak interactions are produced with cross-sections of units of R:

$$\Delta R \approx Q_{EM}^2 \cdot \text{(no. of colors)} \cdot (\frac{1}{4} \text{ for bosons}) \cdot \text{(phase space)}, \qquad (4.4)$$

as long as the center-of-mass energy is sufficiently high that the state in question can be pair-produced. Since the production of ordinary particles is also at the level of units of R, the new states should clearly manifest themselves above the background. All we need, then, to search for these particles, is a machine which can produce a sufficient total rate: 10 R^{-1}/day, at $E_{\rm cm} = 1$ TeV.

5. Technics for 1 TeV e^+-e^- Collisions

I should, however, express the optimistic theoretical conclusion of the previous section in units which put it in better perspective. At 1 TeV, a luminosity of 10 R^{-1}/day corresponds to

$$\mathcal{L} \sim 10^{33} \text{cm}^{-2}\text{sec}^{-1}, \qquad (5.1)$$

a value higher by two orders of magnitude than the luminosities which can now be achieved at high-energy e^+-e^- colliders. The design of e^+-e^- colliders for physics in the TeV energy region must confront these two coupled problems of reaching high energy and high luminosity. In this section, I would like to discuss, on a very basic level, some of the technical challenges which must be met to achieve these goals.

To see the basic problems, consider the following estimates,[56] based on extrapolating the technology of the linear collider to TeV energies. The luminosity of a linear collider is given by

$$\mathcal{L} = \frac{N^2 f}{4\pi \sigma_r^2} \cdot H, \qquad (5.2)$$

where N is the number of particles per bunch, f is the repetition rate, σ_r is the transverse size of the bunch at the interaction point, and H is the enhancement resulting from the pinch effect. A straightforward extrapolation of the SLC technology would give, for a typical set of parameters yielding a luminosity of 10^{33} cm^{-2}sec^{-1}, the following: $N = 1.4 \times 10^{10}$, $f = 2000$ Hz (180 Hz cavities operated at 12 bunches/cycle), $H = 6$, and $\sigma_r = 0.14\mu$ (0.1 · the SLC design value). These parameters require each beam to carry a power

$$P = N \cdot f \cdot (1 \text{ TeV}) \qquad (5.3)$$

equal to 4.7 MWatt/beam. This does not seem completely unreasonable until one recalls that the SLAC linac converts electrical line power into beam power with an efficiency of about 3%. A more general relation, in terms of the parameter

$$D = \sigma_z/(\text{focal length}), \qquad (5.4)$$

is

$$\mathcal{L} = \frac{\gamma N^2 f H}{4\pi \epsilon_n \beta^*} = \frac{3.5 \times 10^{31} \cdot P(\text{MW}) \cdot D \cdot H}{\sigma_z(\text{mm}^2)}. \qquad (5.5)$$

To the above constraints we must add one more. The particle densities in each bunch which are required to achieve high luminosity are sufficiently large that each bunch, as it crosses through the other, emits so much radiation that its overall energy is affected. This effect, called "beamsstrahlung", gives rise to an intrinsic energy spread of each beam which grows according to

$$\sigma_E/E \sim E\mathcal{L}. \qquad (5.6)$$

In terms of our basic set of parameters,

$$\frac{\sigma_E}{E} = \frac{414 E(\text{TeV}) \mathcal{L}(10^{33})}{\sigma_z(\text{mm}^2) f(\text{Hz})}. \qquad (5.7)$$

A reasonable design should at least maintain $\sigma_E/E < 0.1$. We will examine in the next section what values of σ_E/E are required by the physics.

A sample design for a 1 TeV on 1 TeV e^+-e^- collider consistent with these constraints is given in Table 1. I should emphasize that this design represents an existence proof for such a collider; it is not an optimized design. The design is based closely on a scaling up of current SLAC technology: The left-hand column assumes a linear accelerator with a gradient equal to that currently available at SLAC; the right-hand column is based on the use of klystrons of basically the same type operating at double the frequency currently used. The

TABLE 1

An incompletely optimized example of a 2 TeV linear collider at 10^{33} luminosity and 10% σ_{E^*}/E^* for two RF wavelength, from Ref. 56.

β_y^* (cm)		1
D		2
σ_z (mm)		2
ϵ_n (mrad)		4×10^{-6}
N		1.4×10^{10}
f (Hz)		2000
b		12
P_b (MW per beam)		4.7
λ (cm)	10	5
G (MV/m)	20	40
L (km – each linac)	50	25
Number of Klystrons	3500	3500
Total Input AC Power (MW)	390	290

cost of a facility based on this technology would be roughly that envisioned for the Superconducting SuperCollider.

One must, however, remember that the art of designing linear e^+-e^- colliders has hardly begun to develop. In the past few years, a number of new acceleration methods have been proposed—among them, the laser near-field, wake-field, and plasma beat-wave concepts—which are based on accelerating structures of high power density and correspondingly small transverse dimension and which might achieve gradients of several hundred MeV/meter. It seems likely that one of these methods could form the basis of a design which would appear radical compared to that of Table 1 but which would be relatively economical in terms of construction cost and power bill. I hope that those of you who are attending this summer school (and those of you who are reading this lecture) will consider this prospect as a challenge.

6. Physics of 1 TeV e^+-e^- Collisions

What, then, can one hope to discover at very high energy e^+-e^- colliders? The most important effects, as I have emphasized repeatedly earlier in this lecture, involve the production of new types of particles. I have already explained that e^+-e^- colliders are ideal facilities for the production of new particles, if the necessary luminosity can be provided. In this section, I would like to amplify this viewpoint in several ways. First, I will discuss a few technical points relevant to the physics of 1 TeV e^+-e^- reactions. Next, I will explain how to compare the energies of e^+-e^- and p-p or p-$\bar{p}$ colliders. Finally, I will discuss a number of phenomena associated with the case of 'strong' 1 TeV physics which will most probably only be visible in the special environment of e^+e^- annihilation reactions.

Let us first dispose of three technical questions. The first concerns a new background not present in current e^+e^- annihilation experiments. I explained in § 3 that the process $e^+e^- \to W^+W^-$ should be a major component of the e^+e^- annihilation cross-section for center-of-mass energies above a few hundred GeV. At first sight, this seemed a good feature, but at very high energies one might worry that W pair production might be a substantial background to other types of new physics. One might worry a bit more on seeing the magnitude of this cross-section, which I show in Fig. 28 for energies up to 1 TeV. Fortunately,

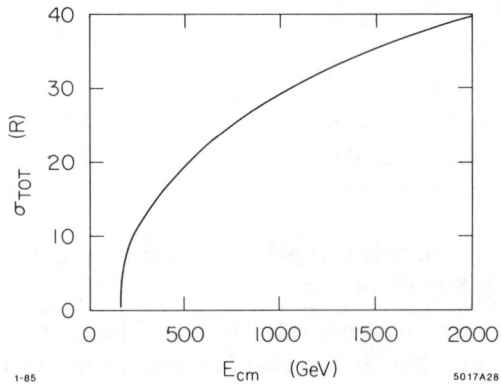

Fig. 28. The total cross-section for the reaction $e^+e^- \to W^+W^-$, in R units, for $E_{\rm cm}$ up to 2 TeV.

since the most important contribution to W pair production comes from the exchange of a neutrino in the t-channel (the first diagram indicated in Fig. 16), this cross-section also becomes increasingly forward-peaked as $E_{\rm cm}$ increases. To assess the sizes of backgrounds, then, we should examine the behavior of the differential cross-section $d\sigma/d\cos\theta$, at angles away from the forward direction. In Fig. 29, I have replotted the three gauge boson production cross-sections shown in Fig. 15 on a linear scale and compared these estimates, made at $E_{\rm cm} = 250$ GeV, to corresponding results at 4 TeV. It is remarkable that the differential cross-section for W pair production, and also, for that matter, the differential cross-sections for Z^0 production, show very little energy dependence except just in the forward direction. Since new-particle production has, as I have argued above, a broad angular distribution, the background from W and Z production turns out to be rather small.[5]

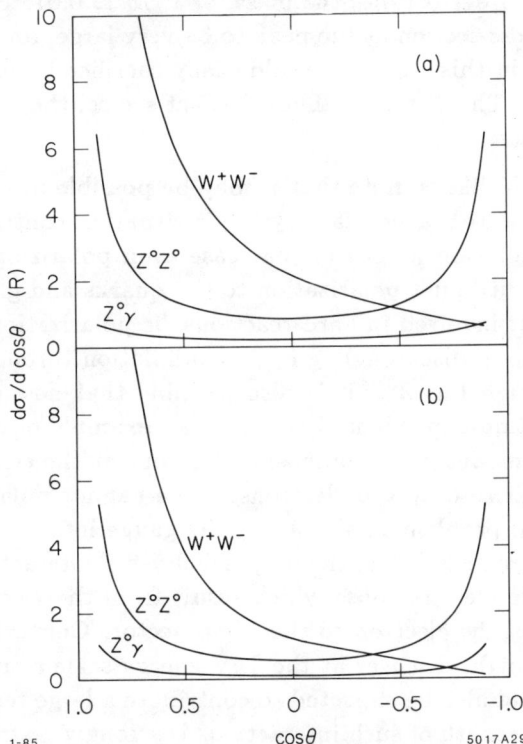

Fig. 29. Angular distributions for the reactions $e^+e^- \to W^+W^-$, $e^+e^- \to Z^0\gamma$, and $e^+e^- \to Z^0Z^0$, for (a) $E_{\rm cm} = 250$ GeV, (b) $E_{\rm cm} = 4$ TeV.

The second technical issue is that of the energy spread of the beams in high-energy e^+-e^- collisions. I noted in the previous section that linear colliders have the property that increasing the luminosity also increases the spread of energies in the beam, as a result of "beamstrahlung" (Eq. (5.7)). Thus, one might well have to live with e^+ and e^- beams with a 10% spread in energy. This decreases the cleanliness of the e^+e^- annihilation environment, but not, I think, significantly. Such a beam spread would have almost no effect on new particle searches away from resonances; it would not, for example, have jeopardized any of the experiments discussed in § 3. I will argue below that, if there are new strong interactions at TeV energies, the new resonances associated with these interactions should be rather broad, with widths of order 10% of their masses. Such resonances could be explored without extremely fine energy resolution. The one case in which a fine energy resolution would be valuable is if there existed a new Z^0, with mass roughly 1 TeV. For such a particle, one would expect a width of order α times the mass, or $\Gamma/M \approx 0.01$. However, one would also expect the cross-section at the peak to be very large, roughly $(M/\Gamma)^2 \approx 10^4$ units of R. Thus, in this case, one could easily sacrifice luminosity for a gain in energy resolution. The "beamstrahlung" effect seems, then, not to be a serious problem for physics.

Finally, I would like to note that it may be possible to polarize at least the electron beam at a TeV e^+-e^- facility. This situation contrasts markedly with high-energy p-p collisions, since in that case even polarizing the protons completely gives very little net polarization to the quarks and gluons which are the elementary objects involved in hard reactions. Is polarization useful? Certainly in would be a help in disentangling e^+e^- annihilation through the photon from annihilation through the Z^0. It is also possible that new phenomena will be strongly polarization-dependent; I will give an example of one such process in my discussion of models with composite electrons at the end of this section. Is it enough to polarize only the electrons if one cannot polarize the positrons? This should be no problem at all. All of the gauge interactions of the electron preserve the electron's helicity; the only helicity-flip interactions of the electron in the standard model are those which result from the electron mass or other direct couplings of the electron to the Higgs sector. Conversely, if new interactions which one might discover at the TeV energy scale could flip an electron's helicity, they would also be expected to contribute a large term to the electron's mass; thus, the strength of such interactions is strongly restricted. Helicity conservation would imply that left-handed electrons annihilate only right-handed positrons, up to minute corrections of order (m_e^2/E_{cm}^2).

Let us now turn from technicalia to more central issues of physics. At the end of § 4, I stressed that the search for new particles is our most promising route to the discovery of new physics which replaces the standard model at energy scales of order 1 TeV. I have already emphasized that, on their own merits, e^+-e^- colliders are ideal instruments for the search for new particles, since they offer low backgrounds and democratic, if not large, production cross-sections. I would like now to assess their capabilities in a somewhat more pointed way—by comparing their capabilities to those of high-energy p-p or p-$\bar{p}$ colliders.

How should one compare the power of e^+-e^- and p-p or p-$\bar{p}$ colliders to produce novel particles with masses in the range of 1 TeV? On the one hand, conventional technology allows much higher center-of-mass energies for proton colliders than for electron colliders. (The design studies for the Superconducting SuperCollider have argued the technical feasibility of a p-p collider with $E_{cm} = 40$ TeV.) On the other hand, new particles are produced from proton-proton collisions only in reactions of elementary constituents—quarks and gluons—which in general carry only a small fraction of the proton's momentum. This means that an e^+-e^- collider operating at a fixed energy will have a power to produce new particles equivalent to that of a p-p collider of a substantially higher energy. We must consider this comparison of energies with some care.

Let us begin with a very simplistic picture. Roughly half the momentum of the proton is carried by gluons, and the rest is divided among the three valence quarks and the sea. Very crudely, then, one might expect that the effective center-of-mass energy of a p-p collider available for producing new particles will be about 1/6 of the total center-of-mass energy, so that colliders with

$$E_{cm}(e^+e^-) \approx \frac{1}{6} \cdot E_{cm}(pp) \qquad (6.1)$$

will have comparable power to produce new particles. It is, in principle, quite straightforward to quantify this estimate, by considering comparable production processes, integrating over realistic parton distributions, and including the (generally rather small) effects of QCD scaling violations.

There is, however, one additional consideration which plays a crucial role. Cross-sections for producing new particles necessarily involve exchanges of virtual states which are off-shell by an amount of order the new particle mass. Thus, by dimensional analysis, such cross-sections decrease with the mass M of the new particle, according to $\sigma(M) \sim M^{-2}$. This behavior is precisely the same as that seen in the precipitous decrease of the R unit in e^+e^- annihilation as E_{cm} increases. In e^+-e^- colliders, however, the total cross-section is also decreasing as E_{cm}^{-2}, so that one can, at least in principle, compensate these small cross-sections by increasing the luminosity. In p-p collisions, however, the total cross-section increases with E_{cm}, and the production of jets with transverse momentum greater than some given value represents an increasing fraction of the total cross-section. Since general purpose detectors can tolerate only a certain total event rate, there is a limit beyond which one cannot usefully raise the luminosity. For the conditions of the SSC, it is generally agreed that this limit corresponds to a luminosity of order $10^{33} \text{cm}^{-2}\text{sec}^{-1}$ for detectors without precision vertex chambers but may be as low as $10^{32} \text{cm}^{-2}\text{sec}^{-1}$ if precision vertex chambers (which might, for example, tag jets with b quarks) are necessary for specific important experiments. If the luminosity is fixed but the cross-sections for interesting events decrease with E_{cm}, the ability of a p-p collider to produce new particles of high mass will be limited: It is not simply that the proportionality constant in Eq. (6.1) will be smaller but, rather, that the equivalent E_{cm} for e^+e^- annihilation will grow as a smaller power of $E_{\text{cm}}(pp)$.

Let me now discuss this constraint more quantitatively. To do this, I will make use of the recent landmark study of Eichten, Hinchliffe, Lane, and Quigg (EHLQ) on the physics capabilities of high-energy p-p colliders.[58] These authors have computed the magnitudes of signal and background, as a function of the center-of-mass energy of the collider, for a variety of new phenomena predicted by theories of 1 TeV physics; this allowed them to formulate criteria for estimating the largest value of the mass for which a new particle should be detectable in experiments done at a given E_{cm}. For clarity, I have selected five representative processes which cover the range of new physics they consider: production of a new W boson, pair production of a new heavy quark Q, production of the supersymmetric partner of the gluon (called the gluino or $\tilde{g}$), production of a new heavy lepton L in association with its neutrino, and deviations from the QCD predictions for high-$p_\perp$ jet production as the result of the presence of a q-q contact interaction (parametrized by a mass Λ) of the type discussed for electrons at the end of § 3. For each of these processes, I have converted the mass limit found by EHLQ into an equivalent center-of-mass energy for discovery of the analogous effect in e^+e^- annihilation in the following way: For W production, I have taken the equivalent E_{cm} to be the mass of the new W; though the W can be produced only in pairs, the corresponding Z boson, normally quite close in mass, can be produced directly. For Q and L production, I have taken the equivalent values of E_{cm} to be $2m_Q$ and $2m_L$. For gluino production, I have taken the equivalent E_{cm} to be $3m_{\tilde{g}}$; the gluino is difficult to produce directly in e^+e^- annihilation, but in models where the mass of the gluino is as large as 1 TeV, the masses of the partners of the quarks and lepton are comparable or perhaps smaller. For the contact interaction, I have taken the equivalent E_{cm} equal to $\Lambda/30$; this reflects the sensitivity of current probes of Λ in Bhabha scattering reported at the end of § 3.

The results of this comparison are shown in Figs. 30 and 31; these figures assume p-p luminosities of 10^{32} and 10^{33}, respectively, integrated over a practical year of 10^7 seconds. The W, Q, and $\tilde{g}$ processes have relatively large cross-sections in p-p collisions; for these, the one-sixth rule (6.1) is not a bad approximation for $E_{\text{cm}} = 10$ TeV, though it rapidly becomes an overestimate as the p-p center-of-mass energy is increased. Over the range of energies shown, the actual relation between the values of E_{cm} for e^+-e^- and p-p is close to

$$E_{\text{cm}}(e^+e^-) \sim \sqrt{E_{\text{cm}}(pp)}, \qquad (6.2)$$

if both energies are expressed in TeV. Except for this last stipulation, this is

Fig. 30. Equivalent energies of e^+-e^- and p-p facilities in searches for a variety of novel particles which might result from new physics at 1 TeV. The five reactions are explained in the text; the capabilities of p-p colliders are taken from the calculations of Ref. 58. This figure assumes a maximum p-p luminosity of $10^{32} \text{cm}^{-2}\text{sec}^{-1}$. The dashed line is a suggested fit to the data: $E_{cm}(e^+e^-) = \sqrt{E_{cm}(pp)/6}$, with all energies in TeV.

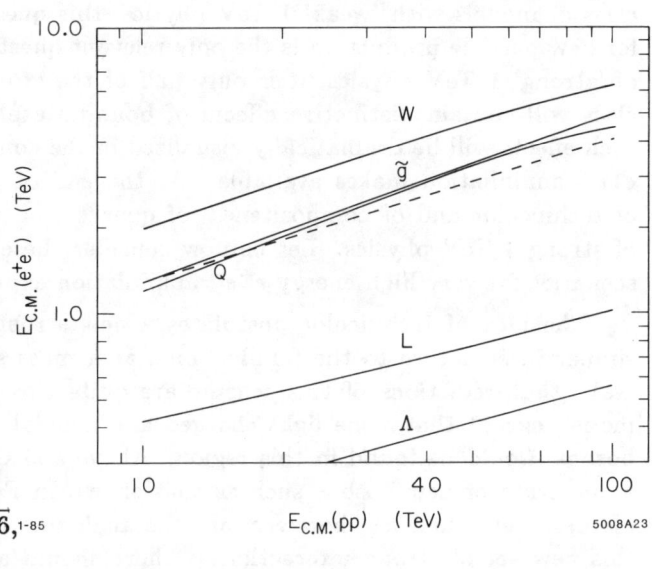

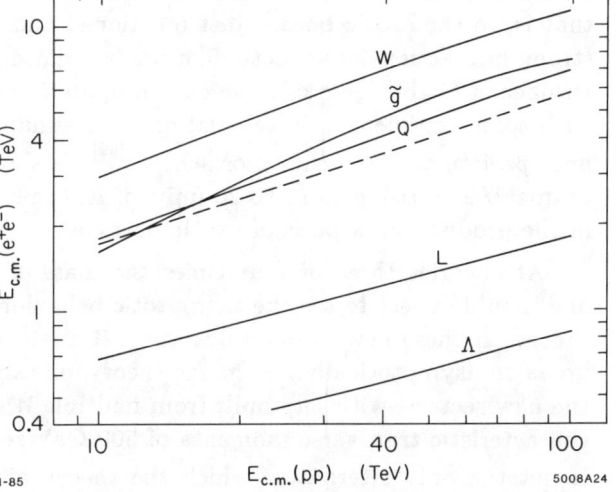

Fig. 31. Equivalent energies of e^+-e^- and p-p facilities, assuming a maximum p-p luminosity of 10^{33} cm^{-2} sec^{-1}. The notation is as in Fig. 30. The dashed line is a suggested fit to the data: $E_{cm}(e^+e^-) = \sqrt{E_{cm}(pp)/3}$.

precisely the relation giving the comparison of energies for colliding-beam and fixed-target experiments.

I have now argued that the advantages of e^+-e^- colliders in producing novel particles do not appear dimmed even by comparison to p-p colliders of much higher energy. If a multi-TeV e^+-e^- collider could be constructed, it would allow searches for new particles as powerful as any which might be conducted at conceivable p-p facilities. In my discussion of § 4, I argued that, for the class of models with 'weak' 1 TeV physics, this question of the reach in mass for new particle production is the only relevant question. However, for the case of 'strong' 1 TeV physics, it is only half of the story. Theories in this latter class will contain distinctive effects of bound state and resonance formation; such effects will be dramatically visualized in the controlled environment which e^+e^- annihilation makes available. At the end of § 4, I described the ideas of technicolor and of compositeness of quarks and leptons as two realizations of strong 1 TeV physics. Let us now consider these implications of these two scenarios for very high energy e^+e^- annihilation experiments.

The idea of technicolor postulates a new strong interaction sector, quite similar in structure to the familiar one, at a mass scale of 1 TeV. Below 100 GeV, the predictions of this scheme are quite close to those of the standard model, except that some light charged and neutral particles resembling Higgs bosons should be found in this region. Above 300 GeV in the center of mass, a spectrum of new bosons such as that shown in Fig. 27 should appear. All of these new particles, however, are the analogues of the π and K mesons of this new set of strong interactions. There should also exist analogues of the ρ—vector bosons bound by the new strong interactions which can be produced as resonances in e^+e^- annihilation. These vector resonances decay to the pions, that is, to the exotic bosons just mentioned and to W and Z pairs. If the new strong interactions are exactly like the familiar ones (except for a change in the number of "light" quarks), one can compute the properties of these resonances with some confidence. A calculation of the shape of the techni-ρ resonance in one particular technicolor scheme[59] is shown in Fig. 32. The peak is a dramatic one, rising to 15 to 20 units of R; the bulk of this cross-section results in the production of pairs of exotic particles.

At energies three or four times the mass of the techni-ρ ($E_{\rm cm} \sim 4$ TeV), one should expect to see the asymptotic behavior of this new strong interaction theory. If these new interactions are just analogous to the familiar ones, built up as an asymptotically free gauge theory, one should see jet-like final states in the new sector—with jets built from multiple W and Z production and having characteristic transverse momenta of 500 GeV relative to the jet axis. But this is not the only alternative which the theory allows. Holdom[60] has argued that technicolor interactions should not exhibit asymptotic freedom but rather should show a more intricate asymptotic behavior; his model predicts that the multi-boson final states are broader in their transverse momentum spectrum:

$$E\frac{d\sigma}{d^3p} \sim \frac{1}{|p_\perp|^\alpha}, \quad (\alpha < 4). \tag{6.3}$$

Alternatively (or in addition), one might find new hard effects, associated with the forces which couple technifermions to ordinary fermions to generate the

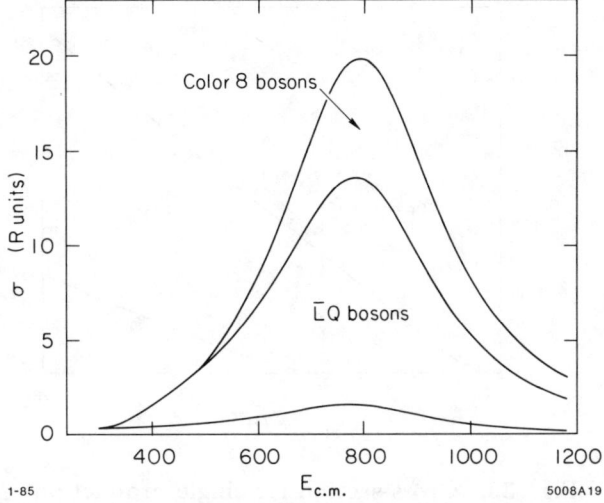

Fig. 32. The behavior predicted in a particular technicolor model (model (a) of Fig. 27) for the cross-section for production of exotic boson pairs in the vicinity of the techni-ρ resonance.

quark and lepton masses. The simplest model of these forces would insist that they are the result of exchanges of new vector bosons, often called ETC bosons. You should recall my comment at the end of § 4 that this model has some serious phenomenological difficulties.[53,54] However, if the scheme makes any sense, it is noteworthy that the scale of the ETC boson masses is not unreasonably high. The mass of the ETC boson which couples to the top quark can be estimated from the relation

$$\frac{m_t}{250 \text{GeV}} \approx \left(\frac{m_T}{m_{\text{ETC}}}\right)^2. \tag{6.4}$$

m_T is the dynamical mass of a technifermion, about 400 GeV; then this ETC boson mass should be roughly 1 TeV. ETC bosons can be produced singly in e^+e^- annihilation, in association with a top quark and a technifermion, producing a final state with a t, a $\bar{t}$, and a multi-W jet. If the ETC bosons carry both color and $(SU(4))$ technicolor, the cross-section for this process is reasonably large; this cross-section is shown in Fig. 33 for two possible values of the ETC boson mass.

Let us now turn to the effects of composite structure within the leptons. e^+e^- annihilation experiments are particularly sensitive to the presence of such composite structure because the total cross-section is dominated by the relatively weak processes of 1-photon and 1-Z^0 exchange. Strong interactions involving the

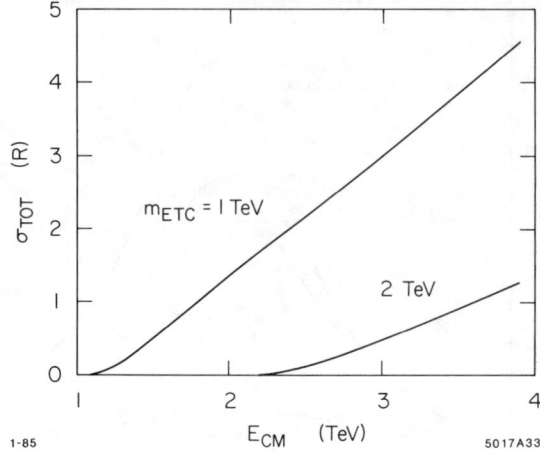

Fig. 33. Cross-section for single production of ETC bosons in e^+e^- annihilation, setting the mass of the ETC boson as 1 TeV and 2 TeV in the two curves, and ignoring the masses of the top quark and technifermion produced in association with this boson.

composite structure of the electron would lead to an annihilation cross-section which we would estimate geometrically to be

$$\sigma \sim \frac{1}{\Lambda^2}, \qquad (6.5)$$

where, as in the discussion below Eq. (3.10), Λ is a mass scale whose inverse gives the size of the composite state. The cross-section can become comparable to the electromagnetic point cross-section (2.1) when Λ is still an order of magnitude less than E_{cm}. We have noted this sensitivity already at the end of section 3 when we described current experiments which probe for the existence of a composite size Λ^{-1}. TeV-energy e^+e^- annihilation experiments should be similarly sensitive. It is worth noting that at energies of order 1 TeV, well above the scale of weak-interaction symmetry breaking, the left- and right-handed components of the electrons are distinct species with different $SU(2) \times U(1)$ quantum numbers and should be expected to have different constituents. This can lead to some remarkable effects. As an example, we might consider a theory in which the left-handed electron and the left-handed muon (the components of these particles which couple to the W) share some constituents and so are coupled by a contact interaction. The helicity-dependence of the interaction leads to structure in the forward-backward asymmetry and in the polarization asymmetry as a function of E_{cm}. The first of these is displayed in Fig. 34; this figure indicates that TeV-energy e^+-e^- colliders will be sensitive to composite structures of extremely small size. The availability of polarized electron beams would be very useful in untangling the precise space-time structure of the interaction and, from it, the quantum numbers of the underlying constituents.

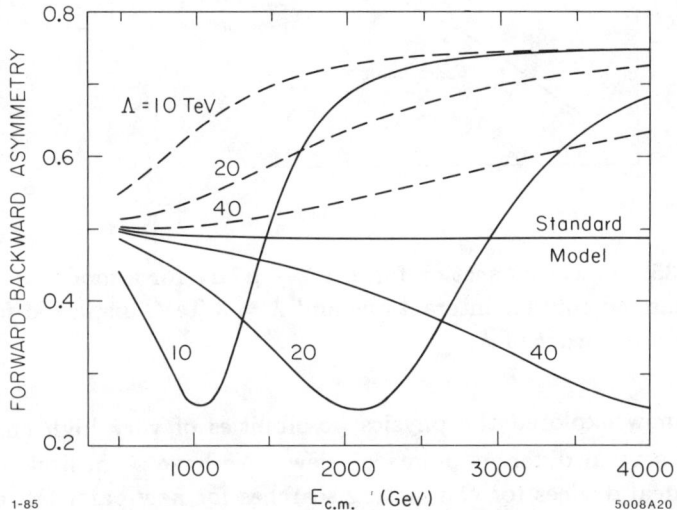

Fig. 34. Effect on the process $e^+e^- \to \mu^+\mu^-$ of a contact interaction linking left-handed electrons and left-handed muons. The graphs show the behavior of (a) the forward-backward asymmetry and (b) the polarization asymmetry for various values of the compositeness scale Λ.

On the other hand, it is possible that the composite structure of electrons and muons will actually appear at a more accessible energy. If $\Lambda = 3$ TeV, a value well above the current experimental bounds, the constituent interactions predict huge effects on the μ pair cross-section for E_{cm} above 300 GeV. In the multi-TeV region, the couplings due to composite structure would dominate all other contributions to e^+e^- annihilation. To display the cross-sections which would result, it is necessary to change our standard of cross-section from the R unit to some absolute level such as the nanobarn. The geometrical cross-section (6.5) associated with a Λ value of 3 TeV is 0.1 nb, a value of the same order as the current PEP and PETRA annihilation cross-sections. The behavior of the μ pair cross-section as a function of E_{cm} is sketched in Fig. 35. The cross-section should exhibit the typical strong interactions effects of resonant structure; it may even show the Regge asymptotic behavior of strong interaction processes.

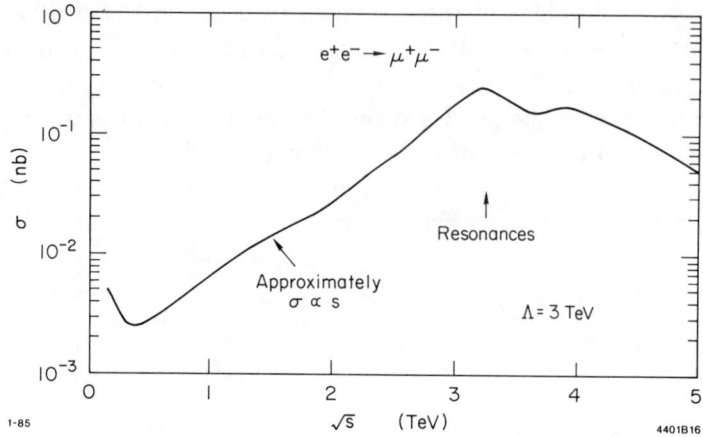

Fig. 35. The cross-section for $e^+e^- \to \mu^+\mu^-$ for a model with left-handed contact interactions and $\Lambda = 3$ TeV, displayed for $E_{\rm cm}$ of the order of Λ.

We have now explored the physics possibilities of very high energy e^+-e^- colliders from several different points of view. We have seen, first, that e^+-e^- colliders are ideal devices for conducting searches for new particles, and I have emphasized that this search will play a central role in our exploration of the TeV energy region. But we have seen also that many pictures of TeV physics predict more remarkable effects, which characterize directly the underlying theory if they can be observed with sufficient clarity. Here, again, the low background levels and the control of the collision energy available in e^+-e^- collisions could be of great value. Can such machines be built? Perhaps, with your thought and effort. I hope some of you can find a way to construct these most elegant devices and to realize the beautiful experiments which they promise.

REFERENCES

1. G. J. Feldman and M. L . Perl, *Phys. Repts.* **33**, 285 (1977).
2. K. Berkelman, *Phys. Repts.* **98**, 145 (1983).
3. S. L. Wu, *Phys. Repts.* **107**, 59 (1983).
4. J. Ellis, in **Proceedings of the 1979 ICFA Workshop**, U. Amaldi, ed. (CERN, 1980).
5. F. Bulos, *et. al.*, in **Proceedings of the 1982 DPF Summer Study on Elementary Particle Physics and Future Facilities**, R. Donaldson, R. Gustafson, and F. Paige, eds. (Fermilab, 1982).
6. J. Ellis, in **Proceedings of the XIV International Symposium on Multiparticle Dynamics**, P. Yager and J. F. Gunion, eds. (World Scientific, Singapore, 1984).
7. W. Panofsky, in **Proceedings of the 1981 International Symposium on Lepton and Photon Interactions at High Energy**, W. Pfeil, ed. (Bonn, 1981).
8. T. Appelquist and H. Georgi, *Phys. Rev.* **D8**, 4000 (1973); A. Zee, *Phys. Rev.* **D8**, 4038 (1973).
9. W. Bartel, *et. al.*, *Phys. Lett.* **129B**, 145 (1983).
10. E. Fernandez, *et. al.*, SLAC-PUB-3479 (1984).
11. R. Hollebeek, in **Proceedings of the 1983 SLAC Summer Institute**, SLAC-Report-2167 (1984).
12. P. Franzini and J. Lee-Franzini, *Ann. Rev. Nucl. Part. Sci.* **33**, 1 (1983).
13. G. Arnison, *et. al.*, *Phys. Lett.* **147B**, 493 (1984).
14. J.-E. Augustin, *et. al.*, *Phys. Rev. Lett.* **33**, 1406 (1974).
15. D. Andrews, *et. al.*, *Phys. Rev. Lett.* **44**, 1108 (1980); T. Böhringer, *et. al.*, *Phys. Rev. Lett.* **44**, 1111 (1980).
16. A. Silverman, in **Proceedings of the 1981 International Symposium on Lepton and Photon Interactions at High Energy**, W. Pfeil, ed. (Bonn, 1981).
17. E. Farhi, *Phys. Rev. Lett.* **39**, 1587 (1977); H. Georgi and M. Machacek, *Phys. Rev. Lett.* **39**, 1237 (1977); A. De Rujula, J. Ellis, E. G. Floratos, and M. K. Gaillard, *Nucl. Phys.* **B138**, 387 (1978).
18. W. Braunsweig, in **Proceedings of the 1981 International Symposium on Lepton and Photon Interactions at High Energy**, W. Pfeil, ed. (Bonn, 1981).
19. H. S. Kaye, Ph. D. thesis, SLAC-Report-262 (1983).

20. G. Arnison, *et. al.*, *Phys. Lett.*, **122B**, 103 (1983); **126B**, 398 (1983).

21. M. Banner, *et. al.*, *Phys. Lett.*, **122B**, 476 (1983); P. Bagnaia, *et. al.*,*Phys. Lett.*, **129B**, 130 (1983).

22. S. L. Wu, Ref. 3.

23. B. W. Lynn and R. G. Stuart, ICTP preprint IC/84/46 (1984), to appear in *Nuclear Physics* **B**.

24. LEP Study Group, CERN/ISR-LEP/79-33 (1979); H. Schopper, in **Proceedings of the 12th International Conference on High-Energy Accelerators**, F. T. Cole and R. Donaldson, eds. (Fermilab, 1984).

25. B. Richter, in **Proceedings of the 11th International Conference on High-Energy Accelerators**, W. S. Newman, ed. (Birkhäuser Verlag, Basel, 1980); H. Wiedemann, in **Proceedings of the 1981 SLAC Summer Institute**, SLAC-Report-245 (1982).

26. O. P. Sushkov, V. V. Flambaum, and I. B. Kriplovich, *Soviet. J. Nucl. Phys.* **20**, 537 (1975).

27. W. Alles, Ch. Boyer, and A. Buras, *Nucl. Phys.* **B119**, 125 (1977).

28. R. W. Brown and K. O. Mikaelian, *Phys. Rev.* **D19**, 922 (1979); I. Hinchliffe, in **Proceedings of the 1982 DPF Summer Study on Elementary Particle Physics and Future Facilities**, R. Donaldson, R. Gustafson, and F. Paige, eds. (Fermilab, 1982).

29. K. H. Lau, in **Proceedings of the 1982 SLAC Summer Institute**, SLAC-Report-259 (1973).

30. S. Yamada, in **Proceedings of the 1983 International Symposium on Lepton and Photon Interactions at High Energies**, D. G. Cassel and D. L. Kreinick, eds. (Cornell, 1983).

31. W. Bartel, *et. al.*, *Phys. Lett.* **114B**, 287 (1982).

32. M. K. Gaillard, I. Hinchliffe, and L. Hall, *Phys. Lett.* **116B**, 279 (1982).

33. L. Gladney, *et. al.*, *Phys. Rev. Lett.* **51**, 2253 (1983).

34. E. Fernandez, *et. al.*, *Phys. Rev. Lett.* **52**, 22 (1984).

35. W. Bartel, *et. al.*, DESY 84/112 (1984).

36. E. Fernandez, *et. al.*, SLAC-PUB-3520 (1984).

37. D. Burke, *et. al.*, SLAC-PROPOSAL-021 (1983).

38. W. Bartel, *et. al.*, *Phys. Lett.* **123B**, 353 (1983).

39. D. P. Barber, *et. al.*, *Phys. Rev. Lett.* **45**, 1904 (1980).

40. B. Adeva, *et. al.*, *Phys. Lett.* **115B**, 345 (1982).

41. M. Althoff, *et. al.*, *Phys. Lett.* **122B**, 95 (1983).

42. F. Wilczek, *Phys. Rev. Lett.* **39**, 1304 (1977).

43. J. D. Bjorken, in **Proceedings of the 1976 SLAC Summer Institute**, SLAC-Report-198 (1976).

44. B. Lee, C. Quigg, and H. Thacker, *Phys. Rev.* **D16**, 1519 (1977); S. Glashow, D. Nanopoulos, and A. Yildiz, *Phys. Rev.* **D18**, 1724 (1978).

45. S. J. Brodsky and S. D. Drell, *Phys. Rev.* **D22**, 2236 (1980); R. Barbieri, L. Maiani, and G. Veneziano, *Phys. Lett.* **96B**, 63 (1980).

46. E. Fernandez, *et. al.*, WIS-EX-243 (1984).

47. D. Bender, *et. al.*, *Phys. Rev.* **D31**, 1 (1985).

48. J. E. Kim, P. Langacker, M. Levine, and H. H. Williams, *Rev. Mod. Phys.* **53**, 211 (1981).

49. S. Weinberg, *Phys. Rev.* **D19**, 1277 (1979).

50. L. Susskind, *Phys. Rev.* **D20**, 2619 (1980).

51. S. Dimopoulos and L. Susskind, *Nucl. Phys.* **B155**, 237 (1979).

52. E. Eichten and K. D. Lane, *Phys. Lett.* **90B**, 125 (1980).

53. E. Eichten, K. D. Lane, and J. P. Preskill, *Phys. Rev. Lett.* **45**, 225 (1980).

54. S. Dimopoulos and J. Ellis, *Nucl. Phys.* **B182**, 505 (1980).

55. M. E. Peskin, *Nucl. Phys.* **B175**, 197 (1980).

56. B. Richter, SLAC-PUB-3371 (1984).

57. The following analysis is based on that of J. Ellis, Ref. 6.

58. E. Eichten, I. Hinchliffe, K. Lane, and C. Quigg, *Rev. Mod. Phys.* **56**, 579 (1984).

59. S. Dimopoulos, *Nucl. Phys.* **B168**, 69 (1980).

60. B. Holdom, *Phys. Rev.* **D24**, 1441 (1981).

HADRON-HADRON COLLIDERS

A. V. Tollestrup
Fermi National Accelerator Laboratory, Batavia, IL 60510

Table of Contents

Introduction 169

The Machines 172

    The FFAG 173
    ISR 178
    S$\bar{p}$pS 179
    TeV I 180
    SSC 181
    The Interaction Region 183
    Phase Space Technology 187
    Beam Stability 190

Detectors for Hadron Colliders 191

    Electrons and Muons 193
    Tracking 193
    Neutrinos 194
    Calorimetry 194
    Calibration 197
    CDF and D0 Detectors 197

Conclusion 203

References 203

HADRON-HADRON COLLIDERS

A. V. Tollestrup
Fermi National Accelerator Laboratory, Batavia, IL 60510[*]

INTRODUCTION

In the last two years there has been a striking change in the perceived role that hadron-hadron colliders will play in the future of high energy physics. This has occurred for several reasons. The first is the obvious push to reach higher and higher center-of-mass energies. For a fixed-target proton machine, $s = 2M_p E$, whereas for a proton-proton collider, $s = 4E^2$, a gain of $2\gamma_p$ over the fixed-target machine. The tests of QCD and its myriad of promised predictions are especially clean at high momentum transfer, where the constituents of the proton, quarks, and gluons can be treated as particles and their interactions studied through their external manifestation as jets. The successful predictions of the electro-weak theory of W and Z production and their decay properties has given a real impetus to probe energies that are still higher in order to understand the underlying structure of the standard theory. The beautiful experiments at CERN confirming the heavy boson predictions have shown the way of the future.

Equally exciting has been the discovery that high energy collisions of hadrons are "clean" and can be interpreted in terms of collisions of the constituent partons. In the past the view has frequently been expressed that such hadron-hadron collision processes are so complicated that little physics could be gleaned from studying them (the "strawberry jam" view of the early 1970s). However, it is now clear that high p_T processes are readily studied through jets, and techniques also have been developed to carry out particle searches using missing p_T, as was successfully done for the W/Z (Ref. 1). The experimental techniques have been developed to the point that it is likely that the existence of the top quark will be confirmed at the CERN $\bar{p}p$ collider - something that looked very difficult and unlikely only five years ago.

Thus we are witnessing a revolution taking place in which the hadron-hadron machine is successfully challenging the e^+e^- colliders which until now have attracted the major attention of physicists using colliding-beam machines. Several factors are at play here. The e^+e^- machine has several inherent advantages.

1. The electrons have radiation damping, which has several consequences for a machine. Since an electron forgets its past history after one damping time constant, resonant effects from magnet imperfections or beam-beam interactions are reduced. Damping leads to an easy mechanism for stacking, which means high currents can be easily achieved. Also, the beam size contracts until it reaches equilibrium with the stochastic quantum excitation

---

[*] Operated by Universities Research Association under Contract with the U.S. Department of Energy.

process. These last two effects lead to high luminosity. Radiation damping, of course, is finally the bête noire in that it limits the energy achievable in an electron synchrotron because of the enormous power radiated.

2. At the energies that have been explored, the momenta of the secondary charged particles from an interaction are relatively easy to measure, as are those of the photons. Thus the total energy, $\sqrt{s}$, has been a powerful kinematic constraint. This remains true even at LEP, where the width of the Z_0 can be directly measured because of this constraint.

3. The intermediate state involving a virtual photon couples to an extraordinarily rich sector of physics in the energy range that has been explored.

In contrast to the e^+e^- machine, the hadron-hadron collider is very sensitive to all kinds of minor error fields which, since there is no damping of the various modes of oscillation of the particles in a bunch, can cause the beams to blow up and thus limit the available luminosity. These effects can come from interactions between a particle and the wall, other particles in the same bunch (intra-beam scattering), or particles in the opposing bunch (beam-beam scattering).

Unlike the e^+e^- case, the pp or $\bar{p}p$ intermediate state is not simple and hence total energy conservation is no longer a useful constraint. Instead, it has turned out that p_T conservation has been the key to the new physics of W/Z and the top quark. Also contributing to this success is the fact that the calorimetry improves in resolution as $\sim 1/\sqrt{E}$. Thus, if the energy scale is large enough, precise measurements can be made. In fact, as the energies expand into the 1 to 10-TeV region, this is the only method available, since magnetic tracking loses accuracy proportional to E and in any case does not measure the neutral component.

In comparing e^+e^- machines with hadron-hadron colliders, one additional factor must be considered and that is synchrotron radiation. At 40 GeV in the center of mass, PETRA radiates 100 MeV/turn of RF. At LEP, to reach the Z_0, 260 MeV/turn is radiated, and at LEP II, which reaches E_{cm} = 200 GeV, 2.56 GeV/turn escapes as synchrotron radiation. Synchrotron radiation per turn changes as E^4/R and when correctly included in the design causes the radius of as electron machine to increase as E^2. However, because of the the large mass of the proton, synchrotron radiation is not important for a hadron collider until one approaches E_{cm} ~ 40 TeV, provided the magnetic field does not exceed 5 to 6 T. (The magnetic field sets the radius and hence affects the radiation of the protons.)

Since a hadron machine can be used in either a fixed-target mode (s = 2 M_pE) or in the pp colliding mode (s = $4E^2$), it is interesting to compare the natures of the experiments that can be carried out in these two different fashions. Consider the Tevatron at 1000 GeV in the fixed-target mode. It has $\sqrt{s}$ = 43.3 GeV. Using a target of beryllium one absorption length long and a beam of 10^{12} particles/sec, one can achieve a luminosity of $\mathcal{L}$ = 7×10^{36}

$cm^{-2} sec^{-1}$. However, in the collider mode, where $\sqrt{s}$ = 1000 GeV, the $\mathcal{L}$ is only 10^{30} $cm^{-2} sec^{-1}$.

The interesting cross sections are scaled by the smallest size structure that can be observed, i.e., $\sigma \sim \pi\lambda^2 \sim C/s$, where s refers to the parton-parton system and C measures the strength of the matrix element. Comparing the collider mode with fixed-target operation, we see that s has changed by a factor of about $2\gamma p$ = 2132, which scales down the structure-dependent cross section by a corresponding amount (i.e., sizes $\sim 10^{-17}$ cm!).* However the $\mathcal{L}$ has not increased proportionally but rather decreased by a factor of 10^6. Whereas the fixed-target experiment has a rate adequate to explore very small cross sections, the collider lacks the luminosity to exploit fully its intrinsic ability to probe very small distances. This illustrates the central problem of colliders, which is to achieve adequate luminosity to measure the small cross sections of interest.

The picture is not as bleak as made to appear above. First, the luminosity in a pp collider can be as large as 10^{33}, which has been shown in a number of recent machine studies. Second, many of the effects involve the production of massive particles through strong interactions with cross sections in the range of 10^{-36} cm^2, making the experiments difficult but not impossible. Nevertheless, the scaling outlined above appears to be intrinsic to physics and will force us to work at ever increasing luminosities as the energy is increased.

A new feature of collider physics is a result of the above considerations, i.e., the detector becomes integrated with the machine. No longer does one have a beam line feeding a series of ever changing experiments. Rather, the modern hadron collider integrates its detectors into its structure as has been the case with e^+e^- machines. The detectors become general purpose facilities where enormous effort and experience is expended for sophisticated instrumentation which must cover 4π solid angle.

The detector must have a very fine granularity, and each element must have good energy resolution. In addition, the ideal detector should provide particle identification for electrons and muons. Since the leptons of interest are frequently immersed in jets, this is a formidable task that has not yet been adequately solved. Finally, one needs accurate charged particle tracking in a magnet field to allow for charge determination of leptons as well as high-resolution vertex detectors in order to find the decays of short-lived particles.

* We should not use the full s of the pp system here since the partons share the protons' momentum. However, even if we use an s reduced by a factor of 10 (Ref. 2), which is indicated by theoretical calculations, the above point remains valid.

The machine design and detector design interact in many ways. For instance:

1. The luminosity determines how many interactions per second the detector must handle, and, if one is using bunched beams, the number of interactions per <u>bunch</u> crossing is inversely proportional to the number of bunches.
2. Scraping off the particles outside the bucket must be provided elsewhere in the machine lattice.
3. The effect of the detector's magnetic field on the circulating beams must be compensated.
4. The machine vacuum affects the detector backgrounds.
5. The bunch length influences the geometry of the detector.
6. Small-angle particles from an interaction may actually pass through low-β quads and machine dipoles before being detected, with the consequence that elements of the detector may have to be incorporated in the machine lattice. This can be a serious complication for a machine using superconducting elements.

The point of the above discussion is that we can no longer construct an accelerator and then propose experiments using it as a facility. The machine and detectors must be designed together and function as a unit. Thus, when the SSC is designed and built, the detectors and experimental areas will be simultaneously planned and constructed. Note also that it takes almost as long to bring a major detector on line (~5 years) as it does to build a modern accelerator, as exemplified at LEP.

The success of the standard theory in predicting the masses of the W and Z as measured in the experiments of UA1 and UA2 has led naturally to the question concerning the structure of the gauge theory: Where is the Higgs that causes the symmetry breaking? It is constrained to lie between 7 GeV and 1 TeV by present knowledge. It is becoming increasingly clear that new discoveries will be made in the 1-TeV region — a subject well documented by Eichten, Hinchliffe, Lane, and Quigg (Ref. 2). The exploration of this energy region is being started with the $\bar{p}p$ collider at CERN and will shortly be supplemented by TeV I at FNAL with the CDF and D0 detectors. The next big step in energy will be the SSC, which will lead to E_{cm} = 40 TeV in the 1990s.

THE MACHINES

After this brief introduction, let us examine the evolution of the hadron-hadron collider and the associated detector. The first large collider proposed was one designed by MURA (Midwestern Universities Research Association) and was a fixed field alternating gradient (FFAG) machine (Ref. 3). This machine was not built, and the next in line was the ISR at CERN in 1970, which was followed by a proposed machine, ISABELLE/CBA, at BNL. The first $\bar{p}p$ bunched beam collider was the SPS at CERN in 1979, to be followed by TeV I at FNAL in 1985. Finally, we hope for the SSC in the early 1990s.

The ISR and CBA were generically related as both involved continuous beams with high currents of 10 to 50 A of circulating protons in two adjacent rings and luminosity $\mathscr{L} > 10^{32}$ cm^{-2} sec^{-1}. The colliders at CERN and TeV I both use only one ring to accelerate protons and antiprotons simultaneously. The beams are bunched, and the development of $\bar{p}$ source technology was necessary before useful luminosities could be attained. The SSC will use bunched beams, but the questions of the cost of pp versus $\bar{p}p$ machines and their relative luminosities are still being studied. The characteristics of the major colliders proposed or built are shown in Table I.

Table I

Machine	Projectiles	E	$\mathscr{L}$
FFAG	pp	15 x 15	~ 10^{32}
ISR	pp, $\bar{p}p$, αα	31 x 31	~ 1.5×10^{32} cm^{-2} sec^{-1}
CBA	pp	400 x 400	$\leq 10^{33}$
S$\bar{p}$pS	$\bar{p}p$	300 x 300	~ 10^{30}
LEAR	$\bar{p}p$	2.2 x 2.2	10^{29}
TeV I	$\bar{p}p$	1000 x 1000	$\geq 10^{30}$
SSC	pp $\bar{p}p$	20 TeV x 20 TeV	10^{33} 10^{32}

The FFAG

In 1958 MURA made a proposal (Ref. 3) to construct a FFAG machine which provided for fixed-target physics at 15 GeV and collider capability between two counter-rotating 15-GeV beams of protons. This machine was developed around a magnet configuration which has not been exploited in high energy physics. Positive and negative magnets were arranged in a ring 180 m in circumference. The equilibrium orbit was such that even though the two opposite polarity magnets had the same length, the orbit had a higher $\int B \, dl$ in the positive bending magnet than in negative benders. This resulted in rather inefficient use of the magnets, which would have weighed 65,000 tons! However, in such a magnet equilibrium orbits for particles rotating in either direction exist, and hence a collider could be realized. Injection was at 50 meV, and the field was fixed and tapered to accommodate orbits between injection and 15 GeV. As a consequence the aperture was 4.9 meters wide. The properties of this machine are shown in Table II, and some drawings of the layout in Figs. 1 through 5.

Table II - FFAG Accelerator Parameters

Magnet System

R_o = 88.6 m = 3,488 in. (radius of 12.5-Bev equilibrium orbit)
R_o^{max} = 88.75 m = 3,494 in. (maximum good field radius)
R_{inj} = 85.53 m = 3,367.3 in. (radius of 200 Mev equilibrium orbit)
R_{min} = 85.33 m = 3,359.4 in. (minimum good field radius)
ΔR = 3.42 m = 134.6 in. (radial aperture)
L_R = 22 in. (length of radial straight sections)
L_m = 61 in. (azimuthal length of magnet)
G_{inj} = 12 in. (vertical magnet aperture at injection)
g_{inj} = 6 in. = 15.2 cm (available aperture inside vacuum tank at injection)
G_{min} = 6 in. (vertical magnet aperture at R_{max})
g_{min} = 3 in. = 7.6 cm (minimum available aperture inside vacuum tank at R_{max})
$B_{ave}(R_o)$ = 5,048.9 gauss
$B_{ave}(R_{max})$ = 5,614.4 gauss
$B_{ave}(R_{inj})$ = 251.3 gauss
Total magnet weight = 24,100 tons
Magnet power = 47.21 megawatts
N = 48 (number of spiral sectors)
M = 264 (number of radial straight sections)
N_s = 24 (number of superperiods per revolution)

Orbit Parameters

ν_x = 9.78 (horizontal betatron oscillation frequency)
ν_y = 6.29 (vertical betatron oscillation frequency)
A_x = ±20.9 cm (horizontal stability limit)
A_y = ±3.7 cm (vertical stability limit)
k = 35
1/w = 548
Angle of spiral with circle = 5.5°

RF System

Three stages, all operating on the 24th harmonic, with 24 drift cavities per stage
    Stage I: 200 Mev to 800 Mev - 30 cycles/sec
    Stage II: 800 Mev to 2 Bev - 30 cycles/sec
    Stage III: 2 Bev to 12.5 Bev - 3 cycles/sec
Injection frequency = 0.3156 Mc/s
Transition energy = 7.7 Bev
Final frequency = 0.5368 Mc/s
Total RF power = 2 megawatts (including beam)

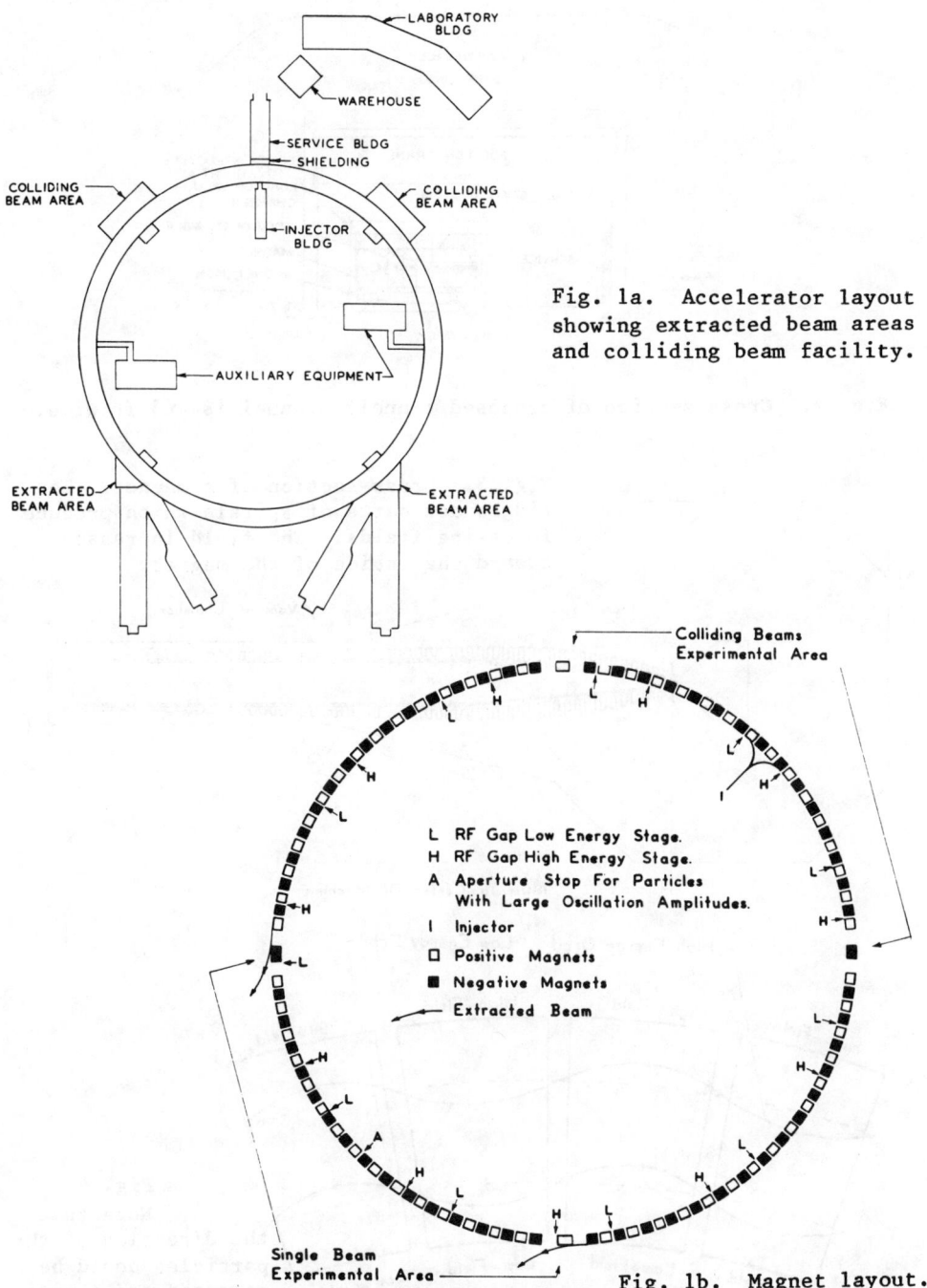

Fig. 1a. Accelerator layout showing extracted beam areas and colliding beam facility.

Fig. 1b. Magnet layout. The magnets reverse sign sequentially around the ring. Fig. 4 shows how positive particles can traverse the machine in either direction.

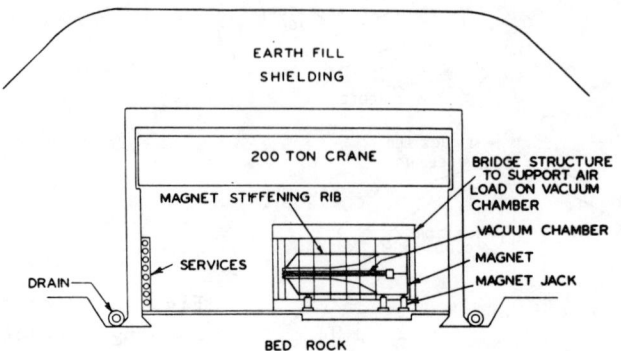

Fig. 2. Cross section of proposed tunnel. Tunnel is √65 ft wide.

Fig. 3. Cross section of a magnet. The ridges are parts of spirals which produce focussing fields. The field increases toward the inside of the magnet.

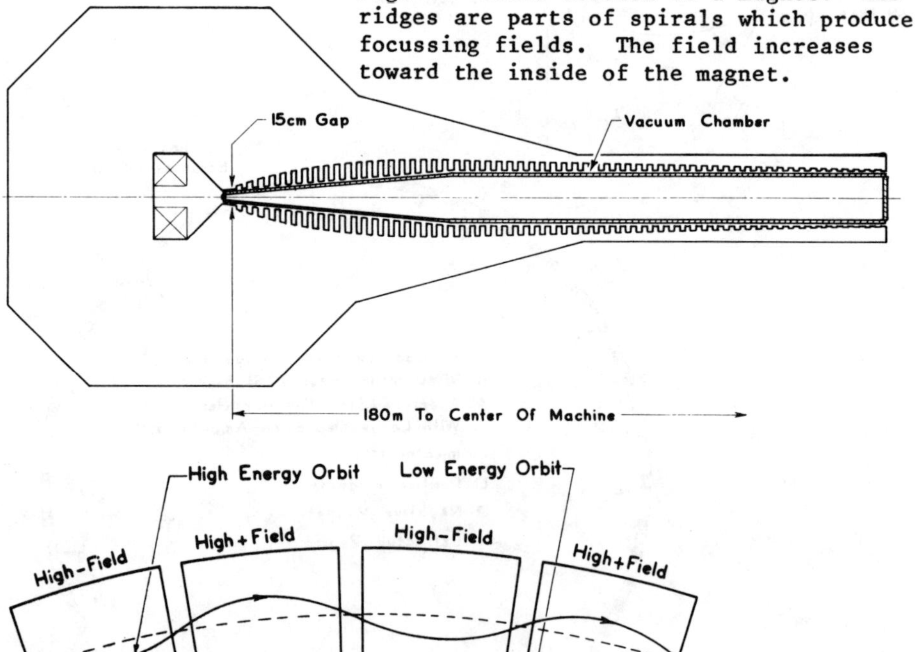

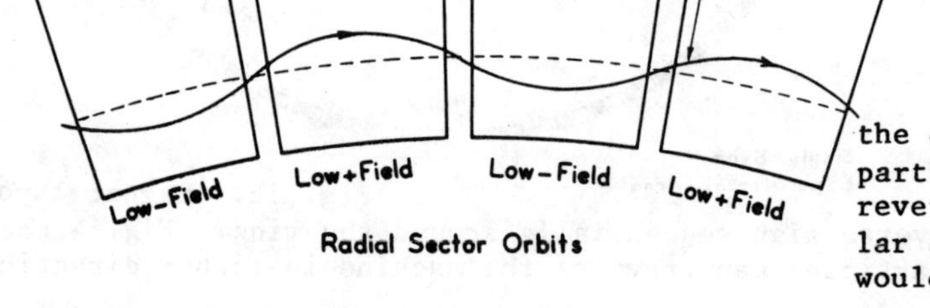

Fig. 4. Note that the direction of the particles could be reversed and a similar set of orbits would exist.

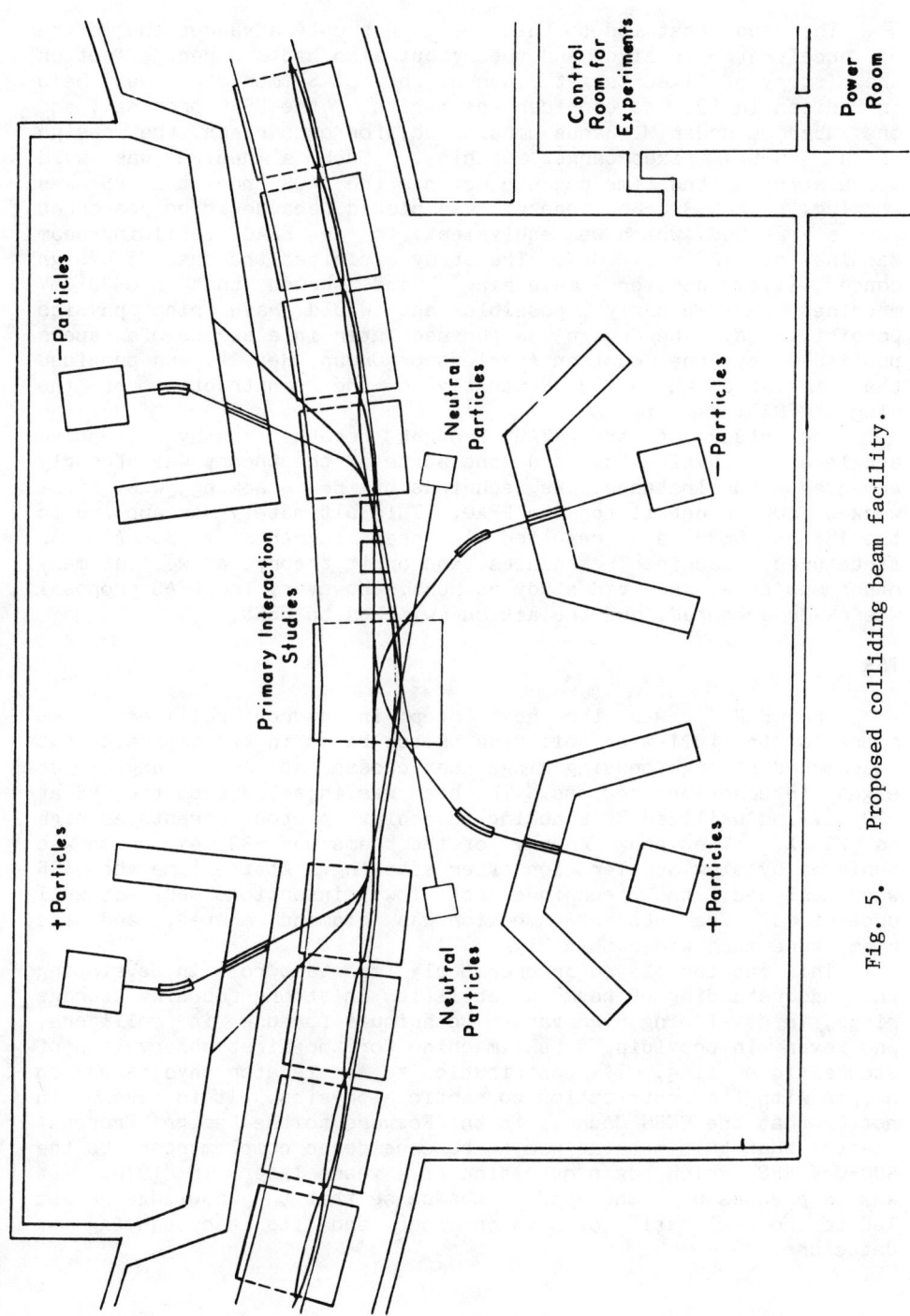

Fig. 5. Proposed colliding beam facility.

177

The study that led to this design not only advanced the state of accelerator design and theory but also had a direct effect on the history of fixed-target synchrotrons. A Summer Study was held in Madison in 1959 to consider the merits of the FFAG proposal, and one subgroup under M. Sands considered, for comparison, the design of a 300-GeV fixed-target machine. Such a design was wild speculation at the time because neither the AGS nor the PS was working. The 300-GeV energy was picked because it corresponded to $\sqrt{s}$ = 24 GeV, which was equivalent to an FFAG colliding-beam machine at 12 x 12 GeV. The study exploited the cascade design concept (i.e., booster → main ring), and showed that a 300-GeV machine was clearly possible and would have rich physics possibilities. The design was pursued later in a series of papers published by the Western Accelerator Group, Ref. 4, and provided the stimulation which led eventually to the construction of the ring at FNAL.

The study of the FFAG brought together many creative accelerator physicists, and the state of the theory was strongly enhanced. For instance, the technique of RF stacking was first worked out in detail for the FFAG. This ultimately was applied to the ISR at CERN and resulted in beam currents > 25 A! RF structures, machine resonances, and orbit theory, as well as many other subjects, received study at MURA. However, the FFAG proposal was never accepted, and the action switches to CERN.

ISR

The CERN ISR was the next step in hadron colliders. It operated in 1971 with colliding proton beams in two separated but interlaced strong-focusing rings that crossed at small angles at eight interaction regions. The beam was injected from the CPS at ~20 GeV and utilized RF stacking to achieve proton currents as high as 22 A. The peak energy of the beams was ~32 GeV, which was achieved by slow acceleration after stacking. At the time the ISR was designed, the techniques of low-β insertions were not well understood. The interaction region was diamond shaped, and the beams were much wider than high.

This machine played an enormously important role in developing the understanding of beam instabilities in strong-focusing storage rings, in developing high vacuum techniques for use in colliders, and even in providing a test machine for the first observation of stochastic cooling. Its contribution to accelerator physics was on a par with its contribution to particle physics. It is also worth noting that the CERN Council in the Forward to the Design Proposal stated that this machine was to be considered complementary to the 300-GeV SPS, which began operation five years later in 1976. It was a precursor of the $S\bar{p}pS$ in the sense that the knowledge gained led to the realization of a $\bar{p}p$ collider and its accompanying 4π detectors.

SppS

The next step was also taken at CERN, where it was realized that a single ring carrying protons and antiprotons in opposite directions could be made to provide luminosities of ~10^{30} cm^{-2} sec^{-1}. In order to achieve this, it is necessary to collect antiprotons from a target hit by protons. However, the phase space density of these antiprotons is much too low for use in a collider. For instance, the transverse phase space for $\bar{p}$ produced at 19 GeV by the CPS is ~100π mm mr, and the density in momentum is about 0.5 $\bar{p}$/eV. For a successful collider the transverse phase space must be compressed by a factor of 10 in each direction, and the momentum density increased to 10^5 $\bar{p}$/eV in order to achieve useful luminosities. The first process is called transverse cooling, and the second is called momentum stacking.

Two methods of accomplishing compaction in phase space have been invented: electron cooling and stochastic cooling. Complete descriptions of these processes are given in Ref. 5a and 5b and will not be repeated here. Suffice it to say that in 24 hours enough $\bar{p}$'s can be collected and processed so that a luminosity of 2×10^{29} cm^{-2} sec^{-1} can be achieved with three bunches of antiprotons (~10^{11}/bunch) colliding with three bunches of antiprotons (~2×10^{10}/bunch). The bunches are injected at low energy into the SPS and then accelerated to an energy that is limited by the power dissipation in the magnets (~300 GeV). The colliding beams are bunched and collinear with a Gaussian overlap σ of about 20 cm. A β^* at the interaction point of 0.75 m x 1.75 m has been obtained simultaneously in two interaction regions.

For elastic scattering experiments, a very large β^* is desirable in order to reduce the angular spread of the beam to allow measurements of the cross section at small t. Such measurements are not only interesting for the physics content, σ_T and dσ/dt, but are necessary for calibrating the luminosity monitor. The elastic scattering experiments at CERN were run with β^* = 100 m.

The characteristics of the CERN AA Source are shown in Table III. In the SppS machine the integrated effective $\mathcal{L}$ is limited by the Source intensity. Three batches of ~2.5×10^{10} antiprotons are injected and then accelerated and used for collisions. However, the $\mathcal{L}$ lifetime is finite because gas scattering and intra-beam scattering increase the transverse beam size, and gas collisions remove particles from the beams. The decay is exponential with a lifetime of 10 to 20 hours. The peak luminosity is determined by how many $\bar{p}$'s can be collected and cooled in a typical lifetime. Thus there is a premium on developing an intense $\bar{p}$ souce for this type collider, and a new source, the ACOL, will be commissioned in 1987. Nevertheless the intensity has been high enough to allow detection of the $W^{\pm}$ and Z^o - a more than modest discovery!

TeV I

Further development of the p̄p collider is being carried out at FNAL with the project called TeV I. A source ten times stronger than the CERN AA is being constructed for operation in early 1986, and s will be much higher since the Tevatron will be operated at 1 TeV. There will be two collision halls, B0 and D0, housing large detectors, as well as smaller experiments in other straight sections. The properties of the TeV I Source are given in Table III.

Table III

	CERN	TeV I
Antiprotons/hour	5×10^9	10^{11}
Lifetime hours	15	20
p bar/bunch	2.5×10^{10}	6×10^{10}
Number of bunches	3	3
$\mathcal{L}(0) \times 10^{29}$ cm^{-2} sec^{-1}	1.5	10

The increased source intensity is obtained for two reasons. First, the flux of antiprotons is much higher since they can be produced by 120-GeV protons from the main ring instead of 19-GeV protons like those from the PS which are used at CERN. Second, this higher flux of antiprotons is cooled in two stages: a debuncher ring and an accumulator ring.

It was realized in the FNAL design that a cascade system in which transverse phase space is cooled first before momentum stacking is attempted would allow a higher frequency RF system to be used. An increase in the band width of the stochastic stacking system results in a higher rate of collecting p̄'s. A second ring (ACOL) utilizing these principles is also being planned at CERN in order to achieve a higher p̄ flux.

Both the Sp̄pS and TeV I will have limited $\mathcal{L}$ because of the beam-beam effects from multiple collinear crossings. They both represent machines that will use existing rings to give a first look at values of s far beyond that which either machine is capable of achieving in the fixed-target mode. Neither represents the ultimate that can be achieved in a machine designed specifically as a collider. Both have had collision regions "grafted on" to an existing machine. However, the striking success at CERN, showing that hadron-hadron colliders can produce beautiful physics, along with the enormous step forward in superconducting magnet technology exhibited by the Tevatron, has pointed the way to the next step.

SSC

This machine represents the culmination of all the knowledge we have gained about machines, detectors, and high energy particle physics. The properties as outlined in the SSC Reference Designs Study carried out at Berkeley in early 1984 are given in Table IV (Ref. 6).

Table IV
Abridged Parameter List - Reference Design A

General Parameters

Maximum energy per ring (TeV)	20
Luminosity - each interaction region ($cm^{-2} sec^{-1}$)	10^{33}
Number of interaction regions	6
Injection energy (TeV)	1
Magnet type	2-in-1, side-by-side
Standard beam separation (horizontal) (cm)	14.0
Peak magnetic field (T)	6.5
Peak current (A)	5400
Bunch spacing of overall design	variable
Bunch spacing used in this list (m)	10
Number of events per crossing (at a cross section of 100 mb)	3.3
Circumference (km)	90
Orbit frequency (kHz)	3.33
Orbit period (μsec)	300
Number of particles per bunch	1.45×10^{10}
Number of bunches per ring	9000
Number of particles per ring	1.30×10^{14}
Average beam current (mA)	69.7
Invariant transverse emittance (μm)	1.0
Beam-beam tune shift per crossing	0.0017

Magnets and Lattice

Amplitude function at interaction point (m)	1
Free space at interaction point (m)	±20
Crossing angle (μrad)	30-100
Length of interaction region insertion (m)	500
Phase advance for interaction region insertions (deg)	360
Length of standard half-cell (m)	100
Phase advance per cell (deg)	80
Magnetic (physical) length of dipole (m)	16.6 (17.5)
Number of dipoles per half-cell	5

Number of standard half-cells per ring	738
Number of half-cells per dispersion suppressor	4
Number of dipoles per dispersion suppressor*	10
Number of dispersion suppressors per ring	18
Number of dipoles per ring*	3870
Number of standard quadrupoles per ring	804
Number of utility insertions (abort, injection, RF)	3
Length of utility insertion (km)	2.0
Phase advance for utility insertion (deg)	200
Maximum beam separation (abort, injection) (cm)	70
Maximum beam separation (RF) (cm)	50
Nominal tune (both planes)	97.76

RF System Related Parameters

Frequency (MHz)	360
Peak voltage per turn (MV)	20
Total cavity length per ring (m)	25
Acceleration period (sec)	1000
Momentum spread at injection, σ_E/E	1.5×10^{-4}
Momentum spread at 20 TeV	5.0×10^{-5}
Bunch length at 20 TeV - rms (cm)	7.0
Longitudinal emittance at injection (95%) (eV/sec)	0.66
Longitudinal emittance at 20 TeV (95%) (eV/sec)	4.4

Injection System

Linac energy (GeV)	1
Length (m)	200
Beam current (mA)	65
Ion species in linac	H
Invariant transverse emittance (μm)	0.22
Low energy booster energy (GeV)	70
Circumference (km)	1.2
RF frequency (MHz)	60
Cycle time (sec)	3
Bunch coalescing frequencies (typical) (MHz)	60/k, k = 1,2,...5
High energy booster energy (TeV)	1
Circumference (km)	6
RF frequency (MHz)	60
Cycle time (sec)	45

*In units of standard length (16.6 m).

The case considered is a double ring with pp collisions. The beams are bunched, cross at a small angle, and achieve a luminosity of 10^{33} cm^{-2} sec^{-1}. At this $\mathcal{L}$ there will be bunches crossing every 33 nsec, with an average of three interactions per crossing - a real challenge for detector technology.

Since the $\bar{p}p$ option is not studied in the report, it is not known exactly what such a machine would cost compared with the two rings necessary for $\bar{p}p$. Attendees at a workshop at the University of Chicago in January 1984 investigated whether there were any overriding physics reasons for selecting pp or $\bar{p}p$, and they found none.

The major tradeoffs between the two options concern the cost and reliability of a $\bar{p}$ source versus a second ring. Not only is a source necessary for $\bar{p}p$, but also the elimination of multiple collisions of the bunches as they pass around the ring. This is accomplished by using electric fields to establish equilibrium orbits that are separated in space for the two different kinds of particles. This operation requires more sophisticated beam technology and better control of the quality of magnet field than does a single beam in a ring. Further studies of these questions are necessary.

Novel ideas to reduce the cost of the accelerator are being explored. The two-in-one magnet design, which places two superconducting coils in one set of iron collars in Design A of the SSC study, is one such possibility. A second involves placing all of the interaction regions in two clusters on opposite sides of the ring. The regions would be separated by some bending magnets to decouple them from each other but would still be clustered together (over several km) for easy access and servicing. The magnet design is being explored over the region of B from 3 T to 6 T. Earlier designs up to 10 T were proposed, but at this moment such a high field seems an imprudent extrapolation of present technology.

After this brief history and description of hadron colliders, we go on to the definition of the pertinent beam parameters and how these parameters are chosen, and to some consideration of detector design.

The Interaction Region

In a successful collider it must be possible to bring the two beams into collision in such a manner that a useful luminosity is achieved:

$$N_{int} = \mathcal{L} \sigma$$

where N_{int} is the interactions per second from a cross section σ. As stated above, a high $\mathcal{L}$ is desirable because the interesting cross sections are becoming smaller as s increases. However, as shown below, present detector technology limits the useful $\mathcal{L}$ to $\sim 10^{32}$ although special detectors may be able to use 10^{33}. At this higher luminosity a total cross section of 50 mb gives an interaction rate of $N_{int} = 7 \times 10^7$. The problems this gives the detector are discussed later. First we investigate the conditions necessary to achieve $\mathcal{L} = 10^{32}$ cm^{-2} sec^{-1}.

Three different experimental arrangements of the beams may be considered:

1. DC beams of protons in two rings arranged to cross at an angle (IRS, CBA).
2. Proton-antiproton bunched beams in a single ring (S$\bar{p}$pS, TeV I.).
3. Bunched proton beams in two separate rings.

In the last case, the beams may be either collinear or at a small angle in the interaction region. To give an idea of the characteristics of the collision region, Table V lists the characteristic parameters of some interaction regions of several machines (real and proposed!).

The interaction region is where the demands of the detector and the physics experiment must be made compatible with the requirements of designing a stable and easily understood machine. As a result the schism that has existed between machine physicists and detector physicists is being forced to close. Let us examine some of these problems.

The following parameters affect the interaction region:
1. $\mathcal{L}$, luminosity.
2. Bunched/unbunched beams.
3. ε, beam emittance.
4. S_B, bunch spacing.
5. σ_z, bunch lengths; σ_t, transverse Gaussian width.
6. α, crossing angle.
7. N_o, total number of particles.
8. I, beam current.
9. N_B, number of bunches.
10. $\Delta\nu$, tune shift.
11. The lattice function at the interaction region.

An excellent discussion of the relation between these variables is given in the SSC Design Study (Ref. 6). Luminosity is the most important variable. Consider first the number of interacions per second,

$$N = \mathcal{L}\sigma.$$

For $\sigma = 100$ mb and $\mathcal{L} = 10^{33}$ the above equation gives 10^8 interaction/sec! At 20 TeV/beam, this corresponds to one interaction per 10 nsec with an average multiplicity of 100! This rate also corresponds to 600 watts of beam interaction energy being dissipated on elements of the detector or striking the magnets of the accelerator. If superconducting magnets are used, special precautions must be taken to keep small-angle secondary particles from hitting the coils and causing them to quench, and the detector must be constructed of radiation resistant elements.

Next consider the effect of bunched beams versus dc beams. To see the effect of bunching, consider two beams crossing at a small

Table V

Machine	$\mathcal{L}$	L	ℓ_{int}	σ_V	σ_H	α	No. Bunches	I	No. Particles/ Beam
1. CBA 400x400	4×10^{32}	60 m	.24 m	.11 mm	.42 mm	9.8 mr	dc	8 amps	6.4×10^{14}
2. TeV I	10^{30}	±7.5 m	±.4 m	.06 mm	.06 mm	0	3x3	1.6 ma	2×10^{11}
3. SSC	10^{33}	±20	±7 cm	6.8 µ	6.8 µ	30-100 µr	9000	70 ma	1.3×10^{14}
4. ISR (IS)	1.8×10^{36}	10 m		1.28 mm		14.77°	dc	29 amps	5.6×10^{14}

angle so that the effective interaction length is ℓ_e. If beam one is bunched and two unbunched, the $\mathcal{L}$ will be the same as for unbunched beams provided there are the same number of particles in beam one in each case. However, it is now obvious that a number of the particles in beam two do not see any particles in beam one at the crossing region and hence do not contribute to $\mathcal{L}$ and could, therefore, be removed. This number is $\sim \ell_e/S_B$. Since $\mathcal{L} \sim N_1 N_2$, the number of particles in each ring could be reduced by $(\ell_e/S_B)^{1/2}$ without changing the $\mathcal{L}$ if both beams are bunched. One, of course, sacrifices in that the duty cycle decreases considerably for the bunched case.

It is now apparent why bunched beams will be required by the SSC since it is desirable to keep the total energy in the beam as small as possible in order to keep the machine-induced radioactivity low as well as to make it easier to construct the beam dump. Reducing the total number of particles in the beam also reduces possible damage to the detector components or the probability of quenching of superconducting beam magnets by stray beam. Note that 10^{14} particles at 20 TeV represents an energy of 320 MJ! Finally, as soon as the energy is high enough for synchrotron radiation to set in, bunched beams are unavoidable since energy must be supplied from RF cavities.

However, bunched beams adversely affect the detector. For a given $\mathcal{L}$, it is clear that the instantaneous $\mathcal{L}$ must be higher for bunched than for unbunched beams. Since the bunches are traveling with $v = c$ there will be c/S_B colliding bunches per second. If the $\mathcal{L}$ is fixed, then the number of interactions/bunch crossing is

$$n = \mathcal{L} \sigma S_B/c.$$

Assuming S_B = 10 M, σ = 100 mb, and $\mathcal{L} = 10^{33}$, we find that n = 3 interactions/crossing. If the bunch length is only a fraction of a meter, multiple interactions may make events which are very difficult for the detector to untangle; therefore, S_B should be small. However, this may lead to multiple bunch crossings within the sensitive volume of the detector. It also leads to undesirable beam-beam effects, as will be discussed shortly.

The integrated luminosity, $\mathcal{L}$, of two bunches with Gaussian cross section characterized by σ_t, crossing head-on, with N_1 and N_2 particles per bunch, is

$$\mathcal{L}_1 = \frac{N_1 N_2}{4\pi \sigma_t^2}$$

and the $\mathcal{L}$ is

$$\mathcal{L} = \mathcal{L}_1 \frac{c}{S_B} = \frac{N_1 N_2}{4\pi \sigma_t^2} \left(\frac{c}{S_B}\right) = \frac{3}{2} \frac{N_1 N_2}{\beta \varepsilon_t} \frac{c}{S_B}.$$

This equation involves the area of the beam. The relation of the beam dimension to the emittance, ε_t, is described next.

Phase Space Technology

This section gives a general description of beam dynamics for hadron colliders. An excellent and extensive reference for these results is provided in Ref. 7, where the equations are derived and the hadron collider lattice is discussed in detail.

We first consider transverse phase space which has as conjugate variables x, x' and y, y'. If there are no skew fields in the magnets, the x and y motion will be uncoupled, and, as the particle moves through the lattice, it will execute independent motion in the two phase space planes xx' and yy'. Although all accelerators have some small degree of coupling between these two coordinates, for many first-order results it is weak enough to be neglected.

The motion of a particle is complicated as it progresses through the lattice. However, an invarient of the motion is given by

$$w = \gamma x^2 + 2\alpha x x' + \beta x'^2,$$

$$\gamma = (1 + \alpha^2)/\beta,$$

$$\alpha = -\frac{1}{2}\frac{\partial \beta(s)}{\partial s}.$$

The functions $\alpha(s)$, $\beta(s)$, and $\gamma(s)$ are the lattice functions for the machine and are determined once the configuration of dipoles and quadrupoles is fixed. As seen above, the invariant function depends only on $\beta(s)$.

Suppose we observe a particle as it passes a fixed point a large number of times, n, as successive rotations. If the tune of the machine is non-integral, the pairs of coordinates x_n, x'_n will lie on the ellipse described by Eq. (1) (see Fig. 6). The maximum x, x_m, is given by $x_m = \sqrt{w\beta_x}$.

Consider an ensemble of particles located around the ellipse. If we allow these particles to move around the ring, the ellipse defined by the set of particular coordinates will change orientation and relative dimension since α, β, γ all change with s, the position around the ring. However, the area of this ellipse will be an invariant, and, since $x_m = \sqrt{w\beta_x}$, the maximum x_m will occur at the position of maximum β. The position of β_{max} and β_{min} in the normal lattice is at the center of the quadrupoles. Thus the quadrupole apertures limit the maximum excursion that a particle in the ring is allowed to make. Note also Eq. (1) becomes $\beta'(s) = 0$ at β_{max} and β_{min} and that this forces $\alpha(s) = 0$ at these same points.

$$w = \frac{1}{\beta} x^2 + \beta x'^2$$

and the ellipse becomes like that shown in Fig. 7.

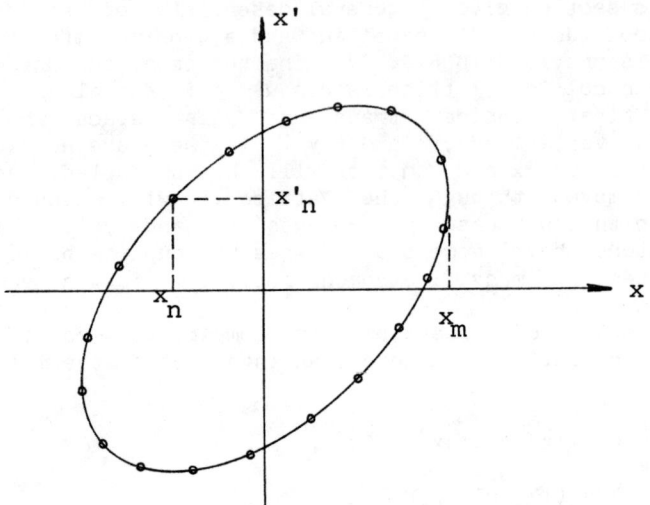

Fig. 6

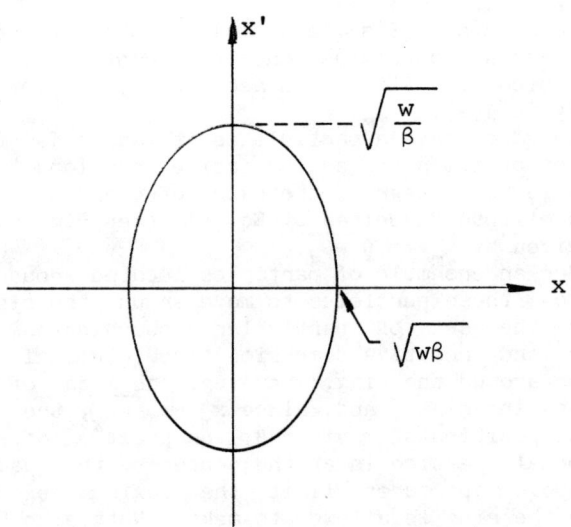

Fig. 7

Next Liouville's theorem may be applied to an ensemble of particles in phase space. Since orbits in phase space do not cross, particles with a given invariant w will always lie on the same ellipse, and ellipses with different w will nest. We can also write that

$$dN = \rho(x,x') \, dx \, dx' \qquad (3)$$

where $\rho(x,x')$ is the density of particles at x,x', and

$$N_0 = \iint \rho(x,x') \, dx \, dx' \qquad (4)$$

is the total number of particles. Now consider a position in the lattice as shown in Fig. 7 where the ellipse is orthogonal to the x,x axes. We may describe the phase space density by

$$\rho(x,x') = \frac{1}{\sqrt{2\pi}\sigma_x} \varepsilon^{-x^2/2\sigma_x^2} \frac{1}{\sqrt{2\pi}\sigma_{x'}} \varepsilon^{-x'^2/2\sigma_{x'}^2},$$

$$\sigma_x^2 = \beta_x^2 \sigma_{x'}^2. \qquad (5)$$

The condition on σ_x and $\sigma_{x'}$ assures that the density is <u>time invariant</u> since the lines of constant density are the invariant ellipses give by Eq. (2). If this condition is true for a beam injected into a machine, then beam emittance is said to be "matched" to the acceptance. The emittance of a beam is defined as the area of the invariant ellipse that contains 95% of the particles. And in the above case of a bi-Gaussian distribution, a simple integration gives

$$\varepsilon = 6\pi\sigma_x^2/\beta_x. \qquad (6)$$

Up to this point the momentum of the beam has been assumed to be fixed. If now the momentum is slowly changed, i.e., the change is small in a betatron period, then it can be proved that the area of the invariant ellipse changes as $1/\gamma\beta$ or inversely with momentum. It has become common practice to define the normalized beam emittance as the value at $\beta\gamma = 1$, and hence

$$\varepsilon = \varepsilon_0/\beta\gamma. \qquad (7)$$

Note that here $\beta = v/c$ and $\gamma = (1 - \beta^2)^{-1/2}$.

A typical modern ion source will provide a beam with transverse normalized emittance in the range of $\varepsilon_0 = 1\pi$ mm mr. However, measurement of the normalized emittance after acceleration frequently gives values 10 to 20 times larger. This dilution in density can be traced generally to the very early stages of acceleration, where space-charge forces are large and Liouville's theorem is no longer applicable.

Beam Stability

The most important consideration from the standpoint of the machine is the stability of the beam. First, of course, it is necessary that the stability be examined from the standpoint of the coherent interaction of the beam from self-induced fields in the wall of the vacuum chamber. Colliders, however, have additional effects. The first is from the field of the colliding bunch or its adjacent bunches that may pass very nearby if the collision angle is small. This is called the beam-beam tune shift.

Consider a particle in beam 1, on axis, passing through a bunch of beam 2. By symmetry, there is no transverse deflection of the particle. However, if it is off axis, it receives a transverse kick from the combined electric and magnetic fields of the oncoming bunch. If the density of the target beam were constant out to a radius a, the kick would be proportional to r out to r = a, whereupon it would fall off as $1/r$. Thus the beam for $r \leq a$ acts in first order like a lens whose focal length is inversely proportional to the number of particles in the bunch and proportional to its effective radius σ_x. This lens induces a shift in the tune of the machine given by

$$\Delta \nu = - \frac{\beta^*}{4\pi f}$$

where f is the first-order focal length of the lens due to the bunch and can be expressed as

$$\Delta \nu = \frac{N_1 \beta^* r_p}{4\pi \sigma_t^2} = \frac{3}{2} \frac{N_1 r_p}{\varepsilon_t}$$

where r_p is the radius of the proton. We see at once a very important result, i.e., both $\mathscr{L}$ and $\Delta \nu$ are $\sim 1/\varepsilon_t$.

It is now thought that, if the tune shift becomes too large, the beam will become unstable. This has not yet been observed for $\bar{p}p$ collisions. However, CERN has operated the collider with bunch-bunch tune shifts as large as 0.003. A great deal of work has been done on trying to understand this problem both analytically (Ref. 8) and experimentaly (Ref. 9). So far the only solid numbers are from CERN, and they are for the so-called weak beam--strong beam collision. Additional results on non-bunched beams are available from the ISR, but these are not useful in understanding the bunched case.

In addition to the collinear crossings described here, bunches may cross non-collinearly. For instance, with small crossing angle, adjacent bunch encounters occur sequentially on each side of the interaction point. These bunches must be sufficiently separated so that they don't become a secondary source of interactions upstream and downstream of the detector. However, more importantly, they must be separated sufficiently so that the fields from the passing bunch do not cause deflections of particles in the oncoming bunch which vary from one side of the bunch to the other. Such a field will "scramble" phase space and cause the emittance to blow up with a resultant decrease in luminosity.

An additional effect that is important in colliders with bunched beams is caused not by the field of crossing bunches but by Coulomb interactions of particles within the same bunch. This is called "intra-beam scattering." Consider a particle with energy higher than the central energy. This particle will perform synchrotron phase oscillations and will slowly move from a forward position in the bunch to a rear position and back again. In the coordinate system of the bunch, one sees particles streaming back and forth through the packet. In this system Coulomb interactions can take place which result in momentum being transferred from the longitudinal to the horizontal direction, as well as diffusion effects in the longitudinal direction. Damping occurs in the vertical direction. As a result the transverse phase space area can grow, with a resultant decrease in $\mathcal{L}$, and bunch length can grow, which may be undesirable from the standpoint of the detector design. Intra-beam scattering has been observed at CERN. The analytical theory has been developed by Bjorken and Mtirgawa (Ref. 10). Lifetimes of 20 hours are anticipated for TeV I, and at CERN lifetimes as short as 10 hours have been seen.

DETECTORS FOR HADRON COLLIDERS

Suitable detectors for hadron colliders were a result of early experiments on the ISR. Indeed most of the physics from this machine is done with detectors of limited solid angle. Thus the high p_T behavior of the π_0 spectrum, the long-range and short-range correlations in rapidity of pions, scaling, and rising σ_T, and many other experiments covered only a restricted region of solid angle. Unlike the e^+e^- experiments, total energy conservation is not a useful kinematic constraint, and much of the energy of the two incoming particles can be carried away along the beam axis by "spectator" partons. Thus extensive evidence for QCD jets was obtained without actually displaying convincing jets as with e^+e^- jet physics. However, as a result of the experience obtained, it became evident that a 4π detector was necessary. At least one such calorimetric detector was proposed by Rubbia et al. 1973 (Ref. 11). This detector was never constructed. However, two of the authors, Rubbia and Darriulat, later constructed the UA1 and UA2 detectors at the $S\bar{p}pS$. Finally Willis et al. (Ref. 12) constructed a highly segmented calorimeter and obtained clear evidence for the same type of jet structure that has recently been exposed so beautifully via the $\bar{p}p$ collision at $\sqrt{s}$ = 540 GeV.

The key feature of the kinematics that has been exploited at the $S\bar{p}pS$ is $\Sigma p_t = 0$. Thus the detector is as complete and "crack free" as possible, and it is highly segmented so that local energy density can be measured and vector momentum sums constructed. The hole for the beams is kept limited to such small angles that particles escaping along the beam axis cannot disturb the p_T balance. Such a calorimeter allows jets to be reconstructed, and hence opens the door to studying parton-parton collisions.

The physics promises to be dominated by highly collimated jets coming from partons materializing. Fig. 8 shows the density in angle, θ, of a 1-TeV jet from a top quark produced at $90°$ in a collision at $\sqrt{s}$ = 40 TeV. The peak density is 10 particles per degree in θ. This was calculated from Isajet. Although jets are very well defined, it becomes increasingly difficult at high energies to employ any type of momentum analysis, and reliance must be placed on calorimetry, where resolution improves as the energy increases.

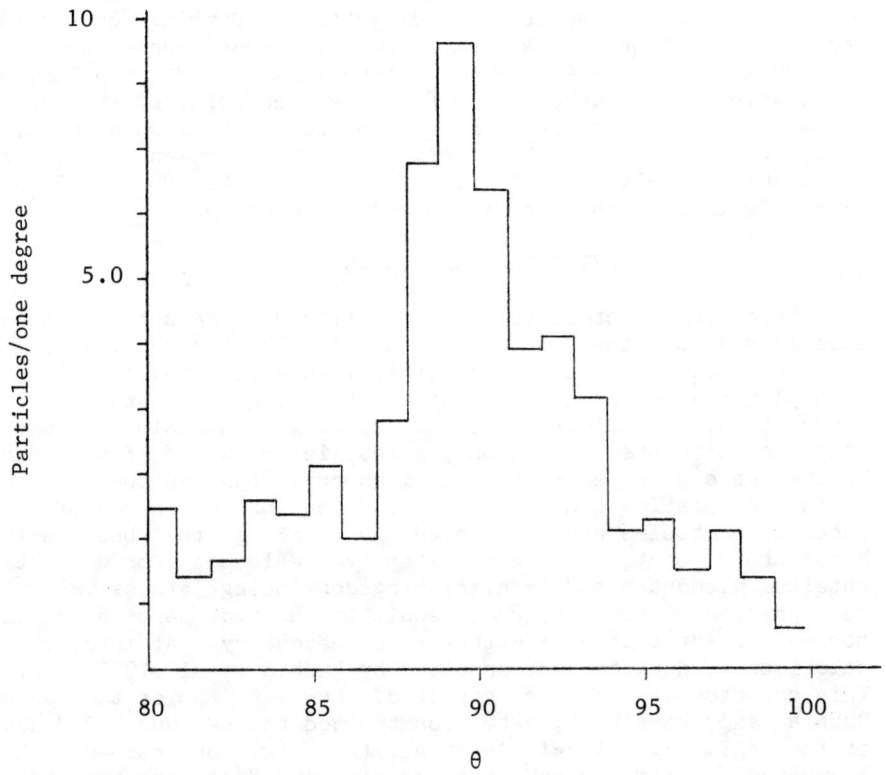

Fig. 8. This figure shows the results of ISAJET for two $90°$ jets of p_T = 1 GeV from pp collisions at $\sqrt{s}$ = 40 TeV.

Electrons and Muons

Equally important is the ability to reconstruct leptons and neutrinos. Electrons must be identified by their characteristic electromagnetic shower. The main difficulty with this technique is that the shower from a photon may overlap the track of a charged pion, thus faking an electron signature. The solution to this problem is high segmentation of the electromagnetic calorimeter so that the shower position and energy can be well determined. Combining this information with the information from a central tracking chamber imbedded in a magnet field allows a match in position and energy to be made in the two different devices. This technique is highly effective for isolated electrons, but becomes very difficult if the electron track becomes embedded in a jet - a situation that can happen if the jets are derived from beauty or charm quarks. A solution to this problem is the use of transition radiation detectors to identify the electron, as is proposed for the TeV I D0 detector.

Muons are identified by their ability to penetrate large masses of iron. Their momentum is determined by measuring their deflection in magnetized iron - a method of limited precision because of multiple scattering. In addition, low momentum muons embedded in jets suffer from a background due to π and $K^{\pm}$ decay in flight as well as "punch-through" from the hadron components. Both of these backgrounds may be reduced by tracking the μ through more than one layer of magnetized iron.

Tracking

Tracking in a central magnetic field serves the purpose of aiding e and μ identification as well as measuring the sign of the lepton. It may also be useful for reconstructing the decays of long-lived particles. So far detectors with field (UA1, CDF) and without field (UA2, D0) have been constructed. As the energy of the collider increases and the energy scale grows, the tracking system encounters serious difficulties. Since one would like to increase the $\mathcal{L}$ as the energy increases, the chamber must handle more events per second. However, as the energy increases, the event multiplicity grows, and, even more importantly the density of tracks and their momentum both increase as one goes to higher s. For magnetic tracking, $\delta p/p \sim Kp$, and it soon becomes difficult to determine even the sign of high energy particles. For instance CDF with a 3-m diameter solenoid and a 1.5-T field has $\delta p/p = 0.003$ p.

The cost and complexity increase rapidly both with size of the magnet and with the magnitude of B. It should be noted that a detector using magnetic momentum analysis for charged particles and an EM calorimeter for γ rays fails miserably at reconstructing the energy of a jet because of interactions of the charged particles in the EM calorimeter. There is no convenient way to eliminate this unknown contribution to the energy registered by the shower counter. Studies show that the energy of a 200-GeV jet can be overestimated by more than 50% when combining charged particle momentum measurements with an EM shower counter.

Neutrinos

Neutrinos are "detected" by the imbalance of visible transverse momentum $p_\nu = -\Sigma \vec{p}_t$, where the sum is over all the cells of the calorimeter. For this procedure to work, the calorimeter must have a large number of cells with good energy resolution so that a precision vector sum can be formed. Also, cracks in the calorimeter must be kept small, since a high energy particle that hits a crack will not be measured in the calorimeter and will give a fake missing momentum. Central tracking in a magnetic field can be used to improve the rejection of this type of false signal since high momentum particles pointed toward cracks are readily identified.

Calorimetry

Improving the calorimeter is a real technical challenge. Requirements can be listed as follows:
1. Good granularity.
2. Good energy resolution.
3. The same response to EM and hadronic showers.
4. Fast pulse response.
5. Ease of calibration.
6. Not sensitive to radiation damage.
7. Homogeneous - no cracks.

Three types of calorimetry are in use or will be in the near future:

EM	Hadron
1. Pb-scintillator	Fe-scintillator
2. Pb-proportional tube	Fe-proportional tube
3. U-liquid argon	U-liquid argon

The first is used in UA1, UA2, and CDF, the second is used in CDF, and the third is proposed for D0. The characteristics of these three detector systems are shown in Table VI. The first two are typical for CDF, and the last is taken from the D0 Design Report. Several important points should be noted.

The response of a uranium calorimeter can be made nearly the same for EM and hadronic showers, whereas using a mixture of Pb and Fe results in a ratio of about 1.3. The primary use of the calorimetry is to measure jets, and this difference in response leads to a decrease in the resolution for jets, since their ratio of hadronic to EM composition can vary. However, it should be noted that some correction to the energy can be made from the observed ratio of energy in the EM and hadron calorimeters.

The question that must be answered is whether or not a given scheme is "good enough," and this criterion is set by the fluctuation due to the neutrino and long-lived K^0 leakage to be expected from a typical jet. Figs. 9 and 10 show the results of an interesting study of CDF (Ref. 13). Fig. 9 shows the results of a hole in the forward and backward direction for the CDF detector. This detector has Pb scint/Fe scint from $40°$ to $140°$ and Pb gas/Fe gas in the end regions $2°$ to $140°$ and $140°$ to $178°$ with

Table VI

	Pb scint/ Fe scint	Pb gas/ Fe gas	LA-U
Energy resolution	$12\%/\sqrt{E}$ $65\%/\sqrt{E}$	$25\%/\sqrt{E}$ $100\%/\sqrt{E}$	$12\%/\sqrt{E}$ $40\%/\sqrt{E}$ hadron
Granularity	Fail	Good	Good
Fast pulse response	Very good	Good	Slow
Calibration	Test beam sources	Test beam sources	Absolute
Response EM/ response hadron	1.3	1.3	1
Cracks	Needs room for shifter-light pipes	Room cables and gas barriers	Cryostat walls
Radiation damage	Scintillator 1000 rad	Solid state preamps	Preamps

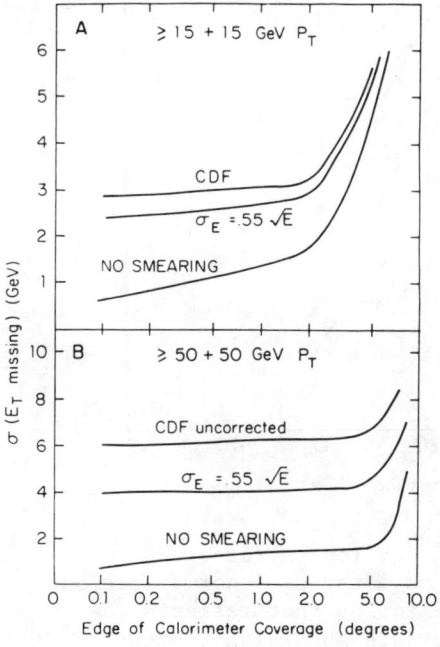

Fig. 9. The rms smearing of missing E_T due to energy lost in a conical hole in the calorimetry along the forward and backward beam directions. Calculations from ISAJET.

characteristics as shown in Table VI. Fig. 9a shows the effect of changing the hole size on missing E_T for parton-parton scattering that results in two jets with $p_T > 15$ GeV and 9b is for $p_T > 50$ GeV. The curve labeled "no smearing" would result from a perfect calorimeter that has losses only due to geometrical losses through the hole. The smaller p_T case is more sensitive to the size of the hole, since it is easier to lose 15-GeV p_T through a 5° hole than 50-GeV p_T. In any case it is apparent that a 2° hole is tolerable. The curve labeled "$\sigma_E = 0.55 \sqrt{E}$" is calculated for a perfect calorimeter with the same response to EM as to hadronic showers. "CDF" refers to a real calorimeter with cracks, punch-throughs, decay ν escaping, and a ratio of 1.35 between EM and hadronic showers.

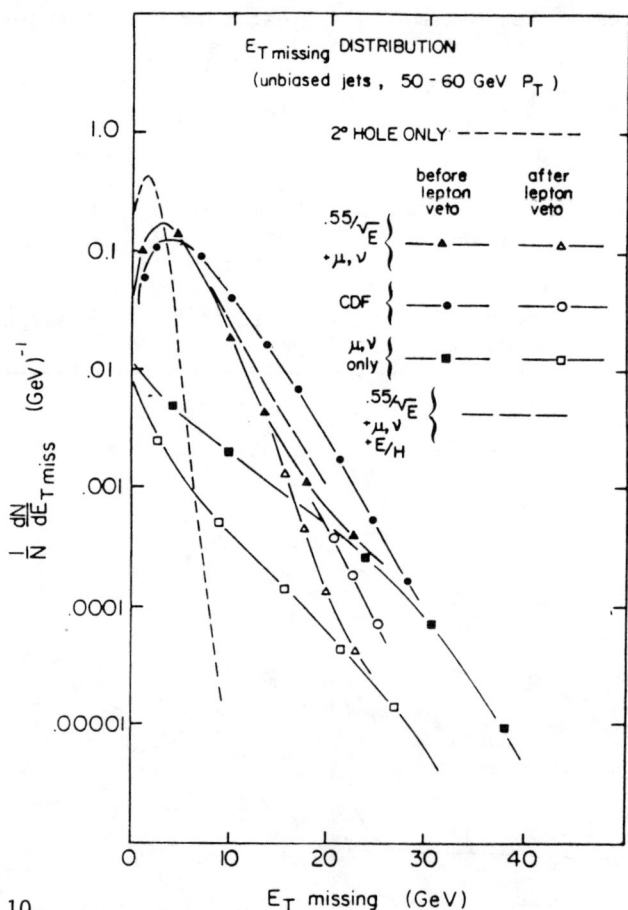

Fig. 10.
Distribution $1/N \cdot dN/dE_{tmissing}$ for CDF and "ideal" detectors with and without 2 GeV lepton veto.
"Before lepton veto" curves have the same normalization. "After" curves have a normalization that reflects the reduction in the number of events surviving the cuts.

Fig. 10 shows the distribution of missing E_T for $\bar{p}p \to 2$ jets with p_T (parton) $\geqslant$ 50 GeV. The dashed curve is for geometrical losses through a $20°$ hole. The curve labeled "$0.55/\sqrt{E}$" is for a $2°$ hole plus effects of missing finite energy resolution and ν and muons escaping the calorimetry. The "CDF uncorrected" curve adds in the CDF calorimeter including cracks, etc. The "CDF after veto" curve shows the offline improvement that can be obtained by examining the tracking data to correct for K decays, μ loss, charged particles hitting cracks, etc. It is clear that, even with the CDF resolution, one is close to the ultimate background obtainable with a better calorimeter because of the effect caused by μ and ν being missed.

Calibration

Detector calibration is a major problem. An enormous advantage of liquid argon is its absolute calibration, and the major disadvantage of gas or scintillation calorimetry is that test beam data for every component of the detector must be obtained. After the initial calibration, sources and light flashers become secondary standards and must be arranged so that the front-end device (phototube or proportional tube) as well as the electronics can be separately monitored. It appears from the results of UA1 and UA2 that it is indeed possible to do this at the 1 to 2% level - a marvelous accomplishment! As the size and complexity of detectors increase for SSC, this will become a dominant concern, and, since tests must be made near full energy, test beams at older machines will not be sufficient. In addition it is difficult and expensive to disassemble a detector in order to calibrate it. Thus in situ calibration techniques and calorimeters with an absolute response will have to be developed.

CDF and D0 Detectors

Two detectors being designed for TeV I are shown in Figs. 11 to 14, which provide a nice comparison between a magnetic and a non-magnetic detector. One could equally well compare UA1 and UA2.

In designing a detector, one must first decide on whether there will be central tracking in a magnetic field. The most important reason for desiring a field is to provide lepton sign determination and to aid in momentum measurement. Once a field is decided upon, one must decide whether to use a solenoid parallel to the beam (CDF) or a dipole at right angles to the beam (UA1). The solenoid gives very uniform coverage for $30° < \theta < 150°$ but misses the end caps. The dipole at right angles gives excellent measurements along the beam and in one direction at right angles, but creates a nasty asymmetry along the field direction. The magnitude of the field is chosen on the basis of the energy scale and tracking chamber resolution.

The radial scale of the calorimetry is set by the tracking chamber since it must surround this region. The transverse angular scale of segmentation is determined by jet structure. The cells

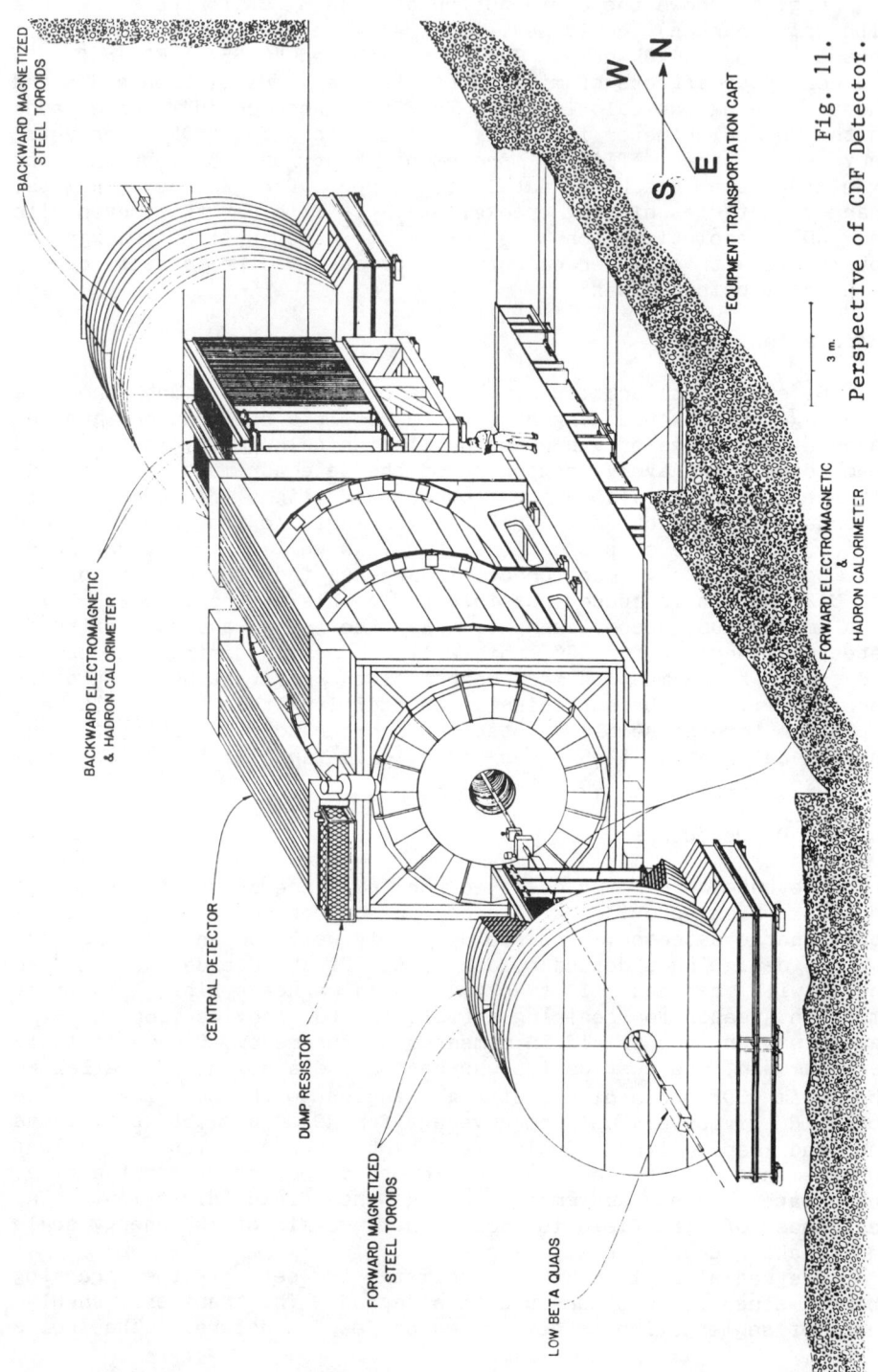

Fig. 11. Perspective of CDF Detector.

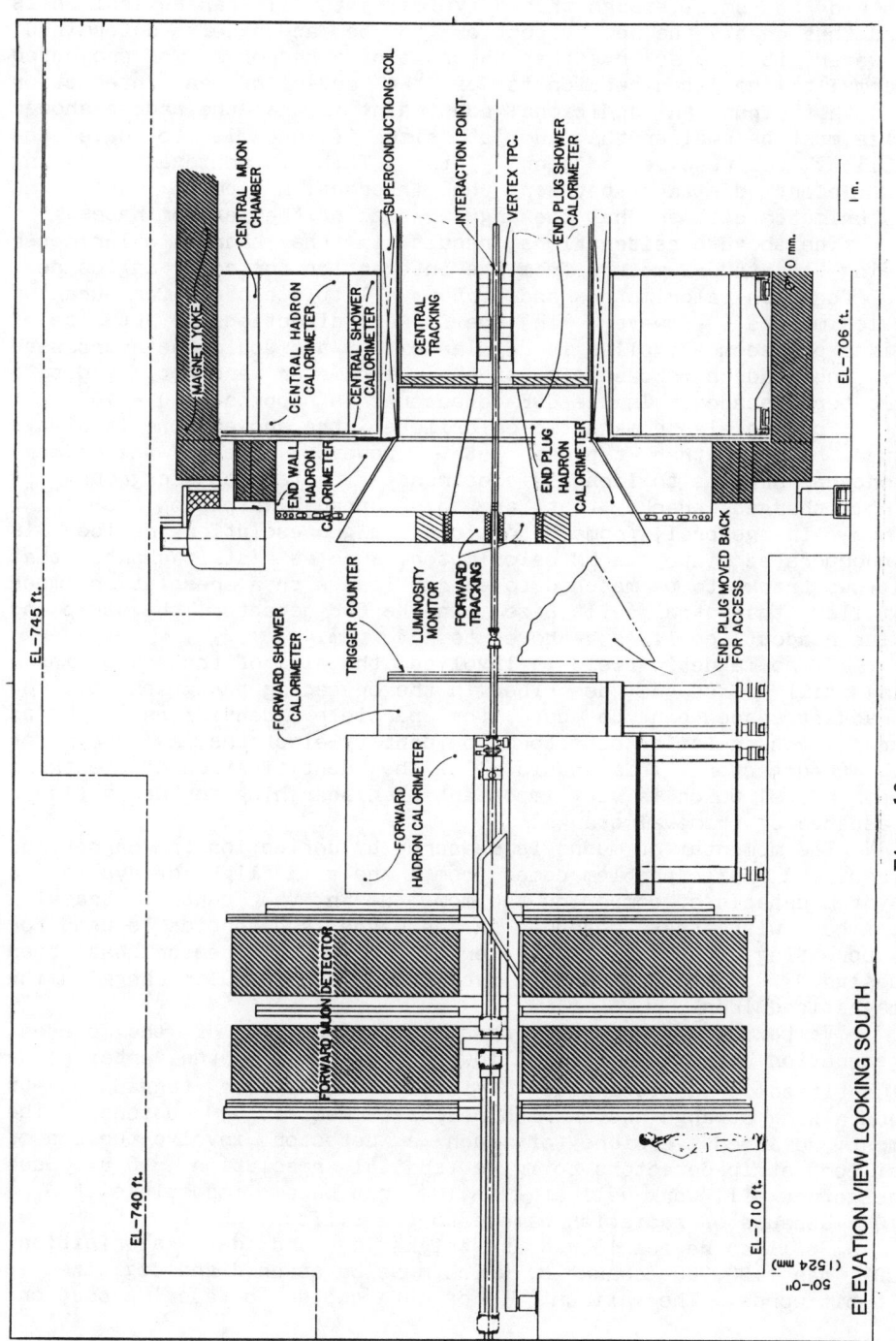

Fig. 12. Cross Section of CDF Detector.

should be small enough that a typical jet will span several cells and thus enable the jet direction to be accurately determined. However, it is also true that the size of a hadron shower projected normal to the jet direction is of the order of an interaction length. Thus an additional constraint is that the hadron shower size must be smaller than the jet size if one is to have the ability to resolve adjacent jets. This is achieved by having sufficient distance between the interaction point and the calorimeter cell so that the angular size of the jet dominates.

The above considerations result in the hadron calorimeter being located ~2 meters from the interaction for polar angles near $90°$ for iron calorimeters and perhaps a little closer for uranium calorimeters. However, in the forward direction the jet size of fixed p_T becomes smaller in angular extent and would be surpassed by the hadron shower size if the calorimeter were not moved to a greater distance. In the CDF detector this happens at $\theta = 10°$.

For the electromagnetic calorimeter, the shower size is always much smaller than typical jets. However, one of the primary concerns here is to identify electrons. The largest background is from charged tracks that are overlapped by a photon from a $\pi°$. Thus, in general, some additional high-resolution device is incorporated into the EM calorimeter, such as a strip chamber that allows tracks to be matched to energy loss with a resolution much smaller than the cell size. In the CDF detector, the strip and wire readout localizes a shower to a few mm.

If no magnetic field is involved, the size of the calorimetry is still fixed as described in the preceding paragraph, but the tracking space can be used for particle identifiers such as transition radiation detectors to identify electrons, as chosen for the D0 detector. This should allow the identification of electrons near jets, which is very important when searching for heavy flavor cascades of primeval quarks.

The momentum of muons is measured by deflection in magnetized iron. A major problem comes from K and π in-flight decays, and a system capable of comparing the momentum in the central tracking chamber with that measured in the magnetized toroids is used for suppressing occurrence of this error. The D0 detector has been designed to measure muons over the whole angular range, using magnetized iron.

Vertex detectors are being planned for all of the present generation detectors. This allows decays close to the vertex to be identified - a technique obviously useful for tagging jets containing strange and charmed mesons as well as τ leptons. The most advanced versions of such a detector involve the use of silicon strip detectors to give spatial resolution ~10 μ. Such detectors will work well at $\mathcal{L} \sim 10^{30}$ but become impossible at $\mathcal{L} = 10^{32}$ because of radiation damage to the silicon.

Still to be considered are triggering and data acquisition. CDF has 100,000 channels, which must be scanned and digitized as 16-bit words. The maximum rate of data copied to tape is about one

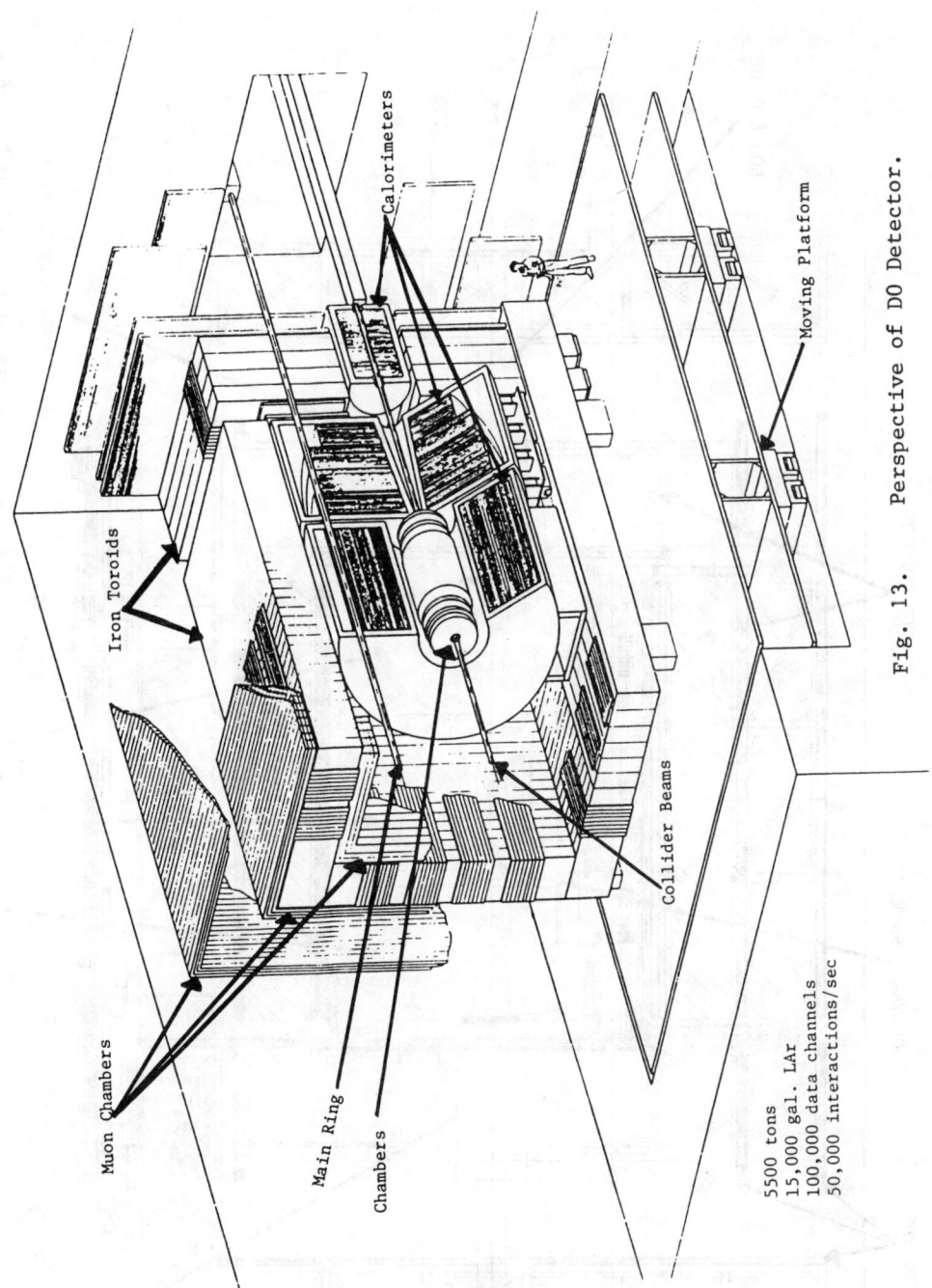

Fig. 13. Perspective of D0 Detector.

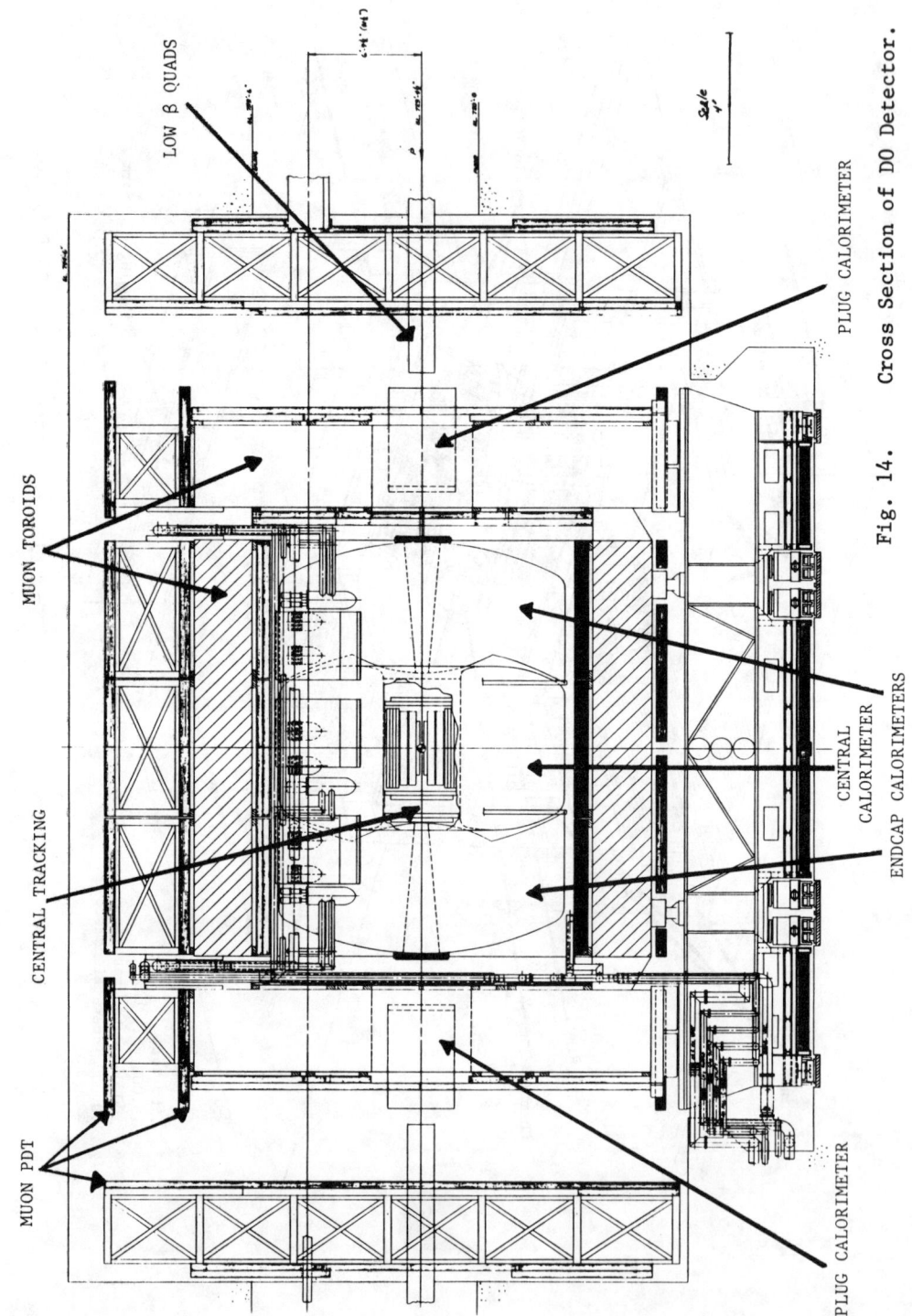

Fig. 14. Cross Section of D0 Detector.

event per second; higher rates become prohibitively expensive in terms of offline computing. The interaction rate at $\mathcal{L} = 10^{30}$ is 5×10^4 events/sec. Thus a reduction of 5×10^4 must be provided by the trigger system. This is achieved first by recognizing clusters of energy and measuring their E_T in order to set a "jet" p_T bias. Stiff muon and electron track candidates can also be recognized by fast electronics. This information can be used to select events such that the rate is reduced to ~100 events/sec without the electronics generating a large dead time.

At the 100-event/sec level, the detector can be completely read out into a buffer, and the data can be passed through special dedicated processors such as the 168E or 3081E or other more modern versions of them. Partial reconstruction of the events in these processors can reduce the data stream to a rate compatible with tape. These level-3 processors may also be used later for aiding the offline data reduction. Indeed this technology is being developed very fast, and considerable progress is being made in developing parallel processors using the Motorola 68000 or equivalent chips. In fact data reduction, although it presents challenges, is probably in better shape than our fundamental detector technology for the next generation of machines and detectors.

CONCLUSION

We are now starting a new era of hadron-hadron colliders. The beautiful experiments at CERN have shown that it is possible to do clean experiments in a world made of leptons and partons. The early studies for the SSC have uncovered no surprises in the accelerator technology that will prohibit scaling present colliders by a factor of 20 in laboratory energy. Luminosities larger than can be accepted by conventional detectors are attainable. Thus the detectors become the focus for the most challenging and difficult technological developments. They must be planned in an integrated fashion with the machine. In addition they will require extensive R&D, which must be started concurrently with that for the machine. Realistic goals must be set because an accelerator with a luminosity higher than can be accepted by realizable detectors will needlessly increase the cost of the machine. Thus the challenge remains of how to measure ever smaller cross sections as the energy increases, and the bottleneck at this point is clearly the detector design!

REFERENCES

1. G. Aruison et al., "Experimental Observation of Isolated Large Transverse Energy Electrons with Associated Missing Energy at s = 540 GeV," Phys. Lett. <u>122B</u> (1983) 103.

   G. Aruison et al., "Experimental Observation of Lepton Pairs of Invariant Mass Around 95 GeV/c^2 at the CERN SPS Collider," Phys. Lett. <u>126B</u> (1983) 398.

2. E. Eichten, I. Hinchliffe, K. Lane, C. Quigg, Super Collider Physics. Am. Phys. Soc., AIP (Oct. 1984).

3.* Proposal for a High Energy Research Facility, Midwestern Universities Research Association, March 14, 1958.

   A Proposal for a High Intensity Accelerator, Midwestern Universities Research Association, submitted: March 30, 1962, revised: September 30, 1963.

4.* A series of about 20 reports on the Design Study for a 300 GeV Proton Synchrotron were prepared by the Western Accelerator Group.

5a. F. T. Cole, F. E. Mills, Annu. Rev. Nucl. Sci. 31, 295 (1981).

5b. A. V. Tollestrup, G. Dugan, Physics of High Energy Particle Accelerators (AIP Conference Proceedings, 105, 1982) p. 954.

6. Superconducting Super Collider Reference Design Study, U. S. Department of Energy, May 8, 1984.

7. M. Month, W. T. Weng, Physics of High Energy Particle Accelerators (AIP Conference Proceedings No. 105, 1982) p. 124.

8. J. F. Schonfeld, Physics of High Energy Particle Accelerators (AIP Conference Proceedings No. 105, 1982) p. 524.

9. L. Evans, Beam-Beam Interactions, CERN 84-15, 1984, p. 319.

10. J. D. Bjorken, S. R. Mtirgawa, Intra-Beam Scattering, Fermilab-Pub-82/47 Thy. (1982).

11. A. Boehm, Z. Bozzo, D. Cheng, G. de Zorzi, R. Ellis, H. Foeth, A. Kernan, F. Muller, B. Naroska, C. Rubbia, G. Sette, A. Straude, P. Strolin, L. Sulak, V. Telegdi, Proposal for a Large 4π Calorimeter to Investigate Multibody Events at the ISR, CERN/ISRC/72-32, November 9, 1972.

12. T. Akesson et al., Phys. Lett. 128B (1983) 354.

13. J. Freeman, R. Perchonok, J. Yoh, Transverse Energy Physics with the CDF Calorimeter, Fermi Note FN399 (1984).

*Copies of these reports are available in the History of Accelerators Collection in the Fermilab Library.

PHYSICS RESULTS OF THE UA1 COLLABORATION AT THE CERN

PROTON-ANITPROTON COLLIDER

C. Rubbia

Harvard University

CERN, Geneva, Switzerland

Reprinted, by permission of the University of Hawaii Press, from Proceedings of the Ninth Hawaii Topical Conference in Particle Physics (1983), edited by R. J. Cence and E. Ma. Published by University of Hawaii at Manoa.

TABLE OF CONTENTS

1 The CERN Super Proton Synchrotron (SPS) as a proton-antiproton collider 206
2 Jets 210
   2.1 Introduction 210
   2.2 The UA1 detector and the trigger conditions 211
   2.3 Definition of jets 212
   2.4 Jet axis and jet energy 215
   2.5 Multiplicities 218
   2.6 Monte Carlo studies 219
   2.7 Jet cross-sections 221
   2.8 Studies on jet fragmentation in charged particles 227
   2.9 Jet fragmentation function 229
   2.10 Transverse momentum with respect to the jet axis 232
   2.11 Structure functions 233
3 Observation of charged intermediate vector bosons 246
   3.1 Introduction 246
   3.2 Event selection for $W \rightarrow e\nu$ decays 248
   3.3 Origin of the electron-neutrino events 250
   3.4 Determination of the invariant mass of the $(e\nu_e)$ system 254
   3.5 Longitudinal motion of the W particle 256
   3.6 Effects related to the sign of the electron charge 259
   3.7 Determination of the parity violation parameters and of the spin of the W particle 261
   3.8 Total cross-section and limits to higher mass W's 263
   3.9 Observation of the decay mode $W \rightarrow \mu + \nu$ 263
4 Observation of the neutral boson Z^0 274
   4.1 Event selection 274
   4.2 Events of type $Z^0 \rightarrow e^+e^-$ 276
   4.3 Events of type $Z^0 \rightarrow \mu^+\mu^-$ 283
   4.4 Backgrounds 286
   4.5 Mass determination 288
5 Comparing theory with experiment 289
   Acknowledgments 292
   References and notes 293

PHYSICS RESULTS OF THE UA1 COLLABORATION AT THE CERN

PROTON-ANITPROTON COLLIDER

C. Rubbia

Harvard University

CERN, Geneva, Switzerland

Reprinted, by permission of the University of Hawaii Press, from Proceedings of the Ninth Hawaii Topical Conference in Particle Physics (1983), edited by R. J. Cence and E. Ma. Published by University of Hawaii at Manoa.

1. THE CERN SUPER PROTON SYNCHROTRON (SPS) AS A PROTON-ANTIPROTON COLLIDER

The conversion of the SPS into a $\bar{p}p$ collider[1] and the associated physics programmes of the UA1 collaboration were motivated by three very specific physics goals, namely :

(i) The observation of jets and a detailed comparison with predictions of QCD.

(ii) The discovery of the charged Intermediate Vector Boson (IVB) $W^{\pm}$ in the electron and muon decay modes and the measurement of its fundamental charge asymmetry in the decay.

(iii) The discovery of the neutral IVB, Z^0, both in the electron and muon decay channels.

These goals are now essentially fulfilled. In addition and perhaps more surprising has been the extreme cleanliness of the events. A modest p_t threshold cut can easily separate the interesting phenomena, removing the background due to spectator quarks.

The first collisions between protons and antiprotons at $\sqrt{s}=540$ GeV in the SPS accelerator operating as a storage ring[2] were observed in the early summer of 1981, not even three years after approval of the project. Two years later, a very large amount of information has become available which will be the subject of these lectures. Before discussing these results, it may be worth recalling that the collider has been the first and so far the only storage ring in which bunched protons and antiprotons collide head-on. Although the $\bar{p}p$-collider uses bunched beams like e^+-e^- colliders, the phase-space damping due to synchrotron radiation is now absent. Furthermore, since antiprotons are scarce, one has to operate the collider in conditions of relatively large tune shift, which is not the case for the continuous proton beams of the ISR. Therefore the machine itself has been breaking new grounds.

One of the most remarkable results of the $\bar{p}p$ collider has been probably the fact that it has operated at such high luminosity, which in turn means a large beam-beam tune shift. Most serious concern had been voiced in the early days of the construction about the stability of the beams, due to beam-beam interactions. The beam-beam force can be approximated as a periodic δ-function of extremely non-linear potential kicks and it is expected to excite a continuum of resonances, in principle with the density of rational numbers. Reduced to bare essentials, we can consider the case of a weak antiproton beam colliding head-on with a strong bunched proton beam. The increment of the action invariant $W = \gamma x^2 + 2\alpha x x' + \beta x'^2$ of an antiproton due to the angular kick $\Delta x'$ is $\Delta W = \beta(\Delta x')^2 + 2(\alpha x + \beta x')\Delta x'$ and can be expressed in terms of the "tune shift" ΔQ as $\Delta x' = 4\pi \Delta Q x/\beta$. If now we assume that the successive kicks are randomized, the second term of ΔW averages to zero and we get

$$\left\langle \frac{\Delta W}{W} \right\rangle = \frac{1}{2}(4\pi \Delta Q)^2.$$

For the design luminosity we need $\Delta Q \sim 0.003$, leading to $(\Delta W/W) = 7.1 \times 10^{-4}$. This is a very large number indeed, giving an e-fold increase of W in only $1/7.1 \times 10^{-4} = 1.41 \times 10^{3}$ kicks! Therefore the only way in which the antiproton motion remains stable is because these strong kicks are not random but periodic and the beam has a long "memory" which allows one to add them coherently rather than at random. Off-resonances, effects of these kicks, then cancel on average, giving a zero overall amplitude growth. The beam-beam effects are very difficult, albeit not impossible to evaluate theoretically, since this a priori purely deterministic problem can exibit stochastic behaviours and irreversible diffusion-like characteristics.

An old measurement at the electron-positron collider SPEAR had further increased the general concern about the viability of the $\bar{p}p$ collider scheme. Reducing the energy of the electron collider (Figure 1) resulted in a smaller value of the maximum allowed tune shift, interpreted as due to the reduced synchrotron radiation damping. Equating the needed beam lifetime for a $\bar{p}p$ collider (where damping is absent) with the extrapolated damping time of an e^+e^- collider gives a maximum allowed tune shift $\Delta Q = 10^{-5} - 10^{-6}$ which is catastrophically low. This bleak prediction did not find itself confirmed by the experience at the collider, where $\Delta Q = 0.003$ per crossing and six crossings is routinely achieved with a beam luminosity lifetime approaching one day. What is then the cause of such a striking contradiction between the experiments with protons and electrons? The difference is related to the presence of synchrotron radiation in the latter case. The emission of synchrotron photons is a main source of quick randomization between crossing and it leads to a rapid deterioration of the beam emittance. Fortunately the same phenomenon provides us also with an effective damping mechanism. The proton-antiproton collider works since <u>both</u> the randomizing and the damping mechanisms are absent. This unusually favourable combination of effects ensures that proton-antiproton colliders become viable devices. They are capable of substantial improvements in the future. Accumulating more antiprotons would permit us to obtain a substantially larger luminosity. A project is on its way at

CERN which is expected to be able to deliver enough antiprotons to accumulate <u>in one single day</u> approximately the integrated luminosity on which the results of these lectures have been based (~ 100 nb^{-1}). On a longer time scale, a $\bar{p}p$ collider built in LEP tunnel with superconducting magnets of high field (10 T) is also conceivable. The luminosity will be further increased because of adiabatic beam damping with

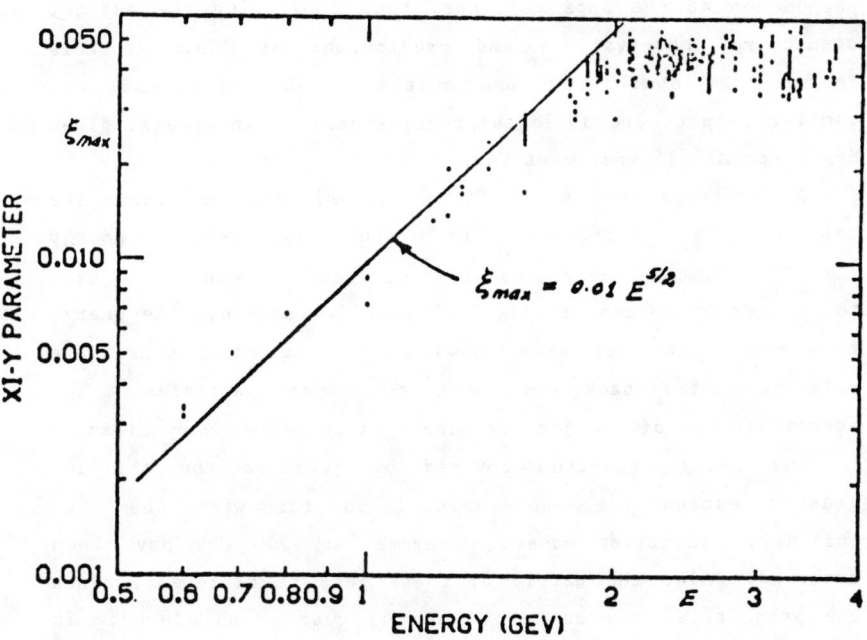

Fig. 1   Maximum allowed tune shift ΔQ at SPEAR as a function of the energy.

energy ($L \sim \gamma/R$). If the proton and antiproton bunches are transferred from the SPS collider to a 10 TeV + 10 TeV collider in the LEP tunnel, a further increase of luminosity of about one order of magnitude is gained. Luminosities of order of 10^{32} cm^{-2} sec^{-1}, which are likely to be at the limit of a general purpose detector, are therefore quite conceivable. A further, important advantage is provided by the emergence of a significant amount of synchrotron damping, which at 10 TeV has an e-folding time of the order of several hours. This could be very helpful in improving even further the beam lifetimes and to increase significantly the attainable luminosities.

2. JETS

2.1. Introduction. Jets appear as the dominant, new phenomenon at the $S\bar{p}pS$ collider, thus confirming the earlier cosmic-ray observations and predictions of QCD. In this lecture we shall make use mostly of UA1 results. Very similar results and analogous conclusions are in general given by the parallel experiment UA2.

As realized very early in the experimentation around the collider, a threshold in the transverse energy $E_T = \Sigma_i E_T^i$ summed over calorimeter cells can be used to trigger on an essentially 100% pure jet sample. The energy flow around the jet axis shows a striking sharp peak on a relatively low background due to other particles. The identification of the jet parameters is therefore very clean.

The energy spectrum covered by jets at the collider greatly exceeds the one explored so far with the e^+e^- collider. Invariant masses in excess of 200 GeV have been observed. Also the nature of these jets is different, since the projectiles now are made both of quarks and gluons. In spite of these differences, however, fragmentation distributions of charged particles appear remarkably similar to the ones measured for e^+e^- jets. A significant fraction of jet events contain more than two jets. For instance for events with $E_T^{(1)} > 20$ GeV, $E_T^{(2)} > 20$ GeV, about 30% have $E_T^{(3)} > 4$ GeV and $\sim$ 10% have $E_T^{(3)} > 7$

GeV. The presence of the third jet strongly suggests the gluon events with bremsstrahlung mechanism (roughly α_s times smaller in cross-section) and very similar to the familiar observation at the e^+e^- colliders. Indeed, the acoplanarity distribution for these events is in excellent agreement with QCD predictions which take precisely this effect into account. Appearance of jets at the collider is interpreted as hard scatterings amongst constituents of the proton and the antiproton. Kinematics of this "elementary" process can be derived from the energies and angles of the jets. There are several processes which can concurrently occur, due to the presence of quarks and gluons :

$$gg \rightarrow gg \quad gq \rightarrow gq$$
$$g\bar{q} \rightarrow g\bar{q} \quad qq \rightarrow qq$$
$$q\bar{q} \rightarrow q\bar{q}$$

Fortunately, in the centre of mass of the parton collision, all processes have almost identical angular distributions. Only cross-sections differ significantly.

2.2. <u>The UA1 detector and the trigger conditions</u>. The UA1 detector has been described in detail elsewhere[3], so only the aspects specifically concerned with this study will be presented. The central part of the detector consists of a large cylindrical tracking chamber centered on the collision point, surrounded by a shell of electromagnetic (e.m.) calorimeters and then by the hadronic calorimeter, which also serves as the return yoke for the 0.7−T dipole magnetic field. There are also tracking chambers and calorimeters in the more forward regions but these were not used in this study. The central detector (CD) and central calorimetry has almost complete geometrical coverage down to 5^0 to the beam axis. In the variables commonly used for such descriptions, this translates to −3.0 to 3.0 in pseudorapidity ($\eta=-\ln[\tan\theta/2]$, where θ is the polar angle from the beam axis), and nearly 2π coverage in azimuth about the beam axis (ϕ).

The central tracking chamber consists of a 5.8-m-long and 2.3-m-diameter cylindrical drift chamber. This chamber provides three-dimensional coordinate information, enabling efficient track reconstruction. This, combined with the 0.7-T magnetic field, results in accurate momentum measurement for nearly all charged tracks.

The central calorimetry consists of lead/scintillator sandwich e.m. shower calorimeters surrounded by iron/scintillator sandwich hadronic calorimeters. These calorimeters are highly segmented in order to obtain position information of the energy deposition. Details of the geometry are given in Table 1.

Making use of the knowledge gained from the previous (1981) collider run [*], a localized transverse energy hardware trigger was implemented to select jet-like events for the 1982 run. This trigger required that the transverse energy (E_T) measured within a calorimeter "block" be greater than 15 GeV. A "block" was defined in the central region as two hadron calorimeter units plus the e.m. calorimeter elements in front of them. A "block" in the end-cap region was defined as the hadronic and e.m. elements comprising one quadrant of an end-cap. With this trigger, a data sample of $\int L \, dt = 14 \text{ nb}^{-1}$ was obtained in the 1982 collider run, which constitutes the sample used for the jet studies reported in this paper. In the 1983 run, approximately 118 nb^{-1} of data were collected. Only the inclusive jet cross-section will include results from the 1983 data sample.

2.3. <u>Definition of jets</u>. Jets are defined as clusters in pseudorapidity/azimuth (η/ϕ) space by the following procedure [*]. An energy vector is associated to each calorimeter cell. For hadronic cells, the vector points from the interaction vertex to the centre of the cell. For electromagnetic cells, the vector points to the energy centroid determined by pulse-height measurements (Gondolas) or by position detectors (Bouchons).

In the subsequent clustering, cells are treated differently depending on their E_T being above or below 2.5

Table 1

UA1 central calorimeters

Calorimeters	Type	Thickness at normal incidence	Rapidity range	Number of cells	Cell size $\Delta\eta \times \Delta\phi$ (approx.)	Comments
Gondolas	Electromag.	26.4 X_0/1.1 λ	$|\eta| < 1.5$	48	$0.11 \times 180°$	Azimuth from light attenuation
C's	Hadronic	5.0 λ	$|\eta| < 1.5$	232	$0.30 \times 15°$	
Bouchons	Electromag.	27.0 X_0/ 1.2 λ	$1.5 \leq |\eta| < 3$	64	$1.5 \times 11°$	Rapidity from position detector
I's	Hadronic	7.1 λ	$1.5 \leq |\eta| < 3$	128	$0.40 \times 15°$	

(X_0 are radiation lengths, λ absorption lengths).

GeV :

- Among the cells with $E_T \geq 2.5$ GeV, the highest E_T cell initiates the first jet. Subsequent cells are considered in order of decreasing E_T. Each cell in turn is added vectorially to the jet closest in (η,ϕ) space, i.e. with the smallest $d \equiv \sqrt{(\Delta\eta)^2 + (\delta\phi)^2}$ (with ϕ in radians), if $d \leq 1.0$. If there is no jet with $d \leq 1.0$, the cell initiates a new jet.

- Cells with $E_T < 2.5$ GeV are finally added vectorially to the jet nearest in (η,ϕ) if their transverse momentum relative to the jet axis is less than 1 GeV and if they are not further than 45^0 in direction from the jet axis.

The cut at $d = 1.0$ has been derived from the jet energy profile (see below).

All triggers selected previously were fully reconstructed in the central detector, and calorimeter depositions were corrected using detailed light attenuation maps and an average response correction factor of 1.13 to allow for the hadronic energy in the electromagnetic calorimeters [*]. After reconstruction of jets according to the procedure above over the full pseudorapidity range $|\eta|<3$, an E_T threshold was applied requiring at least one jet with $E_T > 30$ GeV in the Gondola region. This jet is called the trigger jet. A threshold was applied to the fraction of jet energy deposited in the electromagnetic calorimeters ($\geq 10\%$), to discard signals from beam halo in the hadronic calorimeters. This leaves <1% of events with a potential halo problem, while reducing the good data sample by <2%. To avoid edge effects due to the calorimeter geometry, the trigger jet axis has to be contained in $\eta = \pm 1.2$ and in $\phi = \pm 60^0$ from the horizontal plane. These cuts ensure good energy containment and minimal particle losses due to the narrow dead zones between the calorimeters in the vertical plane.

To understand possible trigger biases, we used a sample of events taken with a global, i.e. $|\eta|<1.5$, $\Sigma|E_T| > 20$ GeV, trigger and studied the efficiency of jet finding. We found that with the localized energy trigger the selection procedure is ~80% efficient for jets with E_T = 30 GeV, and more than 95% efficient for jets with $E_T \geq$ 35 GeV..

2.4. Jet axis and jet energy. The above jet finding procedure results in jets with energy and axis defined from calorimeter depositions. It is therefore necessary to discuss the precision of these parameters.

The definition of a jet axis is rather independent of the algorithm used for finding the jet, as it is mostly dependent on large energy depositions in few calorimeter cells. To assess the precision of the jet axis we have simulated in our detector high E_T jets using a Monte Carlo program with fragmentation according to a cylindrical phase space (CPS) model defined below. The result of this study is that the jet axis given by the jet finding procedure agrees with the vectorial sum of the momenta of fragments from the generated jet to within $\pm 6^0$ in ϕ and ± 0.04 in η (rms). This shows that the granularity of our calorimeters does not introduce any appreciable error in the jet axis definition. If using the same jet finding algorithm on the charged tracks given by the central detector, determined with superior angular resolution, one obtains a charged jet axis which coincides with the calorimetric jet axis to within ± 0.1 in η and $\pm 10^0$ in ϕ (rms). The difference reflects mostly the fluctuations between the charged and neutral parts of jets, and constitutes a lower limit to the precision of the jet axis definition.

The definition of the jet energy, on the other hand, is directly related to the cutoff parameter d in (η,ϕ) space. We use the energy profile and Monte Carlo studies to obtain better understanding of the jet energy.

Given the axis of a jet, the average values per jet of deposited transverse energy and of charged particle transverse momentum can be studied as functions of $\Delta\eta$ and $\Delta\phi$ referred to the jet axis. We restrict ourselves here to the pseudorapidity projection, where the granularity is best. We define an average jet profile by superimposing many jets, leaving out from the average any low-acceptance regions in η

or ϕ. The hemisphere opposite to the jet axis in ϕ (|Δϕ| > π/2) is not included. All jets found in our event sample are included, if their transverse energy is at least 20 GeV and if their axis lies within the same (η,ϕ) limits as used for the trigger jet.

The transverse energy flow as a function of Δη is shown in Fig. 2 (a-c) for three ranges of jet E_T. A clear enhancement is observed on top of a flat energy plateau. The full width of the enhancement at the base is given by Δη = ±1.0, independent of the jet energy (this justifies the cutoff value d ≤ 1.0 in the jet finding algorithm). We distinguish two distinct regions in the average jet, the hard core of the jet (|Δη| < 0.2), and the wings (0.2≤|Δη|<1.0). The relative amount of transverse energy contained in the core increases with the jet E_T. The E_T content of the wings seems rather independent of the jet E_T. Outside the enhancement, for |Δη|>1.0, a constant E_T plateau is observed, whose height is independent of the jet E_T. Its value is substantially higher than the one observed for minimum bias events[5]. When the events with more than two jets are excluded the plateau height lies between the values for minimum bias events and for all events. Hence the origin of the plateau is in two distinct phenomena. The first is due to the debris of the beam particles, namely spectator jets. The second one is the production of multijet events, clearly observed in the present data sample. In this case more than one jet can be emitted in the same azimuthal hemisphere, thus increasing the height of the plateau. It should be noted, though, that an absolute measurement of this plateau is difficult : the particles outside the hard core of the jet are mostly of low momentum, and are not measured with good precision due to the non-linear response of the calorimeters at low energy[6].

We have therefore used the information coming from the central detector for a more quantitative assessment of the jet and plateau composition. The jet profile is now studied with respect to the charged jet axis. Again, particle losses in the small regions of low acceptance in the central detector

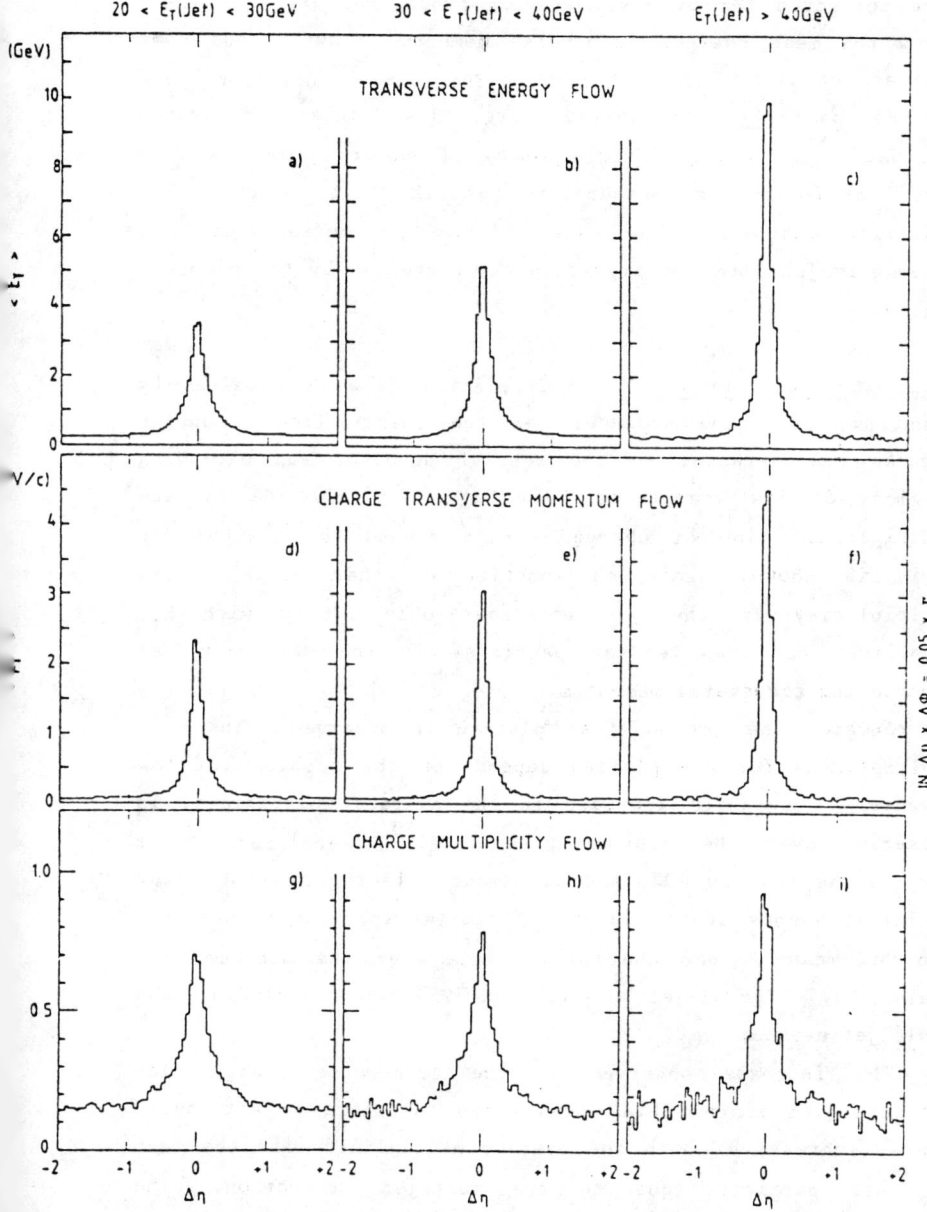

Fig. 2 (a-c): Transverse energy flow as function of Δη, i.e., pseudorapidity distance from the jet axis, for 3 slices of jet E_T. Cells inside $\Delta\phi = \pm 90°$ are used.
(d-e): Charged transverse momentum flow as function of Δη.
(g-i): Charged multiplicity flow as function of Δη.

are corrected for by our averaging procedure. Figs. 2 (d-f) show the mean charged transverse momentum flow in the same slices of jet E_T as before. The same conclusions are reached as for the transverse energy flow. This demonstrates in particular that the granularity of our calorimeters does not influence our measurement of the jet width. The distributions also show that calorimetric measurements for energetic jets are not significantly distorted by the magnetic field.

2.5. Multiplicities. The multiplicity flow is obtained by the same averaging procedure as the energy flow, counting tracks reconstructed in the central detector and excluding regions of low acceptance. In Figs. 2 (g-i) the average multiplicity flow of charged tracks around the charged jet axis is shown, again as function of the jet E_T. The multiplicity in the jet increases only slowly with E_T, implying that the leading particles of the jet carry an increasing transverse momentum.

Outside the jet a flat plateau is observed. The mean multiplicity for the plateau depends on the topology of the events. A definite increase in the average multiplicity is observed over the minimum bias level[7]. For pure 2-jet events the rise is 40%, and is almost a factor of 2 when the multijet events are included. A similar increase is observed in the mean p_t per charged particle over the minimum bias value, 14% for 2-jet events and 25% when including the multijet events.

The increase observed in the transverse energy flow plateau over minimum bias events can therefore be attributed to an increase of both the average multiplicity and the mean p_t per particle, due to the multijet production. The precise determination of charged particle multiplicities inside the jet is difficult. It is not possible to separate low-momentum fragments of the jet, which contribute sizeably to the jet multiplicity, from the multiplicity background arising from spectator or additional jets, on an event-by-event basis. To estimate losses of low-momentum fragments from the jet by the jet finding algorithm, we rely upon a fragmentation model and a full Monte Carlo simulation of the detector.

2.6. Monte Carlo studies. Two Monte Carlo programs with different hadronization models have been compared with the data. The first one is a naive parton-parton hard scattering model without QCD radiative effects, in which the systems of hard scattered partons and spectator jet fragment independently according to cylindrical phase space (CPS). The mean multiplicity of fragments is obtained as a function of s (square of c.m.s. parton-parton energy) from a phenomenological fit to multiplicity measurements in hadronic events from e^+e^- collisions[8] :

$$<n(s)> = 2.0 + 0.027 \exp\{2\sqrt{[\ln(s/\Lambda^2)]}\} \qquad (1)$$

with $\Lambda = 0.3$ GeV.

The second program is ISAJET[9], which incorporates QCD radiated gluons in the final state. The fragmentation model used for the spectators is ISAJET tuned to reproduce minimum bias events. For both models, generated particles are tracked through the magnetic field and the showers in the calorimeters are simulated[10]. In Fig. 3 the experimental transverse energy flow for jets with $E_T > 35$ GeV is compared to the Monte Carlo results. ISAJET gives a better description of the jet shape than CPS: Both programs fail to reproduce the plateau region as they do not include a complete multijet production, in particular through initial state bremsstrahlung. The charged multiplicity flow is shown in Fig. 4, again for jets with $E_T > 35$ GeV. The multiplicity given by ISAJET is low and coincides with the minimum bias level. CPS gives a good overall description of the multiplicity flow.

To measure charged jet multiplicities, we sum up average track multiplicities in the window $\Delta\eta = \pm 1$, apply a flat background subtraction obtained from the region $1 \leq |\Delta\eta| < 2$, and correct for the loss of jet-associated particles outside $\Delta\eta = \pm 1$ according to CPS. As in our data, low acceptance regions are excluded and corrected for.

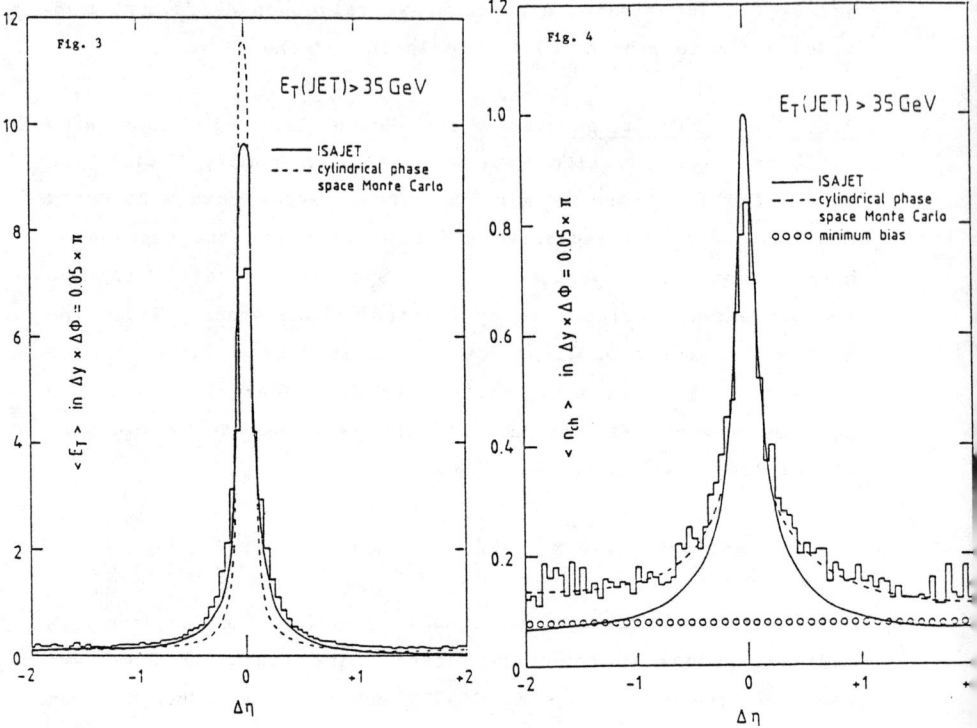

Fig. 3   Transverse energy flow for jets with $E_T > 35$ GeV as function of $\Delta\eta$. Dashed curve: Cylindrical Phase Space model. Full curve: ISAJET model.

Fig. 4   Charged Multiplicity flow for jets with $E_T > 35$ GeV as function of $\Delta\eta$. Dashed curve: Cylindrical Phase Space model. Full curve: ISAJET model.

A further correction is applied for unreconstructed low-momentum particles. The final corrected results on charged track multiplicities in jets are shown in Table 2.

It should be noted that ISAJET when tuned to fit the plateau multiplicity results in larger corrections to the multiplicity inside the jet window, as more fragments are generated far from the jet axis. The 10% systematic error in Table 2 is an attempt to account for this model dependence of the correction.

Using the multiplicity extrapolation (1) from e^+e^- for our energy range would also result in higher multiplicities, of the order of 9.0 charged particles for a jet of 25 GeV. A more detailed comparison with e^+e^- jets can be found in the discussion of jet fragmentation.

2.7. Jet cross-sections. To measure the inclusive jet yield, all events are selected as above, imposing additional constraints on the vertex position (± 40 cm from the detector's centre, for a measured rms spread of ± 12 cm) and on the jet axes (± 1 in pseudorapidity, ± 60° in ϕ from the horizontal plane). All jets with E_T > 35 GeV are considered. Events containing jets with transverse energy above 55 GeV were, in addition, inspected visually on a MEGATEK 3-dimensional display, and a small number of badly reconstructed or background events were rejected, typically events in which cosmic showers or beam halo overlap with collisions.

The inclusive E_T distribution thus obtained is further corrected for geometrical acceptance, experimental procedure, and detector resolution using a Monte Carlo method. Jets are simulated using the models discussed above. The comparison between the reconstructed E_T spectrum and the Monte Carlo input results in a correction factor for each bin of E_T. The global correction is not strongly dependent on E_T and does not exceed a factor 1.25. The inclusive cross-section $d\sigma/dE_T d\eta$ obtained after correction is shown in Fig. 5, together with the data from 1981 published in Ref. 4. In the small region of overlap the measurements agree. All errors

Table 2 Jet population used for the present study
after the cuts mentioned in the text
for different E_T(jet) ranges

N(jets)	E_T(jet) range (GeV)
853	30–35
323	35–40
195	40–50
64	> 50

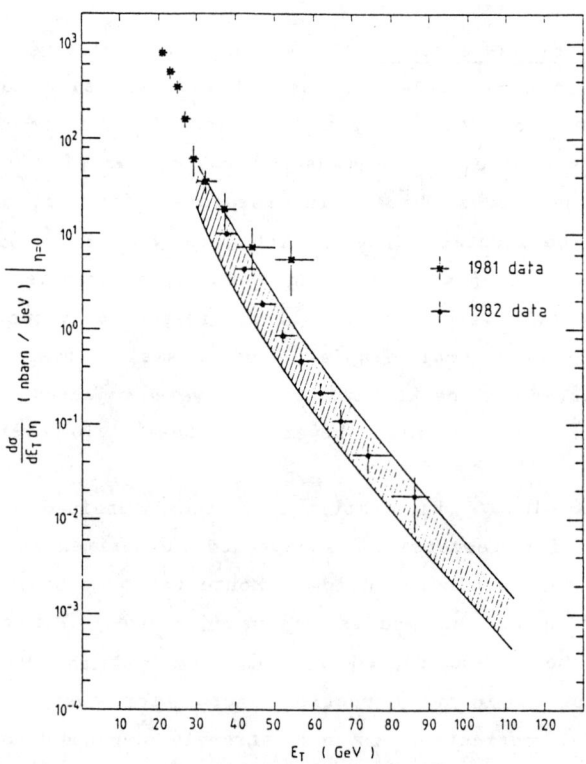

Fig. 5 Inclusive Jet cross section $d\sigma/dE_T d\eta$ ($\eta=0$) as function of E_T. 1981 data (x) and 1982 data (+). The hatched band corresponds to possible QCD predictions [15].

shown are purely statistical. The following systematic errors are not included :

- The uncertainty of the absolute luminosity (± 15%) ;
- Uncertainties in geometrical acceptance due to the choice of the fiducial region, of the jet finding algorithm, and of the fragmentation model used in the Monte Carlo programs (±5%) ;
- An uncertainty in the jet energy which for a given event varies when changing the angular aperture in the jet finding algorithm. This translates into a ±20% uncertainty in cross-section ;
- The dependence of the jet acceptance and jet finding efficiency on the production and fragmentation models used in the Monte Carlo programs (±20%) ;
- The uncertainty arising from the definition of the energy scale : the absolute energy calibration is known to ±5%[5]) ;
- The additional uncertainty in energy scale due to the poorly known particle composition of jets : electromagnetic and hadronic calorimeters exhibit a different response to electromagnetic particles and hadrons, and this response difference is strongly dependent on the particle momenta for $p < 5$ GeV[6]). The uncertainty in the average response correction used is estimated at ±5%. We combine these energy effects into an overall error of ±7.5% of the energy scale, which translates into an uncertainty of a factor 1.5 in the cross-section.

Our overall systematic uncertainty in the cross-section can therefore be given as a factor of 1.65, essentially independent of the jet transverse energy. The measured cross-section is compared in the figure to a band of QCD predictions with a width corresponding to uncertainties in the theory[11]. Note that QCD predicts transverse momentum distributions for partons, while our measurements refer to the sum of the jet fragment momenta. Despite large experimental and theoretical uncertainties, the agreement observed is excellent. The large increase in jet cross-section from ISR to collider energies, by three orders of magnitude, is thus confirmed. In order to obtain the cross-sections for producing several jets in an event, we now consider, in addition to the trigger jet, all jets with $E_T > 15$ GeV reconstructed with a jet axis inside $|\eta| < 2.5$, again attempting to avoid edge effects and jets that might be faked by spectator background.

In Fig. 6 we show the fraction of events with 1, 2, and 3 jets (trigger jets included) as functions of the E_T of the trigger jet. The 2-jet topology dominates over the full range in E_T at a level of $\sim 80-85\%$. The fraction of 1-jet events becomes negligible at high trigger jet E_T, whereas the fraction of 3-jet events rises in the region of low E_T and levels off at $\sim 15\%$. We should stress that our jet finding algorithm, with the window $\Delta\eta = \pm 1$, and the additional requirement $E_T > 15$ GeV, can be expected to have a direct influence on the number of jets found, and that the topological cross-sections as presented here have to be understood in relation to a given jet finding procedure. We also have not corrected these cross-sections in any way for geometrical acceptance.

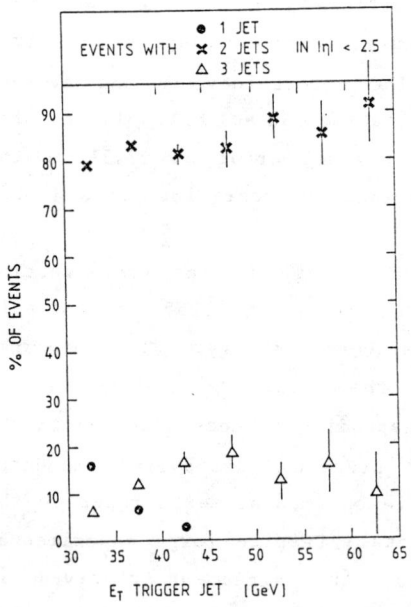

Fig. 6

Fraction of events with 1, 2 and 3 jets of $E_T > 15$ GeV, found by the jet algorithm in a pseudorapidity window $|\eta| < 2.5$, as function of the trigger jet E_T.

The occasional presence of a third jet strongly suggests a gluon bremsstrahlung mechanism similar to what has been observed in hadronic e^+e^- events. QCD predicts multijet events due to quark-gluon and gluon-gluon couplings with rates that are proportional to the products of coupling constants appearing in the bremsstrahlung processes of the original parton. 3-jet events would then, for instance, be produced with a cross-section roughly α_s times the cross-section for 2-jet production. The rate of multijet events can be estimated by measuring the differential cross-section in terms of some suitable parameter describing the non-coplanarity. One such parameter is p^{out}, the momentum perpendicular to the plane defined by the trigger jet and the beam momentum. For large enough p^{out} the 3-jet production rate can be calculated perturbatively from QCD[14],[15] and be compared with the data.

For the study of non-coplanarity we used events with a trigger jet as defined in 2.3. above. In order to minimize any effects coming from problems in jet finding we first calculate p^{out} directly from all E_T-vectors not belonging to the trigger jet. To avoid contamination from the spectator jets we require these E_T-vectors to have $|\eta| < 2.5$. The p^{out} is reconstructed by adding the E_T-vector components perpendicular to the plane defined by the energy axis of the trigger jet and the beam direction, separately on both sides of the plane. If the energies of the jets are balanced, the two $|p^{out}|$ values should be the same. The difference shows a width which is consistent with the experimental resolution. We therefore take as p^{out} the average value of $|p^{out}|_{left}$ and $|p^{out}|_{right}$. The resulting p^{out} distribution is shown in Fig. 7a. This distribution has a contribution from the p_t of the background, i.e. from particles not belonging to the jets, which is difficult to estimate accurately. From the transverse energy flow in minimum bias events one can roughly estimate the contribution to be about 5 GeV. To understand further the size of the background we have repeated the analysis using the jets as found by the jet algorithm. The results are shown in Fig.

7b. One sees that the p^{out} distributions obtained have the same slope but are shifted by about 5 GeV down in p^{out} when compared with the distribution obtained from all E_T-vectors. This shift is consistent with the estimate from minimum bias events. In the same figure is also shown the p^{out} distributions obtained from events with two back-to-back jets generated with our CPS Monte Carlo. In this model the partons are strictly coplanar; the non-zero p^{out} comes from the hadronization model and the resolution of the apparatus. It is clear from the figure that this 2-jet model cannot reproduce the observed large p^{out} tail.

Instead the distribution obtained from a complete QCD calculation for 3-jet production[2] as shown in Fig. 7b agrees much better with the data. The band reflects again the theoretical uncertainties, which are of the same origin as for the inclusive jet cross-section. Note that the QCD calculation is for partons, i.e. there is no broadening due to

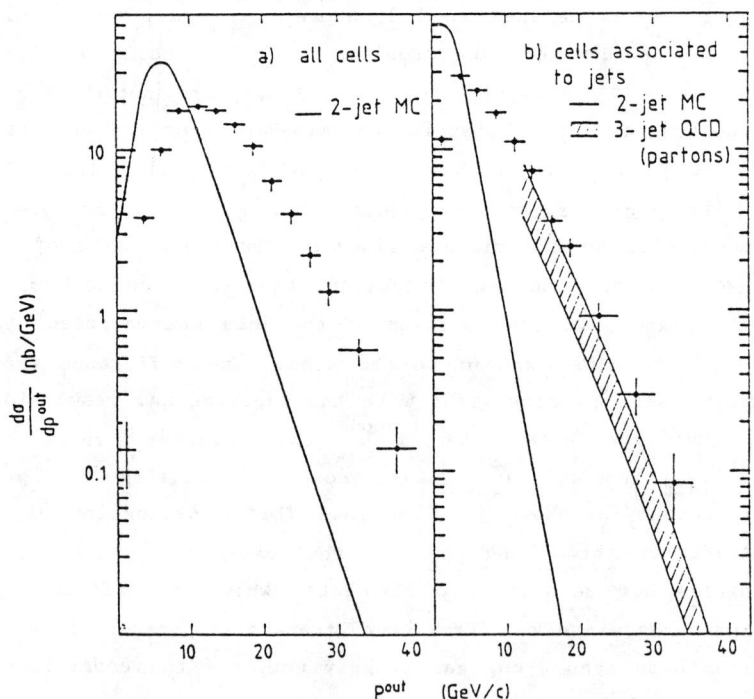

Fig. 7 p^{out} distribution from calorimeter cells (a) and from jets (b), compared to a 2-jet model and to a perturbative QCD 3-jet calculation [16].

fragmentation and no background from spectator partons either. We conclude from our study that we observe multijet events with a rate that is roughly consistent with the expectation from QCD.

2.8. Studies on jet fragmentation in charged particles. So far the jets are found by using exclusively the information coming from the electromagnetic and hadronic calorimeters, via the clustering method explained above and in Refs. 2,6,7. The axis and the total energy of the jet are known from the vectorial sum of all its cell energy vectors. Charged particles are then associated with a particular jet, provided they are confined inside a cone centered around the jet axis. The jet axis and directions of charged particles show obvious correlation up to an angle of $35°$, as shown in Fig. 8. The half aperture of the cone is then fixed to that value. The dashed curve is an estimate of the background of the beam fragments (spectators) not associated with the jet, by a simple Monte Carlo where the particles are generated uncorrelated according to a cylindrical phase-space model.

Owing to the fixed opening angle of the cone, only the central region of the calorimeters is used. This region extends from -1.5 to +1.5 in pseudorapidity η. To minimize edge effects, we restrict the jet-axis direction to be contained in $-1.25 \leq \eta \leq 1.25$. In order to consider only charged particles pointing to the central calorimeter cells, and consequently depositing their energy in those cells, the pseudorapidity range of charged particles is within the limits $-1.5 \leq \eta \leq +1.5$. If the jet axis is close to its pseudorapidity limit of 1.25, charged particles can be emitted in the region $|\eta| > 1.5$. The loss of particles due to this situation is small, of the order of 7%.

Because of the horizontal orientation of the magnetic field and the presence of gaps between the two halves of the calorimeters in the vertical plane, the azimuthal angle about the beam line of the jet axis is restricted to lie within four sectors of a $15°$ half opening angle centered on $±45°$ axes. This restriction in solid angle does not affect any

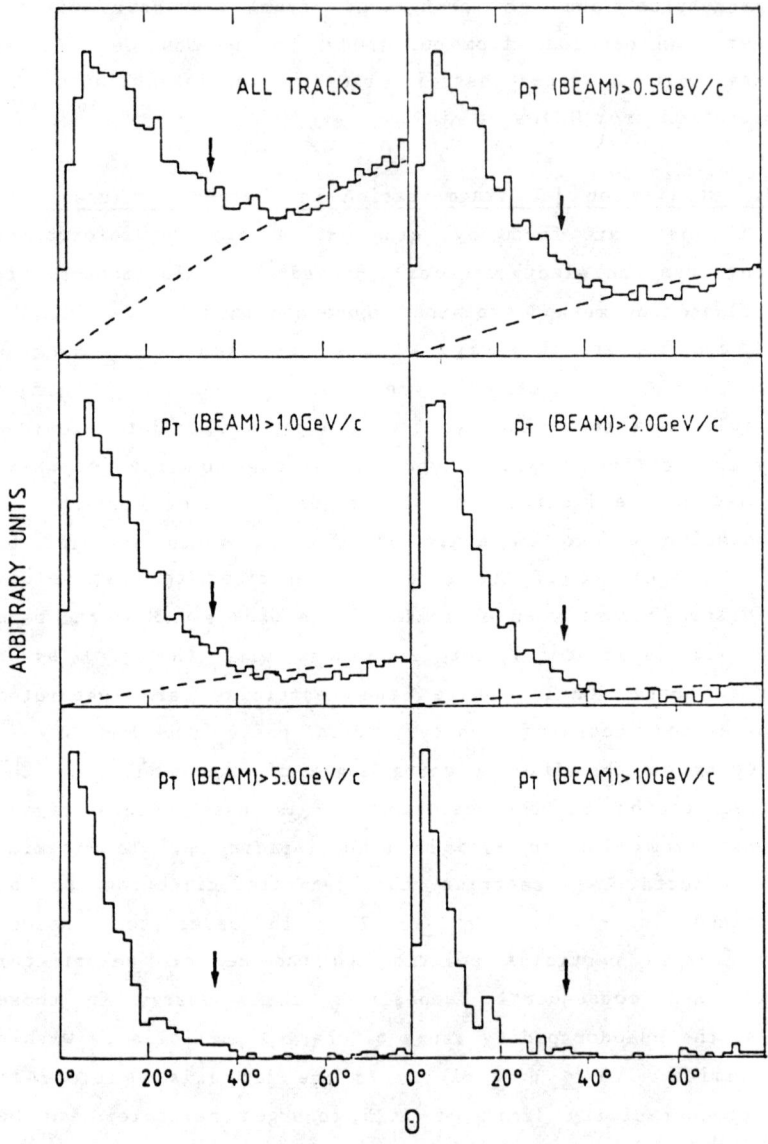

Fig. 8 Distribution of the spatial angle between the jet axis as given by the calorimeters and the directions of charged particles. Arrows indicate the 35⁰ cut. The dashed line is the result of a Monte Carlo assuming no correlation between jet axes and charged particles.

result of the present fragmentation study, as everything is normalized to the final jet population obtained after the above-mentioned cuts.

A loss of 13% is estimated for charged particles emitted outside the sectors. Including the rapidity acceptance, the overall geometrical acceptance is, then, of the order of 80%. A systematic uncertainty of 10% has been added to the statistical errors. No correction has been introduced for track-finding efficiency within jets. However, visual scanning of a reduced population of events did not show any evidence of unfitted straight tracks. The percentage of tracks giving an ionization larger than 1.7 times that of minimum ionizing particles has been measured in the central drift chamber volumes. The small percentage observed, 4.5%, is compatible with the tail of the dE/dx distribution. We deduce that no undetected multi-tracks are contained in the data sample. A small fraction of the tracks within the 35^0 cone (< 5%) has been discarded because of non-association with the main vertex of the interaction. After the acceptance cuts listed above, the jet population used for the present analysis is given in table 2.

2.9. <u>Jet fragmentation function.</u> The inclusive variable used for the jet fragmentation study has been chosen to be

$$z = p_L^{\pm}(\text{jet axis})/E(\text{jet}), \quad 0 \leq z \leq 1,$$

where $p_L^{\pm}(\text{jet axis})$ is the momentum of the charged particle projected on the jet axis. The energy of the jet, E(jet), is defined as the modulus of the vectorial sum of all energy cell vectors belonging to the jet.

The distribution of the variable z – the fragmentation function – is defined as :

$$D(z) = (1/N_{jets})(dN_{ch}/dz).$$

The integral of D(z) is then the mean charged multiplicity contained in the jet :

$$N_{ch}(\text{jet}) = \int_0^1 D(z)dz$$

Several corrections have been applied to $D(z)$. These are discussed below.

The form of the background below $D(z)$ is obtained in the control region $52.5° \leq \theta \leq 70°$, where θ is the spatial angle between the jet axis and the charged particles. Background normalization is given by the Monte Carlo prediction of the number of particles uncorrelated with the jet for $\theta < 35°$ (dashed curve in Fig. 8).

For relatively high z's, all the charged particles are contained in the $35°$ cone around the jet axis. At low values of z this is no longer true. Many soft particles are emitted at large angles. In particular, below 0.02 the emission of particles looks so isotropic that the correlation of the particles with the jet is no longer obvious. Loss of jet particles emitted outside the cone varies from 35% for z between 0.02 and 0.03, to 5% for z around 0.07.

The jets we are dealing with in this paper are very energetic ; they have a total energy above 30 GeV. Associated charged particles with a high z value (therefore a high momentum) have then a large momentum uncertainty. The smearing in z which creates an important deformation of $D(z)$ for values above 0.7 has been removed from the data, assuming an exponential law for $D(z)$ at large z.

Monte Carlo studies with full calorimeter reconstruction indicate that the uncertainty in the measurement of the jet energy is more or less constant between 30 and 50 GeV, and is of the order of 15%. Five percent of the resolution comes strictly from cell association of the jet cluster algorithm, the rest of it being due to the finite resolution and granularity of the calorimeters. No correction has been applied to the data in order to take into account the jet energy smearing.

Figure 9 shows the plot for $D(z)$ with $z > 0.02$. This distribution falls rapidly with z at low z values. At higher z values its form is approximately exponential.

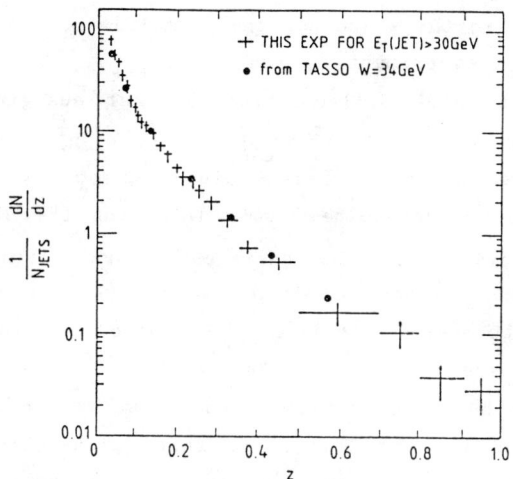

Fig. 9 Charged-Particle fragmentation function for $E_T(\text{jet}) > 30$ GeV, compared with similar results from the TASSO detector at PETRA at $W = 34$ GeV.

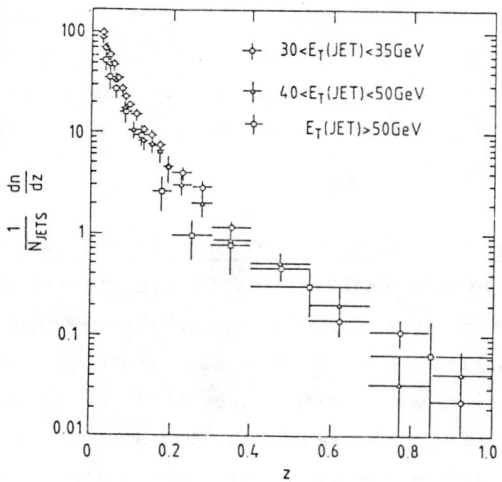

Fig. 10 Charged-particle fragmentation functions for $E_T(\text{jet}) = 30\text{-}35$ GeV, 40-50 GeV, and > 50 GeV.

We can compare the shape and the normalization of D(z) for the present experiment with $(1/\sigma_{tot}) \times (d\sigma/dx_L)$ obtained by the TASSO Collaboration for jet energies of 17 GeV, where $x_L = p_L^{\pm}/p_{beam}$, and $p_L^{\pm}$ is the momentum of the charged particles projected on the jet axis whose direction is determined from minimizing the sphericity of the e^+e^- events[8]. The energies of the jets are of course different for both cases. However, the comparison is meaningful because scaling violations in e^+e^- annihilations are known to be small[13]. No striking differences can be observed between these two sets of data, as can be seen in Fig. 9. This means that quark-dominated and gluon-dominated fragmentation functions are, on the whole, not different from each other, at least for values of z > 0.02.

Within our own data we can look for possible variations of D(z) as a function of the transverse energy of the jet. After background subtractions and corrections, D(z) is plotted in Fig. 10 for three E_T bands : 30-35 GeV, 40-45 GeV, and > 50 GeV ; D(z) is approximately independent of the jet energy. A possible tendency for D(z) to shrink at low z with increasing E_T(jet) cannot be excluded. This is not observed in the high-z region, probably on account of the very large uncertainties in the data introduced by the track momentum smearing, which are difficult to remove entirely owing to the lack of statistics[14].

2.10. Transverse momentum with respect to the jet axis.

The jet axis given by the calorimeters is not precise enough for studying the transverse momentum p_t of the charged particles with respect to the jet axis. For this reason it is replaced by a charged jet axis whose direction is given by the vectorial sum of all charged-particle momenta. The charged particles used to define this axis are inside the cone of 35° half aperture around the calorimeter jet axis. Of course, the charged jet axis is correct only if we assume that the charged and neutral axes are aligned evenly. If this assumption is not valid on an event-by-event basis, it is probably true statistically.

As we have seen before, the association of particles with the jet is questionable at lower z values. For this reason a cut z > 0.1 is applied to select particles unambiguously associated with the jet. Owing to the "seagull effect" discussed below (Fig. 11), this cut will result in a higher mean p_t within the jet, compared with a mean p_t value obtained for all particles belonging to the jet regardless of their z value. For all jets with E_T > 30 GeV, the variation in the average transverse momentum of charged particles measured with respect to the jet axis is plotted in Fig. 11 as a function of z. A "seagull effect" is observed, showing the increase of $\langle p_t \rangle$ from a value of 0.5 GeV/c at a z value around 0.1 to a value approaching 1 GeV/c for z values above 0.5.

The invariant p_t spectrum, $(1/p_t)(dN/dp_t)$, is shown in Fig. 12 together with the results of a fit :

$$(1/p_t)(dN/dp_t) = A/(p_t + p_{t_0})^n$$

for all jets with E_T > 30 GeV. The above function was shown to reproduce well the p_t spectrum of charged particles in minimum bias events[15]. The p_t spectrum is well fitted by the values p_{t_0} = 4 GeV/c, n = 14.8. The observed mean p_t value internal to the jet is $\langle p_t \rangle$ = 600 MeV/c, after having applied the cut z > 0.1 on all particles. A large p_t tail is observed up to p_t = 4 GeV/c. This tail could well be an indication of gluon bremsstrahlung. On the other hand, it could also be due to an experimental misalignment of the jet axis, or to events whose leading particles are neutrals.

Evolution of the mean p_t within the jet has been studied for the following regions of E_T(jet) : 30-35, 40-50, and > 50 GeV. Fig. 13 shows the p_t spectrum obtained for each of these transverse energy bands. The mean p_t increases from 600 MeV/c at E_T = 30 GeV to 700 MeV/c for E_T > 50 GeV.

2.11. *Structure functions.* So far nucleon structure functions were the exclusive domain of lepton-hadron

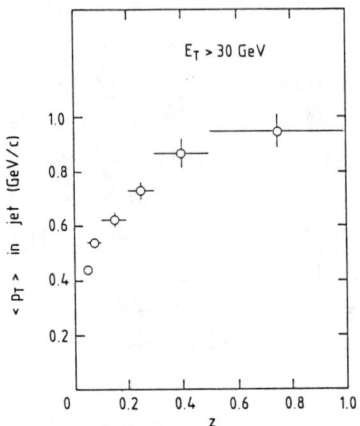

Fig. 11 Variation of $\langle p_t \rangle$ with respect to the jet axis for charged particles as a function of z. Errors are due to statistics only.

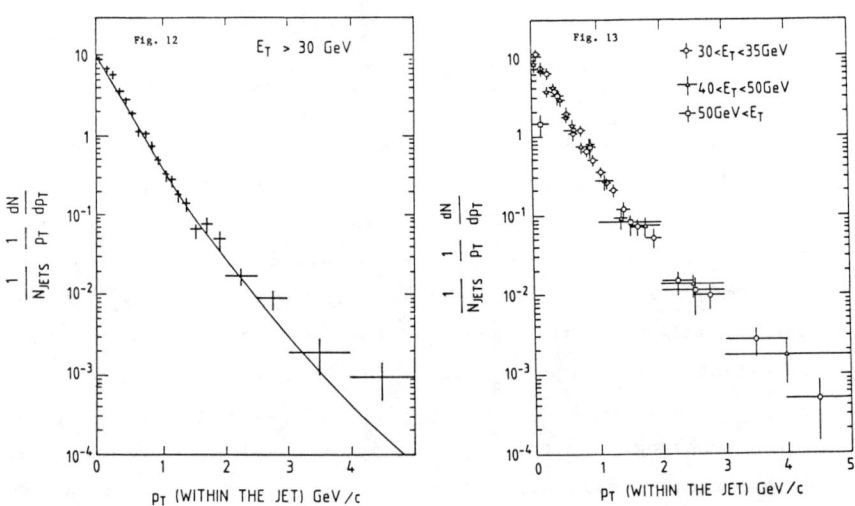

Fig. 12 $(1/p_T)(dN/dp_T)$ spectrum (p_T with respect to the jet axis) for charged particles with $z > 0.1$. The solid line is the result of a fit $1/(p_T + p_T^0)^n$ with $p_T^0 = 4.0$ GeV/c, $n = 14.8$.

Fig. 13 $(1/p_T)(dN/dp_T)$ spectra (p_T with respect to the jet axis) for charged particles with $z > 0.1$ and for three ranges of E_T(jet) = 30-35 GeV, 40-50 GeV, and > 50 GeV.

scattering experiments. The observation of well defined two-jet events in proton-antiproton collisions at high energy opens up the possibility of proton structure function measurements at values of four-momentum transfer squared (Q^2) in excess of 2000 GeV^2, far higher than previously accessible using lepton beams. In the parton model two-jet events result when an incoming parton from the antiproton and incoming parton from the proton interact with each other to produce two outgoing high transverse momentum partons which are observed as jets.

If $d\sigma/d\cos\theta$ is the differential cross-section for a particular parton-parton subprocess as a function of the c.m.s. scattering angle θ, the corresponding contribution to the two-jet cross-section may be written :

$$d^3\sigma/dx_1 dx_2 d\cos\theta = [F(x_1)/x_1][F(x_2)/x_2] d\sigma/d\cos\theta \qquad (1)$$

where $F(x_1)/x_1 [F(x_2)/x_2]$ is a structure function representing the number density of the appropriate partons in the antiproton [proton] as a function of the scaled longitudinal momentum $x_1 [x_2]$ of the partons.

The differential cross-sections for the possible subprocesses have been calculated to leading order in QCD[16]. The elastic scattering subprocesses [gluon-gluon, gluon-quark(antiquark), and quark(antiquark)-quark(antiquark)] have a similar angular dependence and become large as $\cos\theta \to 1$ [like $(1-\cos\theta)^{-2}$] as a consequence of vector gluon exchange. In the approximation that the elastic subprocesses dominate and have a common angular dependence, the total two-jet cross-section may be written[17] in the form of eq. (1). In particular, if $d\sigma/d\cos\theta$ is taken to be the differential cross-section for gluon-gluon elastic scattering :

$$d\sigma/d\cos\theta = (9/8)[\pi\alpha_s^2/2x_1 x_2 s](3+\cos^2\theta)^3 (1-\cos^2\theta)^{-2} \qquad (2)$$

where α_s is the QCD coupling constant and s is the total c.m.s. energy squared, then the structure function $F(x)$ becomes

$$F(x) = G(x) + (4/9)[Q(x) + \overline{Q}(x)] \tag{3}$$

where $G(x)$, $Q(x)$, and $\overline{Q}(x)$ are respectively the gluon, quark, and antiquark structure functions of the proton. In eq. (3) the factor 4/9 reflects the relative strengths of the quark-gluon and gluon-gluon couplings predicted by QCD.

The experimental angular distribution of jet pairs produced in $\bar{p}p$ collisions is analysed as a test of vector gluon exchange and results are presented on the structure function $F(x)$ defined by eqs. (1)-(3).

A set of Monte Carlo events[10] generated using ISAJET[9] has been analysed in parallel with the real data. Isajet generates two jet events and simulates the fragmentation of each jet into hadrons including the effects of QCD bremsstrahlung. The Monte Carlo program simulates in detail the subsequent behaviour of the hadrons in the UA1 apparatus. The Monte Carlo events are used to calculate various corrections which are discussed below, and to estimate the jet energy resolution and the uncertainty in the determination of the jet direction.

After full calorimeter reconstruction, jets are defined using the UA1 jet algorithm[8]. The energy and momentum of each jet is computed by taking respectively the scalar and vector sum over the associated calorimeter cells. A correction is applied to the measured energy ($\sim$ +10%) and momentum ($\sim$ +6%) of each jet, as a function of the pseudorapidity and azimuth for the jet, on the basis of the Monte Carlo analysis, to account for the effect of uninstrumented material and containment losses.

After jet finding, events are selected with $\geq$ 2 jets, within the acceptance of the central calorimetry $|\eta| < 3$. While the majority of these events have a topology consistent with two balanced high E_T jets, some 10-15% of the events have additional jets with $E_T > 15$ GeV[18]. A preceding analysis has shown that multijet events are largely accounted for in terms of initial- and final-state bremsstrahlung processes[19]. For this analysis, in order to compare with theoretical expectations for the two-jet cross-section, additional jets, apart from the two highest E_T jets, are ignored. The r.m.s. transverse momentum of the two-jet system (taking account of resolution) is then $\sim$ 10 GeV. A further correction is then applied, on the basis of the Monte Carlo analysis, to the energy ($\sim$ +12%) and momentum ($\sim$ +7%) of the two highest E_T jets in order to account for final state radiation falling outside the jet, as defined by the jet algorithm. After all corrections, averaged over the full acceptance, the jet energy resolution $\delta E/E \sim \pm 26\%$ and the uncertainty in the jet direction (in pseudorapidity) $\delta \eta \sim \pm 0.05$.

For each event the x_1 and x_2 of the interacting partons are computed as follows :

$$x_1 = [x_F + \sqrt{(x_F^2 + 4\tau)}]/2$$

$$x_2 = [-x_F + \sqrt{(x_F^2 + 4\tau)}]/2 \qquad (4)$$

where

$$x_F = (p_{3L} + p_{4L})/(\sqrt{s}/2)$$

$$\tau = (p_3 + p_4)^2/s . \qquad (5)$$

The c.m.s. scattering angle is computed in the rest frame of the final two jets $[(p_3 + p_4)]$ relative to the axis defined by the interacting partons $[(p_1 - p_2)]$ assumed massless and collinear with the beams[20]:

$$\cos\theta = (p_3-p_4)\cdot(p_1-p_2)/(|p_3-p_4||p_1-p_2|). \qquad (6)$$

In eq. (6), p_3 and p_4 are the 4-momenta of the final two jets and p_{3L} and p_{4L} are the longitudinal momentum components measured along the beam direction in the laboratory frame.

The finite angular acceptance of the apparatus and the trigger E_T threshold requirement discriminates against events with small scattering angles (i.e. large $\cos\theta$), and restricts the range of $\cos\theta$ over which the trigger is fully efficient.

Events which are close to the limits of the acceptance in cosθ are rejected by applying a fiducial cut in cosθ. This fiducial cut ($\cos\theta_{max}$) is defined for each event as the maximum value of cosθ for which both the final two jets would fall in the acceptance region $|\eta| < 2.5$ (with at least one jet having $|\eta| < 1.0$) and for which the mean of the transverse energies of the final two jets would exceed (20)35 GeV [in the (un)filtered data set]. Events with azimuthal angle φ (defined by the final two jets) within $\pm 45^0$ of the vertical, where the two halves of the calorimeter are joined, are also rejected. The final event sample consists of 2432 events satisfying the above cuts, for which the acceptance is essentially uniform.

Figures 14 a-d show the raw histograms in |cosθ| for various x_1, x_2 intervals. In each case a cut has been applied on $\cos\theta_{max}$ as indicated. The curve, which has been normalized to the number of events in each histogram, shows the expected angular dependence [eq. (2)] assuming vector gluon exchange (QCD). The data are consistent with the curve independent of x_1, x_2 in accordance with eq. (2).

Figure 15 shows the angular distribution obtained using all the events in the sample, on the assumption that the angular dependence is independent of x_1 and x_2. The distribution is computed starting from the raw distributions in cosθ, classified in intervals of $\cos\theta_{max}$ (Δ cos θ_{max} = 0.1), normalizing one distribution to another in the region of overlap. The final distribution is divided by the number of events with cos θ<0.1 to obtain $(d\sigma/d\cos\theta)(d\sigma/d\cos\theta)_{\cos\theta=0}$. Monte Carlo studies demonstrate that, for the range of cosθ shown, the angular distribution is not appreciably modified by the effects of resolution smearing. In Fig. 15 the data are compared with theoretical forms for gluon-gluon, gluon-quark(antiquark)-quark(antiquark) elastic scattering in QCD (i.e. assuming vector gluon exchange). Theoretical forms for quark(antiquark)-(quark(antiquark) and gluon-quark(antiquark) scattering in an Abelian scalar gluon theory[21] are also shown. The data are clearly consistent

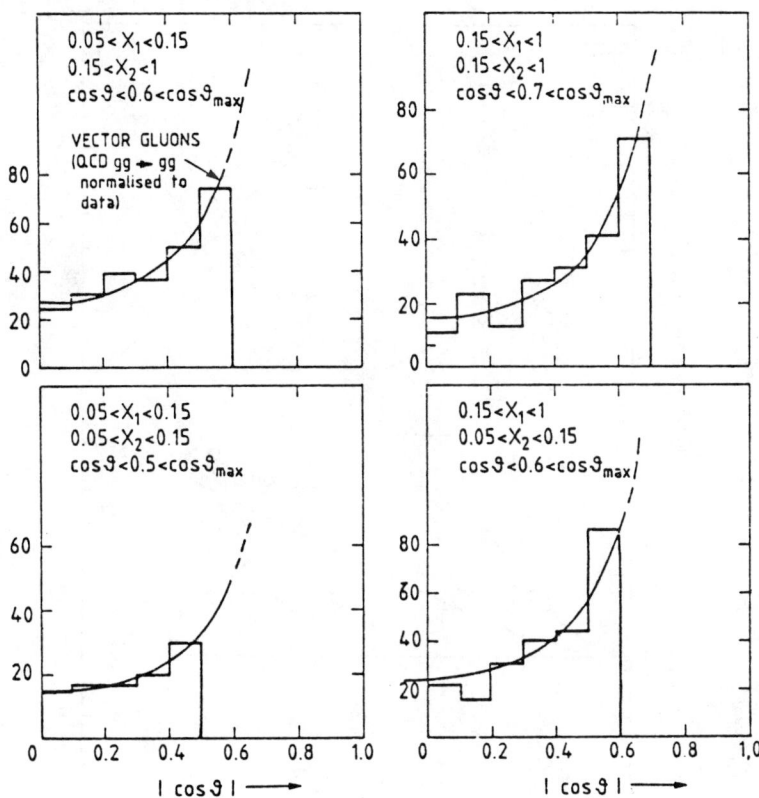

Fig. 14 (a-d): Histograms in cos θ for various x_1, x_2 intervals. A cut on cos θ_{max} has been applied as indicated. The curve, which has been normalized to the number of events in each histogram, is the expected angular dependence [eq. (2)] assuming vector gluon exchange (QCD).

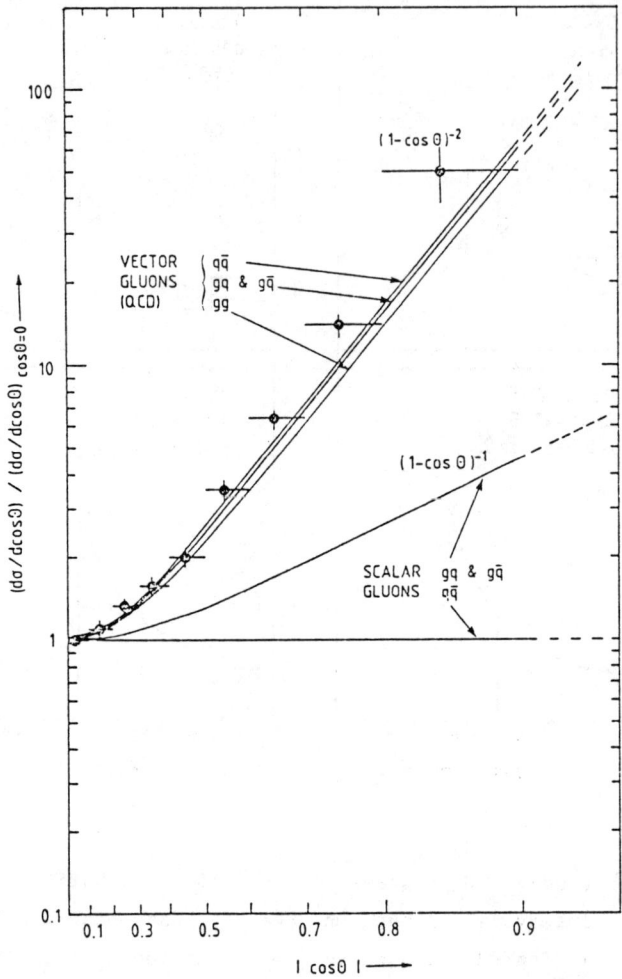

Fig. 15 The angular distribution obtained using all the events in the sample on the assumption that the angular dependence is independent of x_1, x_2. The data are compared with theoretical forms for parton-parton elastic scattering in QCD and in an Abelian scalar gluon theory [9].

with the predictions of the vector gluon theory (QCD), and the Abelian scalar gluon theory is excluded. The data are not yet sufficiently accurate, however, to discriminate between the various subprocesses. A fit to the data of the form $(1-\cos\theta)^{-n}$ for $\cos\theta > 0.4$ yields $n=2.08\pm0.10$. This result is consistent with $n=2$ and, in analogy with the case of Rutherford scattering at low energy, may be regarded as a test of the inverse square law for the interaction between the partons.

In order to extract the structure function $F(x)$ [eq. (3)] the quantity $S(x_1 x_2)$ is defined in terms of the measured differential cross-section $d^2\sigma/dx_1 dx_2$ as follows:

$$S(x_1,x_2) = x_1 x_2 (d^2\sigma/dx_1 dx_2) / \int_0^{\cos\theta_{max}} K(d\sigma/d\cos\theta) d\cos\theta. \qquad (7)$$

If eq. (1) is valid then $S(x_1,x_2) = F(x_1)F(x_2) \cdot S(x_1,x_2)$ is determined from the data by weighting the events individually as a function of x_1, x_2, and $\cos\theta_{max}$. The parton-parton differential cross-section $d\sigma/d\cos\theta$ is taken from eq. (2) with $\alpha_s = 12\pi/[23 \ln(Q^2/\Lambda^2)]$, i.e. assuming five effective quark flavours and $\Lambda = 0.2$ GeV. For this analysis Q^2 is defined by $Q^2 = -t$, where $t = (p_1-p_3)^2$ [or $t = (p_1-p_4)^2$, whichever is numerically smaller] in analogy with deep inelastic scattering. A factor K has been introduced in eq. (7) to allow for the effect of higher-order QCD corrections, which may change the effective Q^2 scale and hence the normalization of the parton-parton cross-section. These corrections have been computed theoretically[22],[23], and are expected to be appreciable even at the energy of the SPS Collider. The results given below have been computed assuming $K = 2$.

The experimental results for $S(x_1,x_2)$ are tabulated in Table 3. The data are symmetric in x_1 and x_2 and have been folded appropriately. The raw event numbers in each bin are given. The data for $\sqrt{\tau} < 0.2$ are based entirely on the unfiltered data set. The data have

Table 3 The experimental result for $S(x_1,x_2)$ (see text). The raw event numbers in each x_1,x_2-bin are also given. The quoted errors are statistical only : the systematic error affecting mainly the overall normalization $\sim \pm 65\%$. The corresponding results for $F(x)$ are also tabulated, together with the mean value of $-\hat{t}$ in each bin.

$S(x_1,x_2)$

x_1 or x_2	0.05-0.10	0.10-0.20	0.20-0.30	0.30-0.40	0.40-0.50	0.50-0.60	0.60-0.80
0.05-0.10	-	-	3.053±0.276 (340)	1.224±0.207 (225)	0.449±0.122 (145)	0.298±0.134 (66)	0.064±0.038 (38)
0.10-0.20	-	2.541±0.247 (263)	0.842±0.118 (392)	0.373±0.061 (202)	0.218±0.047 (115)	0.055±0.027 (30)	0.024±0.015 (16)
0.20-0.30	-	-	0.391±0.080 (75)	0.117±0.044 (28)	0.038±0.029 (13)	0.032±0.038 (11)	0.005±0.011 (1)
0.30-0.40	-	-	-	0.042±0.053 (3)	0.022±0.040 (3)	0.009±0.028 (1)	0.000±0.000 (0)

$F(x)$

	0.05-0.10	0.10-0.20	0.20-0.30	0.30-0.40	0.40-0.50	0.50-0.60	0.60-0.80
$F(x)$	5.497±0.578	1.607±0.082	0.542±0.049	0.222±0.030	0.112±0.027	0.038±0.017	0.012±0.006
$\langle -\hat{t}\rangle$ GeV^2	1500	2200	3000	4100	4100	4600	5100

been corrected for the effects of resolution smearing. This correction varies from − 24% at small x to − 70% at the largest x values. The quoted errors are statistical only: the systematic error affecting mainly the overall normalization is ∼ ±65%, due largely to the uncertainty in the jet energy scale. If factorization holds then $S(x_1,x'_2)/S(x_1,x_2)$ will be independent of x_1 for any choice of x_2 and x'_2. This ratio is plotted in Fig. 16a for $x_2=0.1-0.2$, $x'_2=0.05-0.1$ and in Fig. 16b for $x_2=0.1-0.2$, $x'_2=0.2-0.3$. Since the data are symmetric in x_1 and x_2 the folded data (Table 3) have been used to compute the ratio. In each case the ratio is consistent with a constant (broken line) independent of x_1. The data are therefore consistent with factorization in x_1 and x_2.

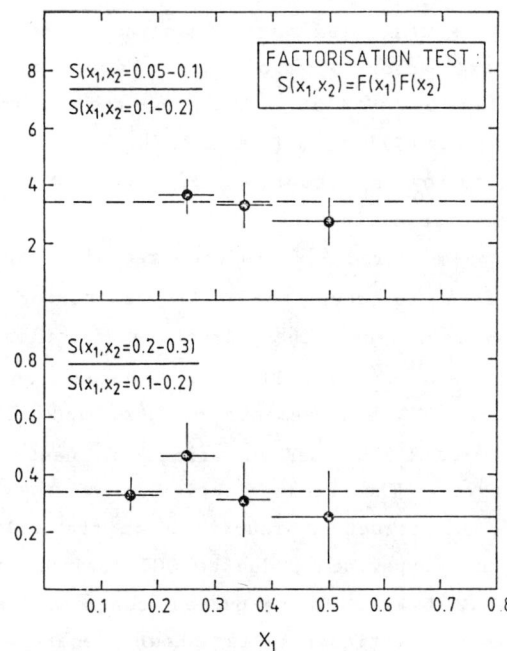

Fig. 16 The ratio $S(x_1,x_2')/S(x_1,x_2)$ for a) $x_2 = 0.1-0.2$, $x_2' = 0.05-0.1$ and b) $x_2 = 0.1-0.2$, $x_2' = 0.2-0.3$. In each case the data are consistent with a constant (broken line) independent of x_1.

Figure 17 shows the structure function $F(x)$ [eq. (3)] obtained from $S(x_1, s_2)$ assuming factorization in x_1 and x_2. The results for $F(x)$ are also tabulated in Table 3, together with the mean value of $-t$ in each x-bin. The measured integral of $F(x)$ over the range $x=0.05-0.80$ is 0.53 ± 0.03. The errors quoted are statistical only. Since the cross-section takes the form of a product of structure functions the errors in the structure function are roughly one half the errors in $S(x_1, x_2)$. In particular, the systematic uncertainty in the overall normalization of the structure function is $\sim \pm30\%$. Over the range $x=0.1-0.8$, the data show an exponential x dependence and may be parametrized by the form $F(x) \sim 6.2\ e^{-9.5x}$ (broken line). Over the same range the data cannot be described by a single term of the form $(1-x)^n$ for any choice of n but may be parametrized as the sum of two such terms e.g. $F(x) \sim 8.0(1-x)^{13} + 1.1(1-x)^4$. Changing the assumed value of Λ by a factor of two (up or down) changes the structure function by $\sim \pm15\%$. Changing the assumed value of the K-factor [eq. (7)] from $K = 2$ to $K = 1$ increases the structure function by a factor of $\sqrt{2}$, i.e. by an amount comparable to the systematic error.

In Fig. 17 the measured $F(x)$ is compared with expectations based on QCD fits to deep inelastic scattering data[24]. The solid curve represents the structure function $G(x) + (4/9)[Q(x)+\bar{Q}(x)]$ at $Q^2 = 20$ GeV^2 based on a QCD parametrization of CDHS measurements of $G(x)$ and $[Q(x)+\bar{Q}(x)]$ at low Q^2 ($Q^2 = 2-200$ GeV^2) using a neutrino beam on an iron target. The broken curve shows the expected modification of the structure function, at the value of Q^2 appropriate in this experiment, due to QCD scaling violations (assuming $\Lambda = 0.2$ GeV). The expected contribution due to the quarks and the antiquarks is shown separately. The two-jet data measure directly, for the first time, the very large flux of gluons in the proton at small x ($x \leq 0.3$) and also demonstrate the existence of QCD scaling violations at large Q^2 relative to the low Q^2 data. The present data suggest somewhat softer parton distributions than would

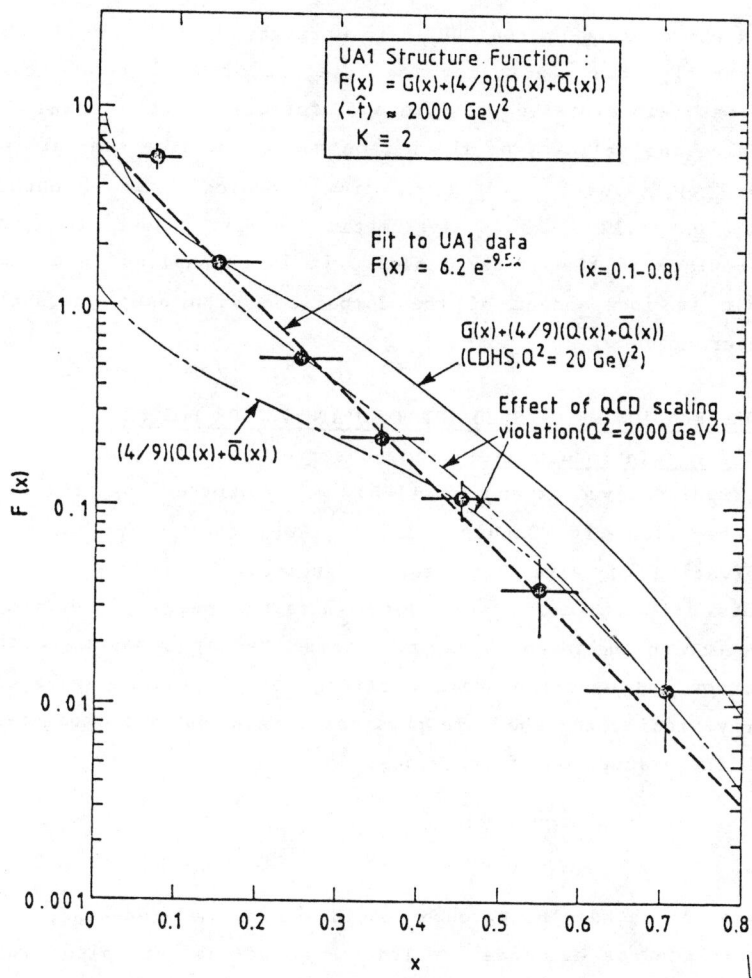

Fig. 17 The structure function of F(x) [eq. (3)]. The errors are statistical only: the systematic error affecting mainly the over-all normalization is ~ ±30%. The broken line represents the parametrization $F(x) = 6.2^{-9.5x}$ (see text). The solid curve represents a QCD parametrization of the structure function $G(x) = (4/9) [Q(x) + \overline{Q}(x)]$ at $Q^2 = 20$ GeV² based on the CDHS [13] measurements at low Q^2 ($Q^2 = 2$–200 GeV²). The broken curves show the expected modification of the structure function at the values of Q^2 appropriate in this experiment. The expected contribution of quarks and antiquarks is shown separately.

be expected based on the CDHS parametrization.

Finally, our result for the integral of F(x) given above may be reinterpreted to yield information on σ_s and K. Assuming the validity of the parton model momentum sum-rule we have $\int_0^1 F(x)dx \leq 1$, from which we obtain $\alpha_s\sqrt{K} \geq 0.12$ (±30% systematic error, due to the uncertainty in the energy scale). It is emphasized that this result is independent of the comparison with deep inelastic scattering data.

3. OBSERVATION OF CHARGED INTERMEDIATE VECTOR BOSONS

3.1. Introduction. The observation of the charged intermediate vector boson (IVB) was reported by the UA1 Collaboration in January 1983[25] on the basis of the observation of five high energy isolated electrons. These events all exibited a very large missing energy, interpreted as neutrino emission. The transverse vector momenta of the electron and neutrino share a strong correlation in angle and energy, indicating the same physical origin and are consistent with the assumption of the process

$$\bar{p}p \to W^\pm + X \qquad (1)$$
$$\hookrightarrow e^\pm \nu_e .$$

Since no other background could be formulated they were interpreted as evidence for IVB. Mass values were also given, $m_W = (80 \pm 5)$ GeV/c^2 (UA1) and $m_W = (80^{+10}_{-6})$ GeV/c^2 (UA2).

Since then the CERN sample has been considerably increased :

UA1[26]	53	$W^\pm \to e^\pm \nu$ events
	14	$W^\pm \to \mu^\pm \nu$ events
UA2[27]	36	$W \to e\nu$ events
Total	103	events

Such a comparatively large sample of events can now be used to proceed further in understanding the phenomenon. In particular the assignment of reaction (1) can now be "proven" rather than "postulated".

(i) The transverse decay kinematics is incompatible with emission of two or more neutrinos and it indicates the two-body decay ($\nu_e e$). A greatly improved value for the mass is also obtained.

(ii) A subset (80% of the sample) of events from UA1 can be completely reconstructed, the two fold ambiguity for the unmeasured longitudinal momentum of neutrino being removed by the overall calorimetric information from the event. Using the well-known relation $x_W = x_p - x_{\bar{p}}$ and $x_p x_{\bar{p}} = m_W^2/s$ we can determine the distribution of the partons participating in the process. One can see that the distributions are in excellent agreement with the assumptions of quarks and antiquarks for the proton and antiproton respectively and exclude "sea" effects and gluons. Therefore the production mechanism is proven to be valence quark-antiquark annihilation.

(iii) A subsample of the fully reconstructed events have the sign of the lepton determined by magnetic curvature. Weak interactions should act as a longitudinal polarizer of the W particles since quarks(antiquarks) are provided by the proton(antiproton) beam. Likewise decay angular distributions from a polarizer are expected to have a large asymmetry, which acts as a polarization analyser. A strong backward-forward asymmetry is therefore expected, in which electrons(positrons) prefer to be emitted in the direction of the proton(antiproton). In order to study this effect independently of W-production mechanism, one has looked at the angular distribution of the emission angle θ^* of the electron(positron) with respect to the proton(antiproton) direction in the W centre of mass. According to expectation of V-A theory the distribution should be of the type $(1+\cos\theta^*)^2$, in excellent agreement with the experimental data. More generally it has been shown by Jacob that for a particle of arbitrary spin J produced with average polarization $<\mu>$ and decaying with an asymmetry parameter $<\lambda>$ one expects

$$<\cos\theta^*> = \frac{<\lambda><\mu>}{J(J+1)} .$$

For V-A theory one expects $<\lambda> = <\mu> = -1$ and $J=1$, leading to the <u>maximal</u> value $<\cos\theta^*> = 0.5$. For $J = 0$ one obviously expects $<\cos\theta^*> = 0$ and for any other spin value $J > 2$, $<\cos\theta> \leq 1/6$. Experimentally, we find $<\cos\theta^*> = 0.5 \pm 0.1$ close to its maximal value, which supports both the J=1 assignment and maximal helicity states at production and decay. Note that the choice of sign $<\mu>=<\lambda>=\pm 1$ cannot be separated, i.e. right- and left-handed currents at production and decay cannot be resolved without a polarization measurement.

(iv) Finally the production cross-section can be determined: $(\sigma.B)_W = 0.53 \pm 0.08 \ (\pm 0.09)$ nb, where the last error takes into account systematic errors. This value is in excellent agreement with the expectations of the V-A theory $(\sigma.B)_W = 0.39$ nb.

All these results can be used now to perform the interesting exercise of <u>deducing</u> classic weak interactions from $W^\pm$ particle observation. Indeed we know that $W^\pm$ must couple to valence quarks at production and to (eν) pairs at decay, which implies the existance of the beta decay processes n → p + e⁻ + ν_e and (p) → (n) + e⁺ + ν_e. The mass value m_W and the cross-section measurement can be then used to calculate G_F, the Fermi coupling constant. The interaction must be Vector since $J = 1$ and parity is maximally violated since $<\mu> = <\lambda> = \pm 1$. The only missing element is the separation between V + A and V - A alternatives. For this purpose a polarization measurement is needed. It may be accomplished in the near future by studying for instance the decay W → τ + ν_τ and using the τ decay as polarization analyser or producing IVB with longitudinally polarized protons. Isn't this a nice way of teaching weak interactions to young students?

3.2. <u>Event selection for W → eν decays</u>. Results are based on an integrated luminosity of 0.136 pb⁻¹, which is corrected for dead-time and other similar losses and which includes the exposure for which results have been initially reported[25]. The trigger selection used throughout the investigation required the presence of an electromagnetic cluster at angles larger than $5°$, with transverse energy in excess of 10 GeV. After on-line filtering and complete off-line reconstruction, about 1.5×10^5 events had at least one electromagnetic (e.m.) cluster with $E_T > 15$ GeV. By requiring the presence of an associated, isolated[25] track with $p_t > 7$ GeV/c in the central detector, we reduce the sample by a factor of about 100. Next, a maximum energy deposition (leakage) of 600 MeV is allowed in the hadron calorimeter cells after the e.m. counters, leading to a sample of 346 events. We then classify events according to whether there is prominent jet activity. We find that in 291 events there is a clearly visible jet within an azimuthal angle cone $|\Delta\phi| < 30°$ opposite to the "electron" track. These events are strongly contaminated by jet-jet events in which one jet fakes the electron signature and must be rejected. We are left with 55 events without any jet or with a jet not back-to-back with the "electron" within $30°$.

The bulk of these events is characterized by the presence of neutrino emission, signalled by a significant missing energy (see Fig. 18). According to the experimental energy

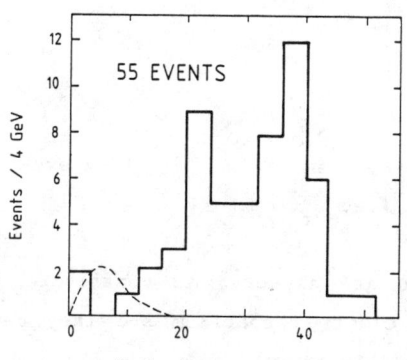

Fig. 18 The distribution of the missing transverse energy for those events in which there is a single electron with $E_T > 15$ GeV, and no co-planar jet activity. The curve represents the resolution function for no missing energy normalized to the three lowest missing-energy events.

resolutions, at most the three lowest missing energy events are compatible with no neutrino emission. They are excluded by the cut $E_T^{miss} > 15$ GeV. We are then left with 52 events. These events have a very clean electron signature (Figs. 19a-c) and a perfect matching between the point of electron incidence and the centroid in the shower detectors, further supporting the absence of composite overlaps of a charged track and neutral π^0's expected from jets.

In order to ensure the best accuracy in the electron energy determination, only events in which the electron track hits the electromagnetic detectors more than $\pm 15°$ away from their top and bottom edges have been retained. The sample is then reduced to 43 events.

We have estimated, in detail, the possible sources of background coming from ordinary hadronic interactions with the help of a sample of isolated hadrons at large transverse momenta and we conclude that they are negligible (< 0.5 events). For more details on background we refer the reader to Ref. 25. We may, however, detect some background events from other decays of the W, namely:

$$W \to \tau \nu_\tau \qquad\qquad (< 0.5 \text{ events})$$
$$\hookrightarrow \pi^\pm (\pi^0) \nu_\tau$$

or

$$W \to \tau \nu_\tau \qquad\qquad (= 2 \text{ events}).$$
$$\hookrightarrow e \nu_e \nu_\tau$$

These events are expected to contribute only at the low P_t part of the electron spectrum and they can be eliminated in a more restrictive sample.

3.3. Origin of the electron-neutrino events.

We proceed to a detailed investigation of the events in order to elucidate their physical origin. The large missing energy observed in all events is interpreted as being due to the emission of one

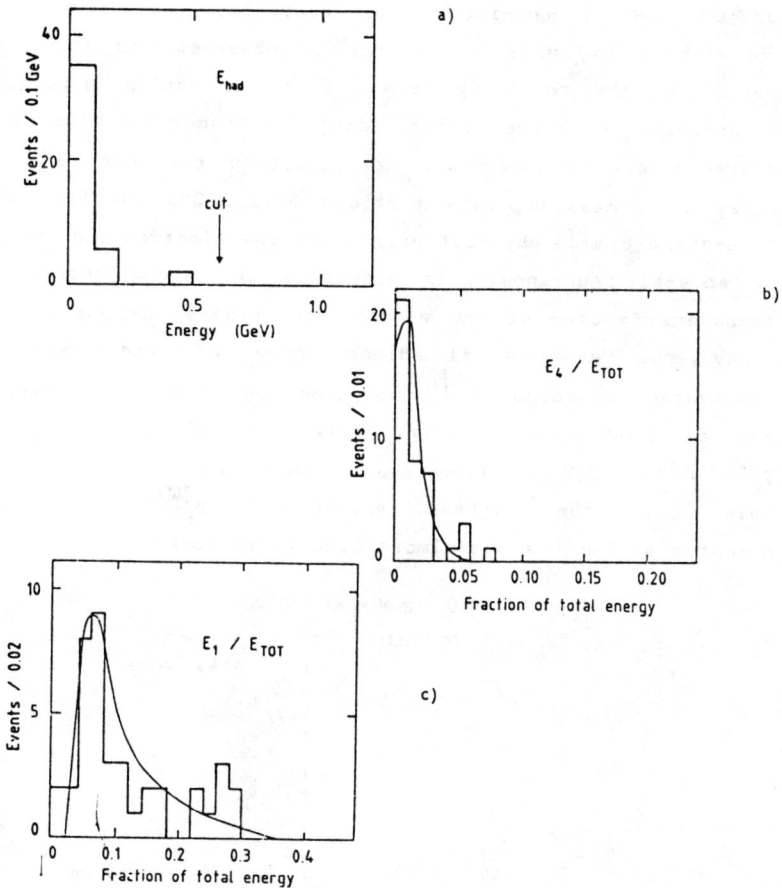

Fig. 19 Distributions showing the quality of the electron signature: a) The energy deposition in the hadron calorimeter cells behind the 27 radiation lengths (r.l.) of the e.m. shower detector. b) The fraction of the electron energy deposited in the fourth sampling (6 r.l. deep, after 18 r.l. convertor) of the e.m. shower detector. The curve is the expected distribution from test-beam data.

or of several non-interacting neutrinos. A very strong correlation in angle and energy is observed (in the plane normal to the colliding beams, where it can be determined accurately), with the corresponding electron quantities, in a characteristic back-to-back configuration expected from the decay of a massive, slow particle (Figs. 20a and b). This suggests a common physical origin for the electron and for one or several neutrinos. In order to understand better the transverse motion of the electron-neutrino(s) system one can study the experimental distribution of the resultant transverse momentum $p_t^{(W)}$ obtained by adding neutrino(s) and electron momenta (Fig. 21). The average value is $p_t^{(W)}$ = 6.3 GeV/c. Five events which have a visible jet have also the highest values of $p_t^{(W)}$. Transverse momentum balance can be almost exactly restored if the vector

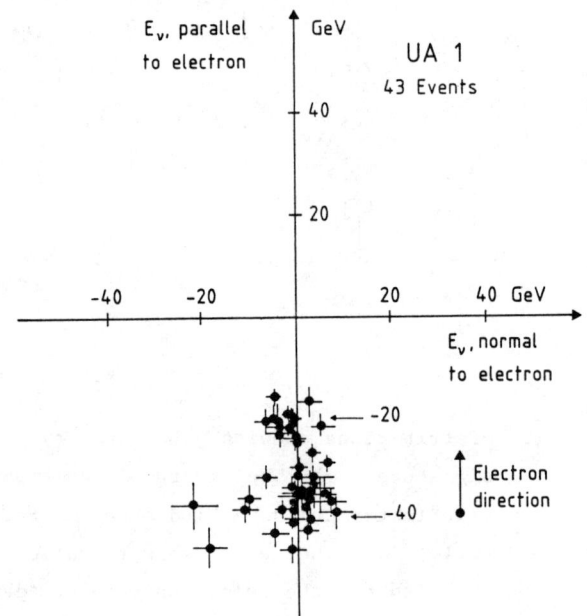

Fig. 20 a) Two-dimensional plot of the transverse components of the missing energy (neutrino momentum). Events have been rotated to bring the electron direction pointing along the vertical axis. The striking back-to-back configuration of the electron-neutrino system is apparent.

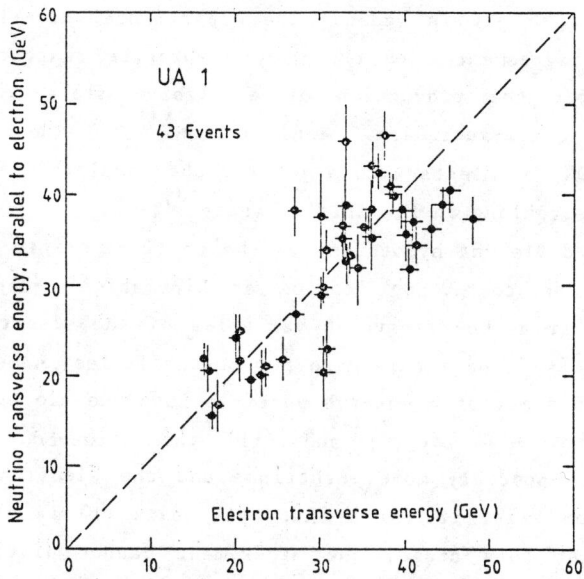

Fig. 20 b) Correlation between the electron and neutrino transverse energies. The neutrino component along the electron direction is plotted against the electron transverse energy.

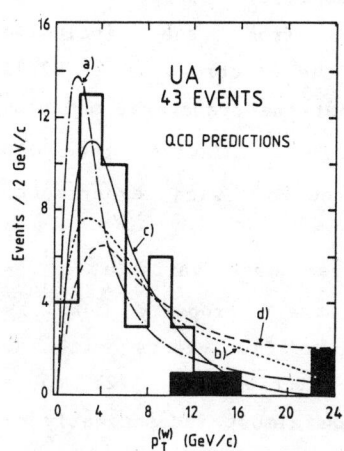

Fig. 21

The transverse momentum distribution of the W derived from our events using the electron and missing transverse-energy vectors. The highest $p_T^{(W)}$ events have a visible jet (shown in black in the figure). The data are compared with the theoretical predictions of Halzen et al. for W production (a) without $[0(\alpha_s)]$ and (b) with QCD smearing; and predictions by (c) Aurenche et al., and (d) Nakamura et al.

momentum of the jet is added. The experimental distribution is in good agreement with the many theoretical expectations from QCD for the production of a massive state via the Drell-Yan quark-antiquark annihilation[28]. The small fraction (10%) of events with a jet are then explained as hard gluon bremsstrahlung in the initial state[29].

Several different hypotheses on the physical origin of the events can be tested by looking at kinematical quantities constructed from the transverse variables of the electron and the neutrino(s). We retain here two possibilities, namely (i) the two-body decay of a massive particle into the electron and one neutrino, $W \to e\nu_e$; and (ii) the three-body decay into two, or possibly more, neutrinos and the electron. One can see from Figs.22a and b that hypothesis (i) is strongly favoured. At this stage, the experiment cannot distinguish between one or several closely spaced massive states.

3.4. Determination of the invariant mass of the $(e\nu_e)$ system.

A (common) value of the mass m_W can be extracted from the data in a number of ways, namely :

i) It can be obtained from the inclusive transverse-momentum distribution of the electrons (Fig. 22a). The drawback of this technique is that the transverse momentum of the W particle must be known. Taking the QCD predictions[28], in reasonable agreement with experiment, we obtain $m_W = (80.5 \pm 0.5)$ GeV/c^2.

ii) We can define a transverse mass variable, $m_T^2 = 2p_t^{(e)} p_t^{(\nu)} (1 - \cos\phi)$, with the property $m_T \leq m_W$, where the equality holds only for events with no longitudinal momentum components. Fitting Fig. 22b to a common value of the mass can be done almost independently of the transverse motion of the W particles, $m_W = (80.3^{+0.4}_{-1.3})$ GeV/c^2. It should be noted that the lower part of the distribution in $m_T^{(W)}$ may be slightly affected by $W \to \tau\nu_\tau$ decays and other backgrounds.

iii) We can define an enhanced transverse mass distribution, selecting only events in which the decay

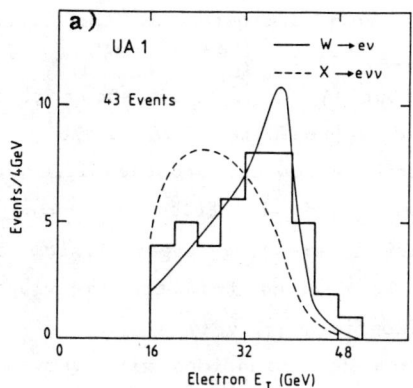

Fig. 22 a) The electron transverse-energy distribution. The two curves show the results of a fit of the enhanced transverse mass distribution to the hypotheses W → eν and X → eνν. The first hypothese is clearly preferred.

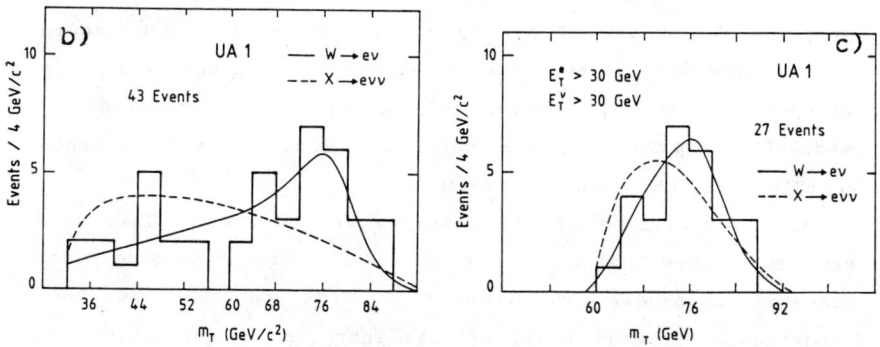

Fig. 22 b) The distribution of the transverse mass derived from the measured electron and neutrino vectors. The two curves show the results of a fit to the hypotheses W → eν and X → eνν.

Fig. 22 c) The enhanced electron-neutrino transverse-mass distribution (see text). The two curves show the results of a fit to the hypotheses W → eν and X → eνν.

kinematics is largely dominated by the transverse variable with the simple cuts $p_t^{(e)}$, $p_t^{(\nu)}$ > 30 GeV/c. The resultant distribution (Fig. 22c) shows then a relatively narrow peak, at approximately 76 GeV/c^2. Model-dependent corrections contribute now only to the difference between this average mass value and the fitted m_W value, m_W = (80.0 ± 1.5) GeV/c^2. An interesting upper limit to the width of the W can also be derived from the distribution, namely Γ_T ≤ 7 GeV/c^2 (90% confidence level).

The three mass determinations give very similar results. We prefer to retain the result of method (iii), since we believe it is the least affected by systematic effects, even if it gives the largest statistical error. Two important contributions must be added to the statistical errors :

i) Counter-to-counter energy calibration differences. They can be estimated indirectly from calibrations of several units in a beam of electrons; or, and more reliably, by comparing the average energy deposited by minimum bias events recorded periodically during the experiment. From these measurements we find that the r.m.s. spread does not exceed 4%. In the determination of the W mass this effect is greatly attenuated, to the point of being small compared to statistical errors, since many different counter elements contribute to the event sample.

ii) Calibration of the absolute energy scale. This has been performed using a strong ^{60}Co source in order to transfer test-beam measurements to the counters in the experiment. Several small effects introduce uncertainties in such a procedure, some of which are still under investigation. At the present stage we quote an overall error of ±3% on the energy scale of the experiment. Of course this uncertainty influences both the $W^\pm$ and Z^0 mass determinations by the same multiplicative correction factor.

3.5. Longitudinal motion of the W particle. Once the decay reaction W → eν_e has been established, the longitudinal momentum of the electron-neutrino system can be determined with a twofold ambiguity for the unmeasured longitudinal component of the neutrino momentum. The overall information of the event can be used to establish momentum and energy conservation bounds in order to resolve this ambiguity in 70% of the cases. Most of the remaining events have solutions which are quite close, and the physical conclusions are nearly the same for both solutions. The fractional beam energy x_W carried by the W particle is shown in Fig. 23a and it appears to be in excellent agreement with the hypothesis of W production in $q\bar{q}$ annihilation[30]. Using the well-known relations $x_W = x_p - x_{\bar{p}}$ and $x_p \cdot x_{\bar{p}} = m_W^2/s$, we can determine the relevant parton distributions in the proton and antiproton. One can see that the distributions are in excellent agreement with the expected x distributions for quarks and antiquarks respectively in the proton and antiproton (Fig. 23b and c). Contributions of the u and d quarks can also be neatly separated, by looking at the charges of produced W events, since $(u\bar{d}) \to W^+$ and $(\bar{u}d) \to W^-$ (Figs. 23d and e).

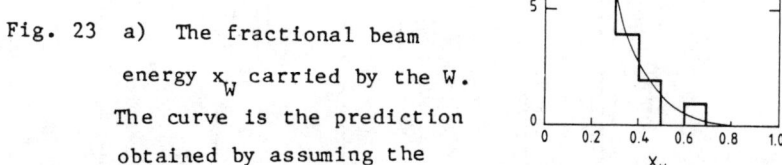

Fig. 23 a) The fractional beam energy x_W carried by the W. The curve is the prediction obtained by assuming the W has been produced by $\bar{q}q$ fusion. Note that in general there are two kinematic solutions for x_W (see text), which are resolved in 70% of the events by consideration of the energy flow in the rest of the event. Where this ambiguity has been resolved the preferred kinematic solution has been the one with the lowest x_W. In the 30% of the events where the ambiguity is not resolved the lowest x_W solution has therefore been chosen.

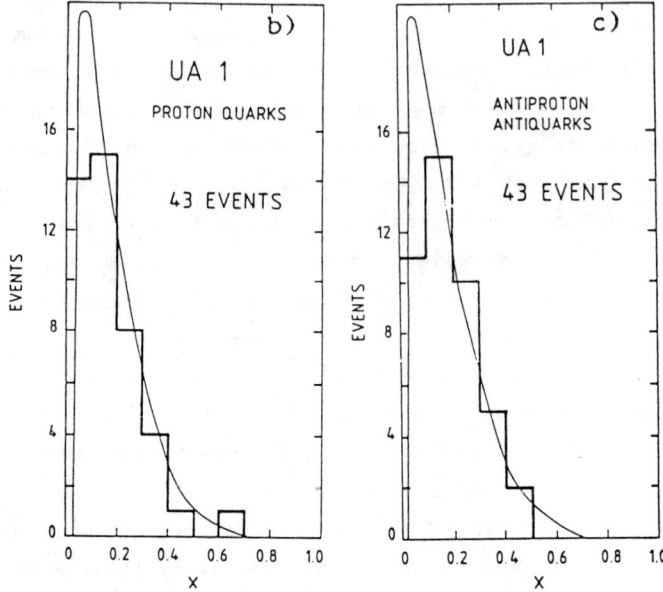

Fig. 23 b) The x-distribution of the proton quarks producing the W by $q\bar{q}$ fusion. The curve is the prediction assuming $q\bar{q}$ fusion.

c) The same as (b) for the antiproton quarks.

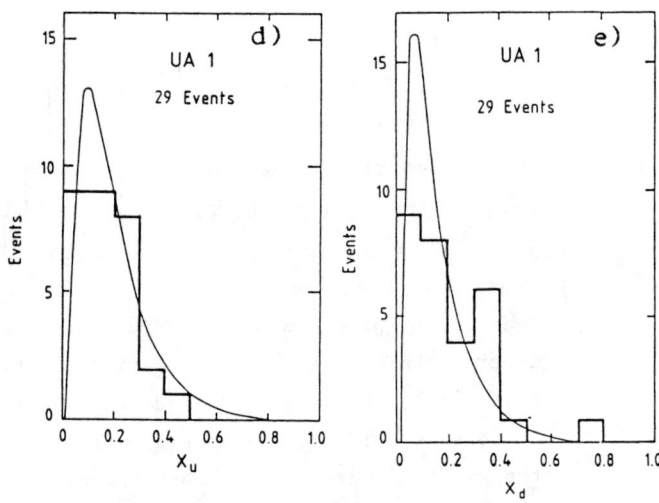

Fig. 23 d) The same as Fig. 23b but for $u(\bar{u})$ quarks in the proton (antiproton).

e) The same as Fig. 23b but for $d(\bar{d})$ quarks in the proton (antiproton).

3.6. Effects related to the sign of the electron charge. The momentum of the electron is measured by its curvature in the magnetic field of the central detector. Out of the 52 events, 24 (14) have a negative (positive) charge assignment; 14 events have a track topology which makes charge determination uncertain. Energy determinations by calorimetry and momentum measurements are compared in Fig. 24a, and they are, in general, in quite reasonable agreement with what is expected from isolated high-energy electrons. A closer examination can be performed, looking at the difference between curvature observed and expected from the calorimeter energy determination, normalized to the expected errors (Fig. 24b). One can observe a significant deviation from symmetry (corresponding to p < E), which can be well understood once the presence of radiative losses of the electron track (internal and external bremsstrahlung), is taken into account[31].

Weak interactions should act as a longitudinal polarizer of the W particles since quarks(antiquarks) are provided by the proton(antiproton) beam. Likewise decay angular distributions from a polarizer are expected to have a large asymmetry, which acts as a polarization analyser. A strong backward-forward asymmetry is therefore expected, in which electrons(positrons) prefer to be emitted in the direction of the proton(antiproton). In order to study this effect independently of W-production mechanisms, we have looked at the angular distribution of the emission angle θ^* of the electron(positron) with respect to the proton(antiproton) direction in the W centre of mass. Only events with no reconstruction ambiguity can be used. It has been verified that this does not bias the distribution in the variable $\cos\theta^*$. According to the expectations of V-A theory the distribution should be of the type $(1 + \cos\theta^*)^2$, in excellent agreement with the experimental data (Fig. 25).

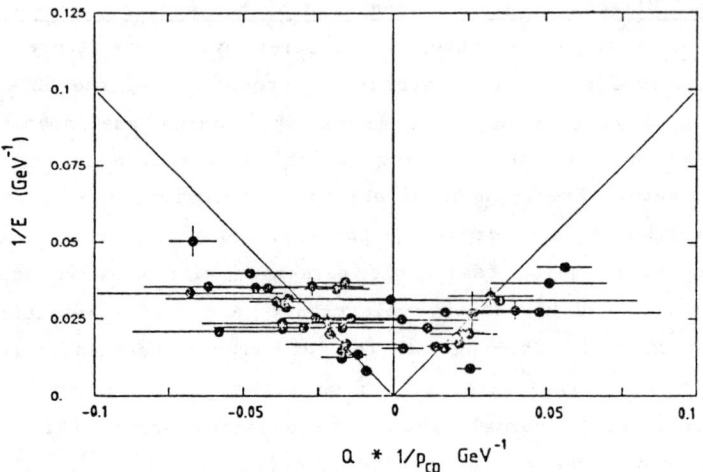

Fig. 24 a) $1/E$ plotted against Q/P_{CD} where E is the electron energy determined by the calorimeter, P_{CD} the momentum determined from the curvature of the central detector track, and Q the charge of the track.

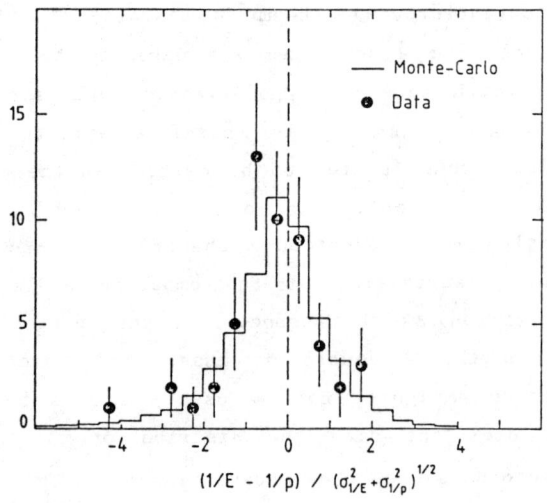

Fig. 24 b) $(1/E - 1/p)$ normalized by the error on the determination of this quantity. The curve is a Monte Carlo calculation, in which radiative losses due to internal and external bremsstrahlung have been folded with the experimental resolution.

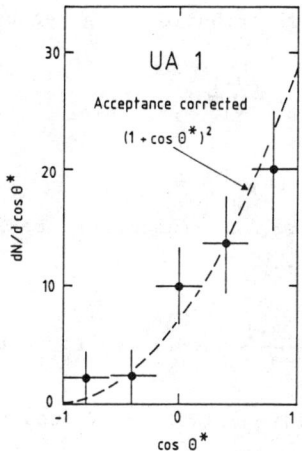

Fig. 25 The angular distribution of the electron emission angle θ^* in the rest frame of the W after correction for experimental acceptance. Ony those events in which the electron charge is determined and the kinematic ambiguity (see text) has been resolved have been used. The latter requirement has been corrected for in the acceptance calculation.

3.7. <u>Determination of the parity violation parameters and of the spin of the W particle.</u> It has been shown by Jacob[32] that for a particle of arbitrary spin J one expects

$$\langle\cos\theta^*\rangle = \frac{\langle\lambda\rangle\langle\mu\rangle}{J(J+1)} \quad ,$$

where $\langle\mu\rangle$ and $\langle\lambda\rangle$ are, respectively, the global helicity of the production system ($u\bar{d}$) and of the decay system ($e\nu$).

The detailed derivation follows closely the paper of reference. Let θ be the angle between the direction of the electron and the spin of the W particle in the rest system of the W-particle. The decay amplitude of W into a ν is proportional to

$$D^{J*}_{\mu\lambda}(\varphi,\theta,-\varphi) \quad .$$

The decay angular distribution is given by

$$I(\theta,\varphi) = \sum_{\mu\mu'} \rho_{\mu\mu'} \frac{2J+1}{4\pi} \sum_{\lambda_1\lambda_2} |c_{\lambda_1\lambda_2}|^2 D^{J*}_{\mu\lambda}(\varphi,\theta,-\varphi) D^{J}_{\mu'\lambda}(\varphi,\theta,-\varphi)$$

with $\lambda = \lambda_1 - \lambda_2$ and $\rho_{\mu\mu'} = \delta_{\mu\mu'}$.

Combining the two D functions making use of the Clebsch-Gordon series :

$$I(\theta,\varphi) = \frac{2J+1}{4\pi} \sum_{\mu} (-1)^{\mu} P_{\mu} \sum_{\lambda_1\lambda_2} (-1)^{\lambda_1\lambda_2} |c_{\lambda_1\lambda_2}|^2$$

$$\sum_{\ell=0}^{2J} c(JJ\ell|\mu-\mu) c(JJ\ell|\lambda-\lambda) P_{\ell}(\cos\theta)$$

with
P_{μ} the matrix density for the W in absence of polarization of the p and $\bar{p}$ beams ; P_{μ} is the probability that the W is in the state of helicity μ.

We now take the average value of $\cos\theta$. This gives a particularly simple expression :

$$<\cos\theta> = \frac{1}{J(J+1)} (\sum_{\mu} \mu p_{\mu})(\sum_{\lambda_1\lambda_2} \lambda |c_{\lambda_1\lambda_2}|^2).$$

The two terms within parenthesis are the average helicities, therefore

$$<\cos\theta> = \frac{<\mu><\lambda>}{J(J+1)}.$$

For V-A one then has $<\lambda> = <\mu> = -1$, $J = 1$, leading to the maximal value $<\cos\theta^*> = 0.5$. For $J = 0$ one obviously expects $<\cos\theta^*> = 0$ and for any other spin value $J \geq 2$, $<\cos\theta^*> \leq 1/6$. Experimentally, we find $<\cos\theta^*> = 0.5 \pm 0.1$, which supports <u>both</u> the $J = 1$ assignment <u>and</u> maximal helicity states at production and decay. Note that the choice of sign $<\mu> = <\lambda> = \pm 1$ cannot be separated, i.e. right- and left-handed currents both at production and decay cannot be resolved without a polarization measurement.

3.8. Total cross-section and limits to higher mass W's. The integrated luminosity of the experiment was 136 nb^{-1} and it is known to about ±15% uncertainty. In order to get a clean W → eν_e sample we select 47 events with $p_t^{(e)}$ > 20 GeV/c. The W → $\tau\nu_\tau$ contamination in the sample is estimated to be 2 ± 2 events. The event acceptance is estimated to be 0.65, due primarily to (i) the $p_t^{(e)}$ > 20 GeV/c cut (0.80) ; (ii) the jet veto requirement within $\Delta\phi = \pm 30°$ (0.96 ± 0.02) ; (iii) the electron-track isolation requirement (0.90 ± 0.07) ; and (iv) the acceptance of events due to geometry (0.94 ± 0.03). The cross-section is then

$$(\sigma.B)_W = 0.53 \pm 0.08 \, (\pm 0.09) \text{ nb},$$

where the last error takes into account systematic errors. This value is in excellent agreement with the expectations for the Standard Model[30] $(\sigma.B)_W$= 0.39 nb.

No event with $p_t^{(e)}$ or $p_t^{(\nu)}$ in excess of the expected distribution for W → eν events has been observed. This result can be used in order to set a limit to the possible existence of very massive W-like objects (W') decaying into electron-neutrino pairs. We find $(\sigma.B)_{W'}$ ≤ 30 pb at 90% confidence level, corresponding to $m_{W'}$ > 170 GeV/c^2, if standard couplings and quark distributions are used to evaluate the cross-sections.

3.9. Observation of the decay mode W → μ + ν. Muon-electron universality predicts an equal number of events in which the electron is replaced by its heavy counterpart, the muon :

$$p\bar{p} \rightarrow W^\pm X \; ; \; W^\pm \rightarrow \mu^\pm \nu_\mu .$$

Although almost identical to decays with electrons [eq. (1)] in theory, the muonic decay has a completely different experimental signature. Whereas an electron produces an electromagnetic shower (detected in the electromagnetic calorimeters), a high momentum muon traverses the whole detector with almost minimum energy loss. Muons are identified by their ability to penetrate many absorption lengths of material. Thus potential backgrounds for muons are radically different from those for electrons. The observation of the same rate for processes $e\nu$ and $\mu\nu$ is therefore not only the most direct confirmation of muon-electron universality in charged-current interactions, but it also provides an important experimental verification of the previous results.

We now briefly describe the muon detection. A fast muon, emerging from the $\bar{p}p$ interaction region, will pass in turn through the central detector, the electromagnetic calorimeter, and the hadron calorimeter, which consists of the instrumented magnet return yoke. After 60 cm of additional iron shielding (except in the forward region), it will then enter the muon chambers, having traversed about $8/\sin\theta$ nuclear interaction lengths, where θ is its emission angle with respect to the beam axis. The number of hadrons penetrating this much material is negligible ; however, there are two sources of hadron-induced background :

i) stray radiation leaking through gaps and holes ;

ii) genuine muons from hadron decays, such as $\pi \to \mu\nu$, $K \to \mu\nu$, etc.

It is therefore essential to follow the behaviour of all muon candidates throughout the whole apparatus. Tracks are recorded in the central detector. The momenta of muons are determined by their deflection in the central dipole magnet, which generates a field of 0.7 T over a volume of 0.7 x 3.5 x 3.5 m^3. The momentum accuracy for high-momentum tracks is limited by the localization error inherent in the system ($\leq$ 100 μm) and by the diffusion of electrons drifting in the gas, which is proportional to $\sqrt{\ell}$ and amounts to about 350 μm after the maximum drift length of ℓ = 19.2 cm. This results in a momentum accuracy of about ±20% for a 1-m-long track at p = 40 GeV/c, in the best direction with respect to the field. In general, the precision depends greatly on the length and orientation of the track. For the muon sample under discussion, the typical error is around ±30%.

In the present investigation the calorimeters have a fourfold purpose : (i) they provide enough material to attenuate hadrons, and constitute a threshold for muon detection of p_t > 2 GeV/c ; (ii) they identify hadronic interactions and/or accompanying neutral particles by an excess in the energy deposition ; (iii) they ensure a continuous tracking of the muon over six segments in depth ; (iv) they provide an almost hermetically closed energy flow measurement around the collision point, which makes possible the determination of the transverse components of the neutrino momentum by transverse energy conservation.

Fifty muon chambers[33], nearly 4 m x 6 m in size, surround the whole detector, covering an area of almost 500 m^2. A graphical display of a W → μν event is shown in Fig. 26, with an expanded view of the muon chambers shown as an insert. Each chamber consists of four layers of drift tubes, two for each projection. The tubes in adjacent parallel layers are staggered. This resolves the left-right drift time ambiguity and reduces the inefficiency from the intervening dead spaces. The extruded aluminium drift tubes have a cross-section of 45 mm x 150 mm, giving a maximum drift length of 70 mm. An average spatial resolution of 300 μm has been achieved through the sensitive volume of the tubes[34]. In order to obtain good angular resolution on the muon tracks, two chambers of four planes each, separated by 60 cm, are placed on five sides of the detector. This long lever-arm was chosen in order to reach an angular resolution of a few milliradians, comparable to the average multiple scattering angle of high-energy muons (3 mrad at 40 GeV/c). Because of space limitations, the remaining side, beneath the detector, was closed with special chambers consisting of four parallel layers of drift tubes.

The track position and angle measurements in the muon chambers permit a second, essentially independent, measurement of momentum. The statistical and systematic errors in this second momentum determination were carefully checked with

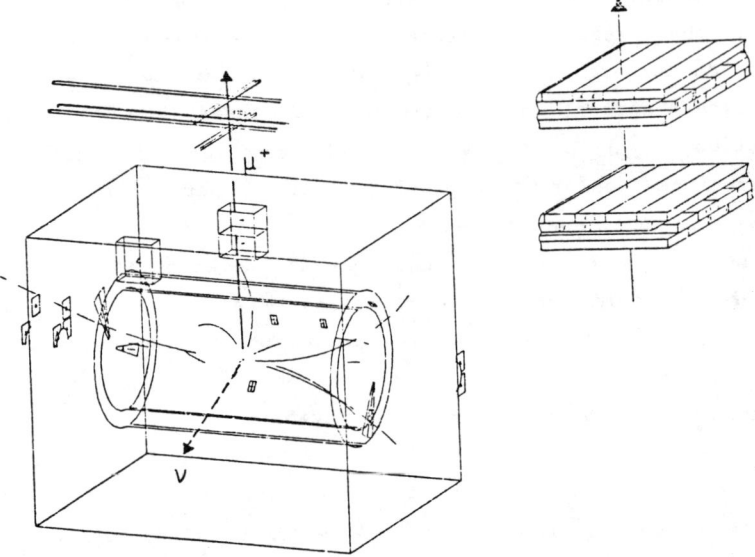

Fig. 26 A graphical display of a $W^+ \to \mu^+ \nu$ event. The vertical arrow shows the trajectory of the 25-GeV/c μ^+ up to the muon chamber while the other arrow shows the transverse direction of the neutrino. The curved lines from the vertex are the charged tracks seen by the central detector, and the petals and boxes illustrate the electromagnetic and hadronic energy depositions. An expanded view of a muon module is shown as an insert.

high-momentum cosmic-ray muons ; Fig. 27 compares the momentum measurements in the central detector and muon chambers. Because of the long lever-arm to the muon chambers, a significant increase in precision is achieved by combining the two measurements.

The presence of neutrino emission is signalled by an apparent transverse energy imbalance when the calorimeter measurement of missing transverse energy is combined with the muon momentum measurement. This determines the neutrino transverse momentum error perpendicular to the muon p_t whereas the error parallel to the muon p_t is dominated by the track momentum accuracy.

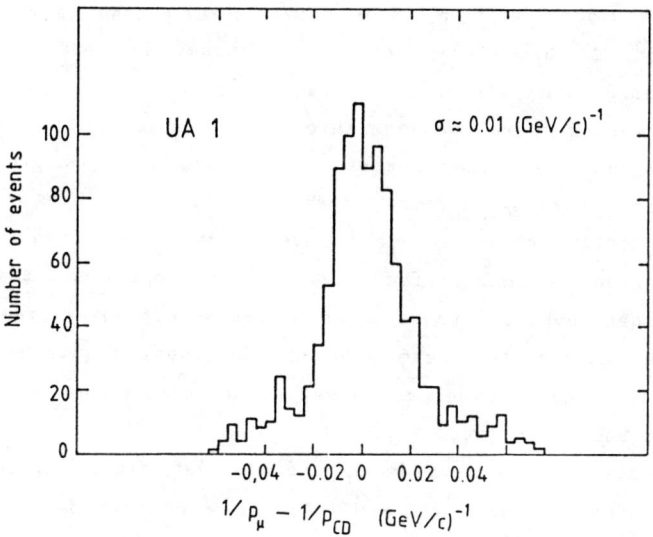

Fig. 27 Distribution of $1/p_\mu - 1/p_{CD}$ for vertical cosmic-ray muons with $p_{CD} > 10$ GeV/c, where p_μ and p_{CD} are the momenta measured in the muon chambers and central detector respectively.

The muon sample is contaminated by several background sources such as leakage through the absorber, beam halo, meson decays, and cosmic rays. Some of the background can be eliminated by requiring a matching central detector track with sufficiently high momentum to penetrate to the muon chambers. All events were therefore passed through a fast filter program which selected muon candidates with $p_t > 3$ GeV/c or $p > 6$ GeV/c. This filter program reconstructed tracks in the muon chambers. For each track pointing roughly towards the interaction region, the central detector information was decoded along a path from the muon chamber track to the interaction region. Track finding and fitting were performed in this path. Events were kept if a central detector track satisfied the above momentum cut and matched the muon chamber track within generous limits. The filter program selected about 72000 events. Since only limited regions of the central detector were considered, the program took about 10% of the average reconstruction time of a full event.

The 17326 events from the fast filter which contained a

muon candidate with $p_t > 5$ GeV/c were passed through the standard UA1 processing chain. Of these, 713 events had a muon candidate with $p_t > 15$ GeV/c or $p > 30$ GeV/c. These events were passed through an automatic selection program which eliminated most of the remaining background by applying strict track quality and matching cuts. Independently of this, all events were examined on an interactive scanning facility. This confirmed that no W-candidate events were rejected by the selection program.

The selection program imposed additional requirements on event topology in order to reject events with muons in jets or back-to-back with jets.

Events were also rejected if the jet algorithm found a calorimeter jet with $E_T > 10$ GeV or a central detector jet with $p_t > 7.5$ GeV/c back-to-back with the muon to within $\pm 30°$ in the plane perpendicular to the beam. Thirty-six events survived these cuts, and were carefully rescanned. After eliminating additional cosmics and probable $K \to \mu\nu$ decays, 18 events remained. The final W-sample of 14 events was obtained after the additional requirement that the neutrino transverse energy exceed 15 GeV. The effects of the different cuts are shown in table 4.

The most dangerous background to the $W \to \mu\nu$ sample comes from the decay of medium-energy kaons into muons within the volume of the central detector such that the transverse momentum kick from the decay balances the deflection of the particle in the magnetic field. This simulates at the same time a high-momentum muon track and, in order to preserve momentum balance in the transverse plane, a recoiling "neutrino". Most of these events are rejected by the selection program. We have performed a Monte Carlo calculation to estimate the residual background. Charged kaons with $3 < p_t < 15$ GeV/c and decaying in the central detector were generated according to a parametrization of the transverse momentum distribution of charged particles[35], assuming a ratio of kaons to all charged particles of 0.25[30]. A full simulation of the UA1 detector was performed, and each track was subjected to the same

Table 4

Selection of $W \to \mu\nu$ candidates. The number of events is indicated at each stage

Events with a $p_t > 5$ GeV/c muon selected by the fast filter program	17,326
Fully reconstructed events with muon $p_t > 15$ GeV/c or $p > 30$ GeV/c	713
Events with a good quality CD track which matches the muon chambers well	285
Events remaining after rejection of cosmic rays	247
Events remaining after a tight cut on the χ^2 of the CD track fit to remove decays	144
Events where the muon is isolated both in the CD and in the calorimeters	53
Events with no jet activity opposite the muon in the transverse plane	36
Events remaining as W candidates after scanning (see text)	18
Events with a neutrino transverse momentum > 15 GeV/c	14

reconstruction and selection procedures as the experimental data, including the scanning of these events. Normalizing to the integrated luminosity of 108 nb^{-1}, we found 4 events in which the K decay was recognized and simulated a muon with P_t > 15 GeV/c. Imposing the additional requirement of P_t > 15 GeV/c for the accompanying neutrino leaves less than one event as an upper limit to the background to W → μν from this source.

In addition, we expect about 5 events in our data sample with muons from decays of pions or kaons with p_t > 15 GeV/c. These will be similarly suppressed by the reconstruction and selection procedures ; in particular such events will be characterized by jets which transversely balance the high-p_t hadrons and are therefore rejected by our topological cuts. The momentum measurements in the central detector and in the iron agree very well (Fig. 28), as a good check of our procedure.

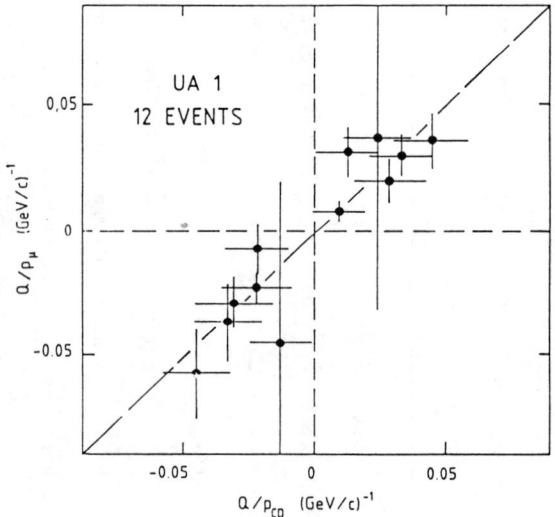

Fig. 28 Two-dimensional plot of Q/p_μ versus Q/p_{CD} for the W → μν events, where p_μ and p_{CD} are the momenta measured in the muon chambers and central detector respectively, and Q is the charge of the muon. The events with tracks in the bottom chambers are not shown.

Eighteen events survive our selection criteria and contain a muon with $p_t > 15$ GeV/c. The muons are isolated, and there is no visible structure to compensate their transverse momenta, in contrast with what might be expected for background events from heavy-flavour decays. Including the muon in the transverse energy balance, all events exhibit a large missing transverse energy of more than 10 GeV, attributed to an emitted neutrino. For the final $W \to \mu\nu$ sample, we consider only those 14 events with a neutrino transverse momentum $p_t > 15$ GeV/c. As in the electron case, the transverse momentum of the neutrino is strongly correlated, both in magnitude and in direction, with the transverse momentum of the muon. Fig. 29a shows this correlation in the direction parallel to the muon p_t. Similarly the component of the neutrino p_t perpendicular to the muon p_t is small. The characteristic back-to-back configuration and the high momenta of both leptons, well above the threshold, are very suggestive of a two-body decay of a massive, slow particle. The large errors in the momentum determination of the muons smear the expected Jacobian peak of a two-body decay. However, the transverse momentum distribution agrees well with that expected from a $W^{\pm}$ decay, once it is smeared with the experimental errors (Fig. 29b).

The transverse momentum $p_t^{(W)}$ of the decaying particle is well measured, because the muon momentum does not enter into its determination. In fact, $p_t^{(W)}$ is simply energy measured in the calorimeters, after subtraction of the muon deposition. The measured distribution is given in Fig. 30a and agrees well with our previous measurement from the $W \to e\nu$ sample, shown in Fig. 30b. Each of the two events with the highest $p_t^{(W)}$ has a jet which locally balances the transverse momentum of the W.

In order to determine the mass of the muon-neutrino system, we have used in a maximum likelihood fit the eight measured quantities for each event (momentum determination of the muon in the CD and in the muon chambers, angles of the muon, four-vector of the energy for the rest of the event) and their relevant resolution functions. We have taken account of the cuts imposed on the measured muon and neutrino transverse momenta[31]. We obtain a fitted W mass of $m_W = 81^{+6}_{-7}$ GeV/c^2, in excellent agreement with the measured value from W → eν. This result is insensitive to

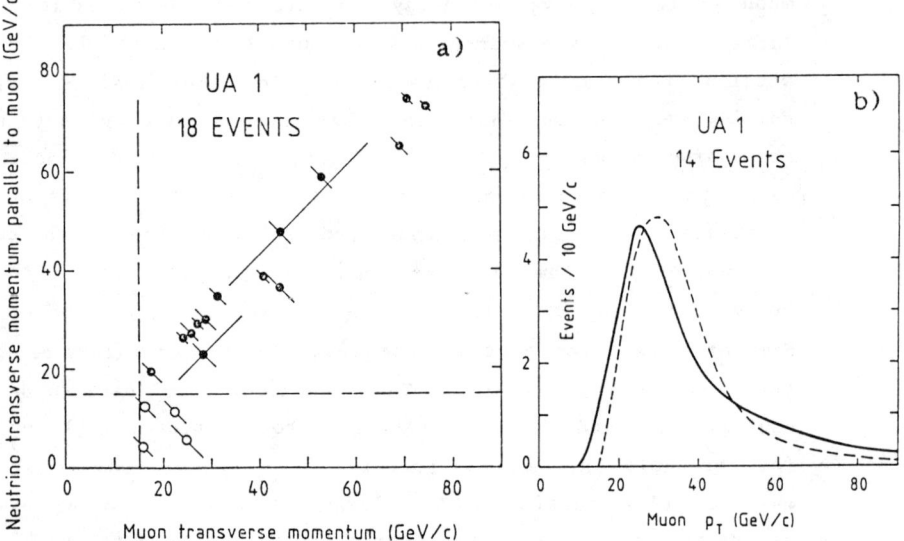

Fig. 29 a) Transverse energy of the neutrino parallel to the muon versus transverse momentum of the muon. Since the two quantities are correlated, error bars are shown for the difference and the sum. The difference in the transverse energy of the W parallel to the muon which is measured in the calorimetry and is therefore not correlated with the transverse momentum of the muon. For the errors in the sum only two error bars are shown for typical events. The filled circles correspond to the final sample of 14 W events, and the open circles to the 4 events with neutrino p_t < 15 GeV/c.

Fig. 29 b) The solid curve is an ideogram of the transverse momentum distribution of the muons in the final sample of 14 W → μν events. The dashed curve is a Monte Carlo prediction, based on the W production spectra measured in W → eν decays and a W mass of 80.9 GeV/c^2 smeared with errors.

the assumed decay angular distribution of the W. If the mass is fixed at the electron value of 80.9 GeV/c^2, a fit of the decay asymmetry gives $<\cos\theta^*> = 0.3 \pm 0.2$, fully consistent with our result from W → eν and with the expected V-A coupling. The asymmetry measurement is not very significant since the ambiguity due to the two possible solutions for the longitudinal momentum of the W could be resolved in only a few cases. This is due to the large momentum errors and the limited acceptance in pseudorapidity ($|\eta| < 1.3$) for the muons.

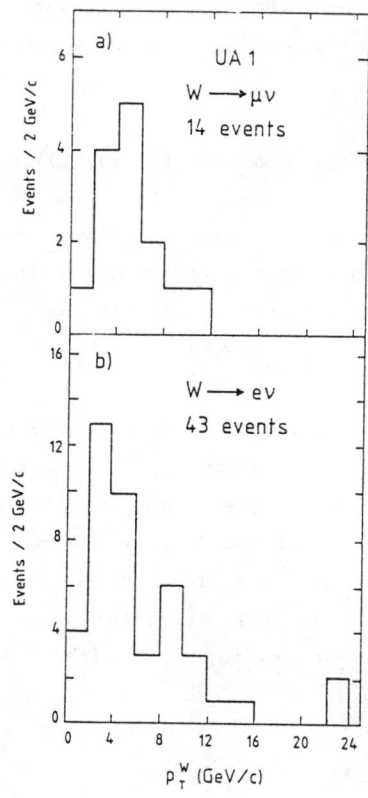

Fig. 30 a) The transverse momentum distribution of the W derived from the energy imbalance measured in the calorimetry.
b) The corresponding distribution from the W → eν data is shown for comparison.

The overall acceptance for the final sample of 14 W →
μν events is limited by two main factors, namely the
geometrical acceptance of the muon trigger system for muons
with p_t > 15 GeV/c (49%) and the influence of the track
quality cuts applied to the muon. The latter has been
estimated by applying identical cuts to an equivalent sample
of 46 W → eν events from the 1983 data sample. 21 events
remain, giving an acceptance of (46 ± 7)%. A further
correction of (87 ± 7)% is included to account for the jet
veto and track isolation requirement. These three factors
give an overall acceptance of (20 ± 3.5)%.

The integrated luminosity for the present data sample is
108 nb^{-1}, with an estimated uncertainty of ±15%. The
cross-section is then

$$(\sigma \cdot B)_\mu = 0.67 \pm 0.17 \, (\pm 0.15) \text{ nb}$$

where the last error includes the systematics from both
acceptance and luminosity. This value is in good agreement
both with the standard model predictions[30] and with our
result for W → eν, namely $(\sigma \cdot B)_e$ = 0.53 ± 0.08
(±0.09) nb.

A direct comparison between the electron and muon results
has been made by selecting those W → eν events which are
within the acceptance of the muon trigger system. Twelve
events remain from the 21 which pass the muon track quality
cuts. After correction for the difference in integrated
luminosity (118 nb^{-1} in the electron case) this gives the
following cross-section ratio, in which systematic errors
approximately cancel :

$$R = (\sigma \cdot B)_\mu / (\sigma \cdot B)_e = 1.24^{+0.6}_{-0.4}$$

4. OBSERVATION OF THE NEUTRAL BOSON Z^0

4.1. Event selection.
We now extend our search to the
neutral partner Z^0, responsible for neutral currents. As
in our previous work, production of intermediate vector bosons

is achieved with proton-antiproton collisions at $\sqrt{s} = 540$ GeV in the UA1 detector, except that now we search for electron and muon pairs rather than for electron-neutrino coincidence. The process is then

$$\bar{p} + p \rightarrow Z^0 + X$$
$$\hookrightarrow e^+ + e^- \text{ or } \mu^+ + \mu^-.$$

The reaction is then approximately a factor of 10 less frequent than the corresponding $W^{\pm}$ leptonic decay channels. A few events of this type are therefore expected in our muon or electron samples. Evidence for the existence of the Z^0 in the range of masses accessible to the UA1 experiment can also be drawn from weak-electromagnetic interference experiments at the highest PETRA energies, where deviations from point-like expectations have been reported.

Events for the present paper were selected by the so-called "express line", consisting of a set of four 168E computers[38] operated independently in real time during the data-taking. A subsample of events with $E_T \geq 12$ GeV in the electromagnetic calorimeters and dimuons are selected and written on a dedicated magnetic tape. These events have been fully processed off-line and further subdivided into four main classes : (i) single, isolated electromagnetic clusters with $E_T > 15$ GeV and missing energy events with $E_{miss} > 15$ GeV, in order to extract $W^{\pm} \rightarrow e^{\pm} \nu$ events ; (ii) two or more isolated electromagnetic clusters with $E_T > 25$ GeV for $Z^0 \rightarrow e^+ e^-$ candidates ; (iii) muon pair selection to find $Z^0 \rightarrow \mu^+ \mu^-$ events ; and (iv) events with a track reconstructed in the central detector, of transverse momentum within one standard deviation, $p_t \geq 25$ GeV/c, in order to evaluate some of the background contributions. We will discuss these different categories in more detail.

4.2. Events of type $Z^0 \to e^+e^-$. An electron signature is defined as a localized energy deposition in two contiguous cells of the electromagnetic detectors with $E_T > 25$ GeV, and a small (or no) energy deposition (≤ 800 MeV) in the hadron calorimeters immediately behind them. The isolation requirement is defined as the absence of charged tracks with momenta adding up to more than 3 GeV/c of transverse momentum and pointing towards the electron cluster cells. The effects of the successive cuts on the invariant electron-electron mass are shown in Fig. 31. Four e^+e^- events survive cuts, consistent with a common value of

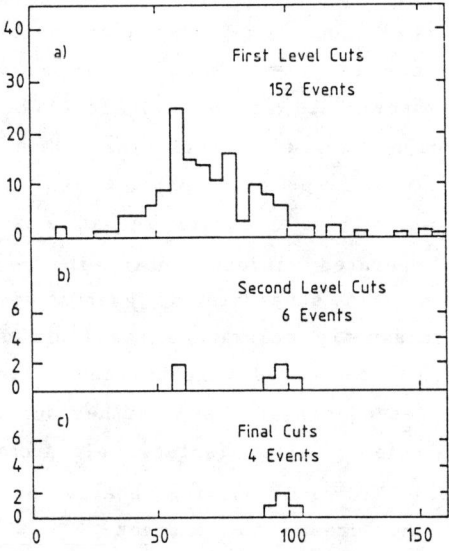

Fig. 31 Invariant mass distribution (uncorrected) of two electromagnetic clusters: a) with $E_T > 25$ GeV; b) as above and a track with $p_t > 7$ GeV/c and projection length > 40 cm pointing to the cluster. In addition, a small energy deposition in the hadron calorimeters immediately behind (< 0.8 GeV) ensures the electron signature. Isolation is required with $\Sigma\ p_t < 3$ GeV/c for all other tracks pointing to the cluster. c) The second cluster also has an isolated track.

(e^+e^-) invariant mass. They have been carefully studied using the interactive event display facility MEGATEK. One of these events is shown in Figs. 32a and 32b. The main parameters of the four events are listed in tables 5 and 7. As one can see from the energy deposition plots (Fig. 33), their dominant feature is two very prominent electromagnetic energy depositions. All events appear to balance the visible total transverse energy components ; namely, there is no evidence for the emission of energetic neutrinos. Except for one track of event D which travels at less than $15°$ parallel to the magnetic field, all tracks are shown in Fig. 34, where the momenta measured in the central detector are compared with the energy deposition in the electromagnetic calorimeters. All tracks but one have consistent energy and momentum measurements. The negative track of event C shows a value of (9 ± 1) GeV/c, much smaller than the corresponding calorimeter deposition of (48 ± 2) GeV. One can interpret this event as the likely emission of a hard "photon" accompanying the electron. Subsequent calibrations of the electromagnetic (e.m.) counters indicated that the centroid of the energy deposition in the calorimeters was significantly displaced with respect to the incident electron track, indicating an angle of $\Delta\phi = (14 \pm 4)$ degrees between the charged and neutral components. This excluded the possibility of an external bremsstrahlung in the vacuum pipe and detector window. The estimated probability of an internal bremsstrahlumg exceeding the angle and energy observed is, according to Berends, about 0.005, or 2% for the sample of four events. A rather similar event has been reported by the UA2 Collaboration. In this case the photon and electron hit separate cells, thus directly indicating a finite e-γ separation. More recently, an event of the type $\mu^+\mu^-\gamma$ has been observed in the UA1 study of $Z^0 \to \mu^+\mu^-$ decays. Also, in this event a small but finite angle ($\sim 8°$) is observed between the muon and the hard photon (E$\sim$30 GeV). Event parameters are summarized in table 8. It is certainly premature to draw any conclusion about the origin of such events. It shows however how

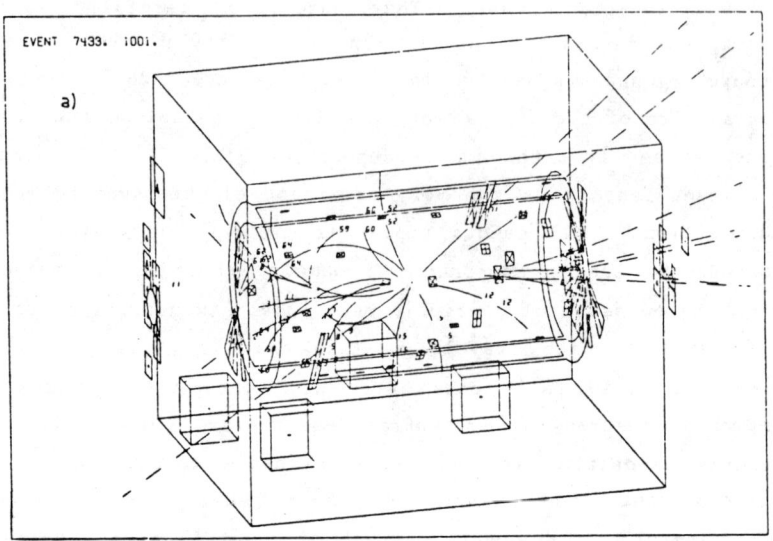

Fig. 32 a) Event display. All reconstructed vertex associated tracks and all calorimeter hits are displayed.

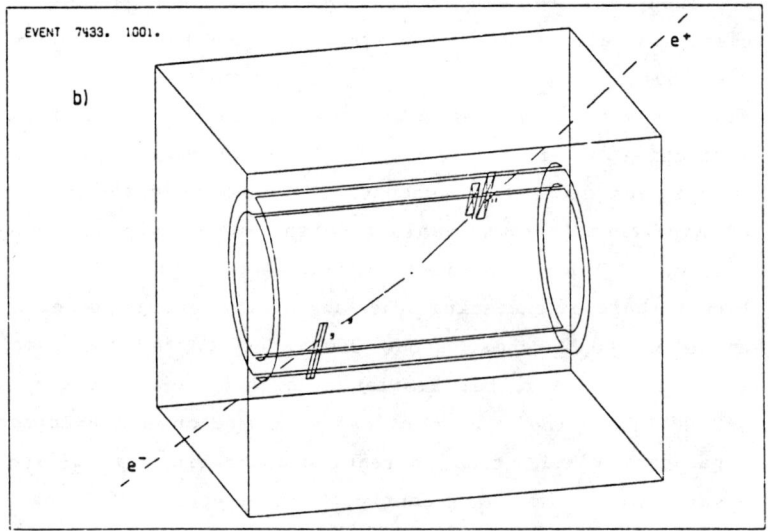

Fig. 32 b) The same, but thresholds are raised to $p_t > 2$ GeV/c for charged tracks and $E_T > 2$ GeV for calorimeter hits. We remark that only the electron pair survives these mild cuts.

Table 3 Properties of the individual electrons of the pair events

Run, event		Drift chamber measurement						Shower counter measurement				
		p (GeV)	Δp a) (GeV)	Q	dE/dx b)	y c)	φ (deg)	E_{tot} (GeV)	Electromagnetic samples (GeV)			Had. energy
									S_1	S_2	S_3	S_4	H
A	7433 1001	33	+9 −6	+	1.8±0.3	1.01	144	44	14	27	3	0.0	0.0
		63	+23 −13	−	1.7±0.2	−1.19	−31	48	6	37	4	0.2	0.0
B	7434 746	27	+19 −8	+	1.6±0.3	−0.36	131	42	2	18	20	1.3	0.1
		93	+66 −28	−	1.8±0.2	−1.45	−60	102	42	56	4	0.2	0.0
C	6059 1010	32	+11 −6	+	1.3±0.2	0.64	67	61	1	37	22	0.6	0.0
		9	+1 −1	−	1.4±0.1	0.24	−121	48	1	23	23	1.3	0.0
D	7739 1279	d)	d)	d)	d)	−0.19	169	51	1	13	34	2.4	0.0
		50	+50 −17	−	1.5±0.2	−0.79	−9	55	8	38	9	0.0	0.1

a) ±1σ including systematic errors.
b) Ionization loss normalized to minimum ionizing pion.
c) The rapidity y is defined as positive in the direction of outgoing $\bar{p}$.
d) Unmeasured owing to large dip angle.

Table 6 Properties of the muons of the dimuon event 6600-222

Q	Track parameters				Normalized ionization I/I_0		abs	μ/CD matching [d]	
	p (GeV/c)	ℓ (cm)	y	φ (°)	e.m. calorimeter	Hadron calorimeter		Position (cm)	Angle (mrad)
+	58.8^{+8}_{-6} [a]								
	60.3 ± 10.8 [b]								
	$59.2^{+6.4}_{-5.2}$ [c]	170	1.19	-27.6	0.8 ± 0.5	1.2 ± 0.5	10.2	$\Delta X_1 = -1.3 \pm 1.9$ $\Delta X_2 = 11.6 \pm 10.7$	$\Delta\phi = -2 \pm 6$ $\Delta\Lambda = 11 \pm 14$
-	63.6^{+30}_{-15} [a]								
	43.1 ± 6.2 [b]								
	$46.1^{+6.1}_{-5.7}$ [c]	80	-0.28	119.1	1.2 ± 0.9	1.6 ± 0.8	11.1	$\Delta X_1 = -0.1 \pm 8.0$ $\Delta X_2 = -8.0 \pm 8.5$	$\Delta\phi = -6 \pm 3$ $\Delta\Lambda = -9 \pm 14$

Momentum determination : a) Central detector and μ chamber (statistical errors only) ;
b) Transverse momentum balance ;
c) Weighted average of (a) and (b) ;
d) Difference between the extrapolated CD track and the track measured in the μ chambers (see Fig. 6).

μ-CD matching :

Remarks: The acceptance of the single muon trigger starts at a transverse momentum of about 2.5 GeV/c and reaches its full efficiency of 97% at 5.5 GeV/c. The geometrical acceptance of the dimuon trigger used in this analysis, reduced the acceptance for Z^0 events to about 30%.

Table 7

Mass and energy properties of lepton pair events

	Run, event	Lepton pair properties			General event properties			
		Mass a) (GeV/c^2)	p_t (GeV/c)	x_F b)	E_{tot} (GeV)	$\Sigma\|E_T\|$ (GeV)	Missing E_T (GeV)	Charged tracks
A	7433 1001	91 ± 5	2.9 ± 0.9	0.02 ± 0.01	274	82	2.1 ± 3.6	27
B	7434 746	97 ± 5	7.9 ± 1.2	0.39 ± 0.01	494	149	9.3 ± 5.0	67
C	6059 1010	98 ± 5	8.0 ± 1.5	0.17 ± 0.01	412	143	3.3 ± 4.8	38
D	7339 1279	95 ± 5	8.4 ± 1.4	0.17 ± 0.01	493	157	0.8 ± 5.0	54
E	6600 222 (μμ)	95 ± 8	24 ± 5	0.14 ± 0.02	278^c)	128^c)	3.4 ± 5.9^c)	28

a) These errors have been scaled up arbitrarily to 5 GeV to represent the present level of uncertainty in the overall calibration of the e.m. calorimeter which will be recalibrated completely at the end of the present run. This scale factor is not included in the error bars plotted in Fig. 8.
b) x_F is defined as the longitudinal momentum of the dilepton divided by beam energy.
c) Includes the muon energies.

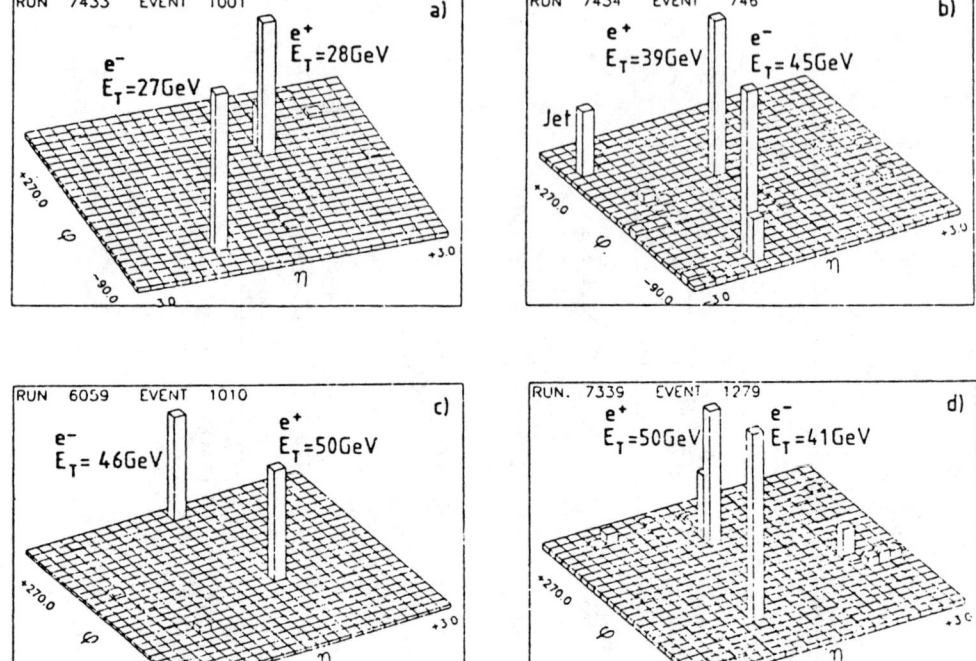

Fig. 33 Electromagnetic energy depositions at angles $> 5^0$ with respect to the beam direction for the four electron pairs.

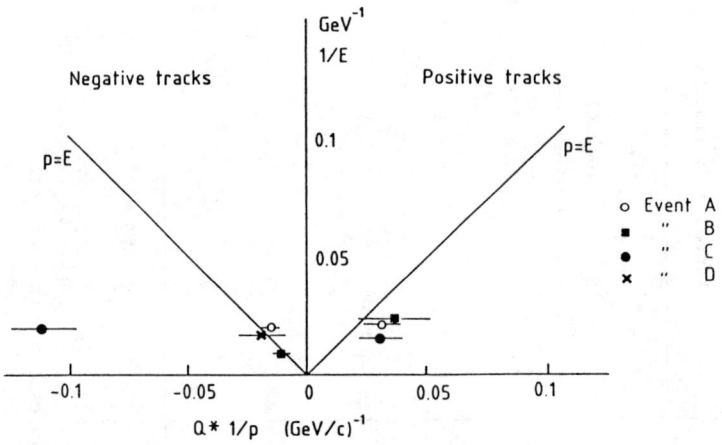

Fig. 34

Magnetic deflection in 1/p units compared to the inverse of the energy deposited in the electromagnetic calorimeters. Ideally, all electrons should lie on the 1/E = 1/p line.

Table 8   Energy, angle, and mass properties of the eeγ events

	UA1	UA2
E_γ (GeV)	38.8 ± 1.5	24.4 ± 1.4
E_{e_1} (GeV)	61.0 ± 1.2	68.5 ± 1.6
E_{e_2} (GeV)	9 ± 1	11.4 ± 0.9
$\Delta\phi(e_2\gamma)$ (°)	14.4 ± 4.0	31.8
$m(e_1 e_2)$ (GeV/c^2)	42.7 ± 2.4	49.8
$m(e_1 e_2 \gamma)$ (GeV/c^2)	98.7 ± 5.0	89.7 ± 2.8
$m(e_1\gamma)$ (GeV/c^2)	88.8 ± 2.5	74.1
$m(e_2\gamma)$ (GeV/c^2)	4.6 ± 1.0	9.1

Notes: For the UA1 event $e_1 = e^+$ and $e_2 = e^-$.
The UA2 numbers are calculated from ref. [4].

entirely new phenomena may be occurring in the collider energy range.

The average invariant mass of the pairs, combining the four consistent values, is (95.2 ± 2.5) GeV/c^2 (table 7).

4.3. Events of type $Z^0 \to \mu^+\mu^-$. Events from the dimuon trigger flag have been submitted to the additional requirement that there is at least one muon track reconstructed off-line in the muon chambers, and with one track in the central detector of reasonable projected length ($\geq$ 40 cm) and $p_t \geq$ 7 GeV/c. Only 42 events survive these selection criteria. Careful scanning of these events has led to only one clean dimuon event, with two "isolated" tracks (Fig. 35). Most of the events are due to cosmics. Parameters are given in tables 6 and 7. Energy losses in the calorimeters traversed by the two muon tracks are well within expectations of ionization losses of high-energy muons (Fig. 36a). The position in the coordinate and the angles at the exit of the iron absorber (Fig. 36b) are in agreement with the extrapolated track from the central detector, once multiple scattering and other instrumental effects have been calibrated with p > 50 GeV cosmic-ray muons traversing the same area of the apparatus. There are two ways of measuring momenta, either in the central detector or using the muon detector. Both measurements give consistent results. Furthermore, if no

284

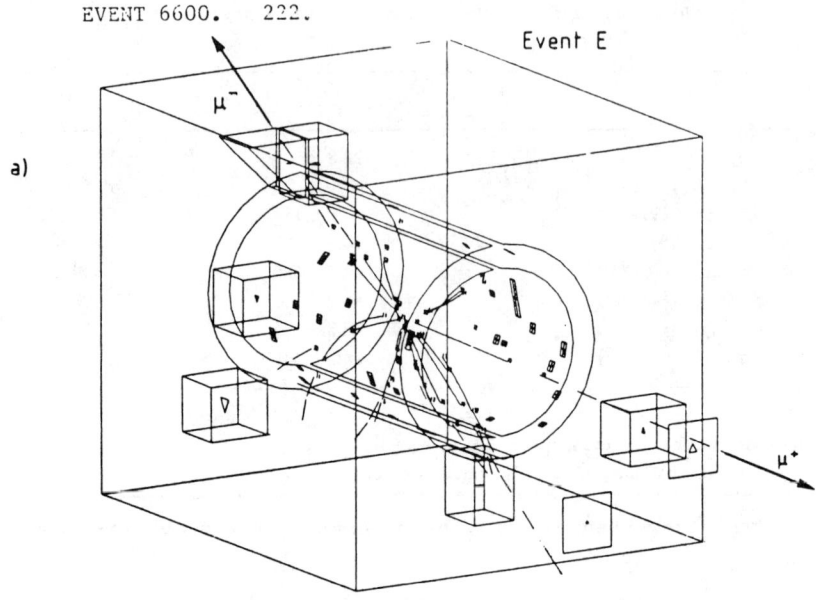

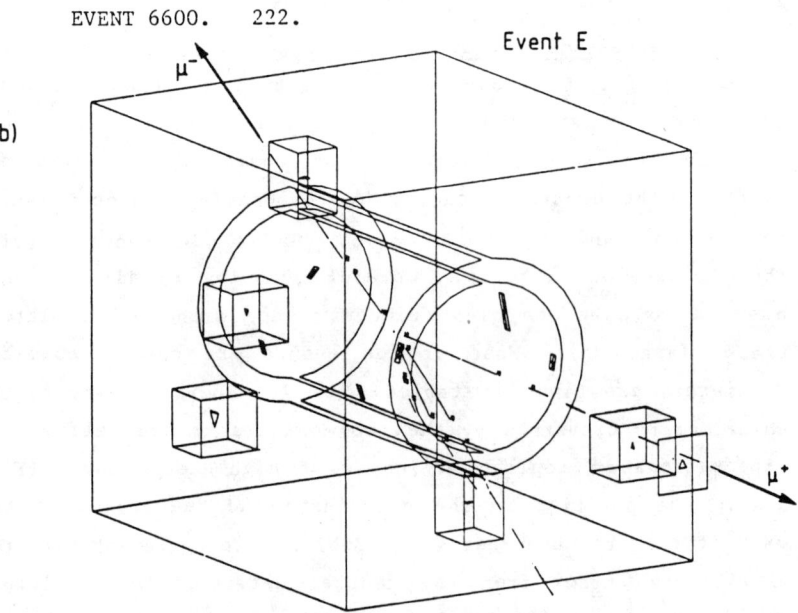

Fig. 35 Display for the high-invariant-mass muon pair event:
a) without cuts and b) with $p_t > 1$ GeV thresholds
for tracks and $E_T > 0.5$ GeV for calorimeter hits.

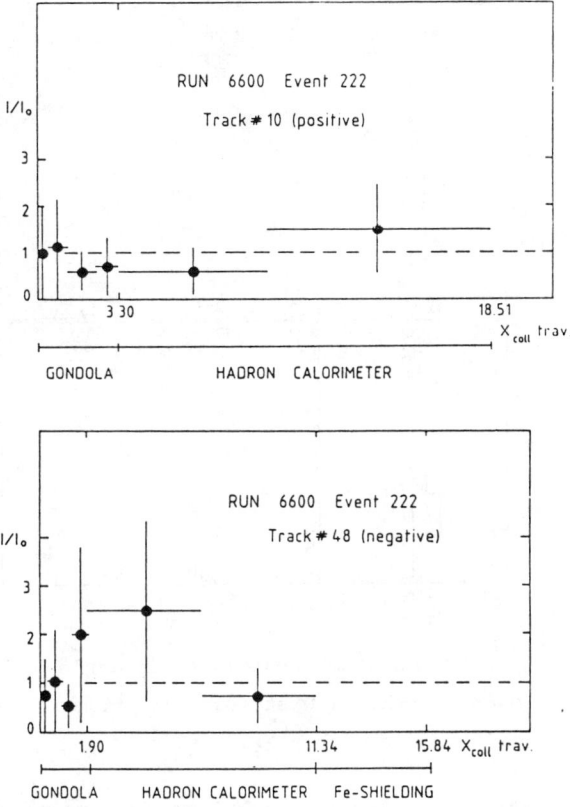

Fig. 36 a) Normalized energy losses in calorimeter cells traversed by the two muon tracks.

neutrino is emitted (as suggested by the electron events which exhibit no missing energy), the recoil of the hadronic debris, which is significant for this event, must be equal to the transverse momentum of the ($\mu^+\mu^-$) pair by momentum conservation. The directions of the two muons then suffice to calculate the momenta of the two tracks. Uncertainties of muon parameters are then dominated by the errors of calorimetry. As shown in table 6 this determination is in agreement with magnetic deflection measurements. The invariant mass of the ($\mu^+\mu^-$) pair is found to be $m_{\mu\mu}$ = 95.5 ± 7.3) GeV/c^2, in excellent agreement with that of the four electron pairs (see table 7).

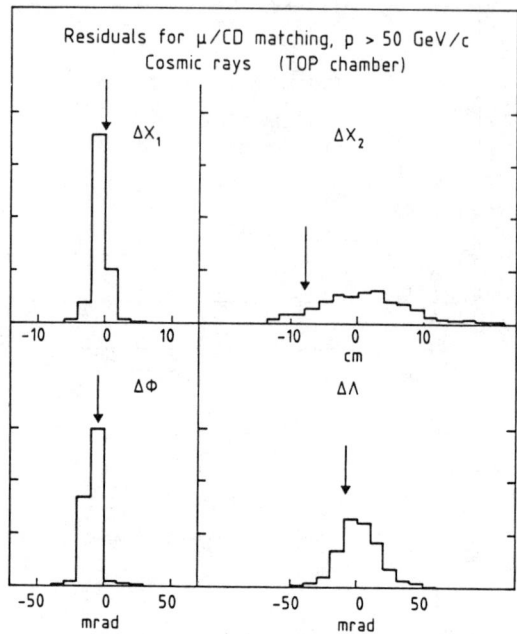

Fig. 36 b) Arrows show residuals in angle and position for muon track. Distributions come from cosmic-ray calibration with p > 50 GeV/c.

4.4. Backgrounds. The most striking feature of the events is their common value of the invariant mass (Fig. 37) ; values agree within a few percent and with expectations from experimental resolution. Detection efficiency is determined by the energy thresholds in the track selection, 15 GeV/c for $e^{\pm}$ and 7 GeV/c for $\mu^{\pm}$. Most "trivial" sources of background are not expected to exhibit such a clustering at high masses. Also, most backgrounds would have an equal probability for (eμ) pairs, which are not observed. Nevertheless, we have considered several possible spurious sources of events :

i) Ordinary large transverse momentum jets which fragment into two apparently isolated, high-momentum tracks, both simulating either muons or electrons. To evaluate this effect, events with (hadronic) tracks of momenta compatible with p_t > 25 GeV/c were also selected in the express line. After requiring that the track is isolated, one finds

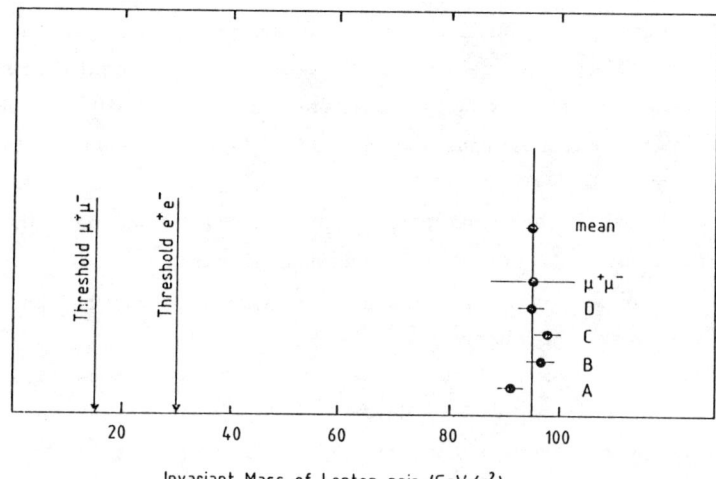

Fig. 37 Invariant masses of lepton pairs.

one surviving event with transverse energy $\sim$ 25 GeV in a sample corresponding to 30 nb^{-1}. Including the probability that this track simulates either a muon ($\sim 2 \times 10^{-3}$) or an electron ($\sim 6 \times 10^{-3}$), we conclude that this effect is negligible[39]. Note that two tracks (rather than one) are needed to simulate our events (probabilities must be squared!) and that the invariant mass of the events is much higher than the background. The background is expected to fall approximately like m^{-5} according to the observed jet-jet mass distributions[14].

ii) Heavy-flavoured jets with subsequent decay into leading muons or electrons. In the 1982 event sample (11 nb^{-1}), two events have been observed with a single isolated muon of $p_t > 15$ GeV and one electron event with $p_t > 25$ GeV/c. Some jet activity in the opposite hemisphere is required. One event exhibits also a significant missing energy. Once this is taken into account they all have a total (jet+jet+lepton+neutrino) transverse mass of around 80 GeV/c^2, which indicates that they are most likely due to heavy-flavour decay of W particles. This background will be kinematically suppressed at the mass of our five events.

Nevertheless, if the fragmentation of the other jet is also required to give a leading lepton and no other visible debris, this background contributes at most to 10^{-4} events. Monte Carlo calculations using ISAJET lead to essentially the same conclusion[30].

iii) Drell-Yan continuum. The estimated number and the invariant mass distribution make it negligible[40].

iv) W^+W^- pair production is expected to be entirely negligible at our energy[41].

v) Onium decay from a new quark, of mass compatible with the observation (~ 95 GeV/c^2). Cross-sections for this process have been estimated by different authors[42], and they appear much too small to account for the desired effect.

In conclusion, none of the effects listed above can produce either the number or the features of the observed events.

4.5. <u>Mass determination.</u> All the observations are in agreement with the hypothesis that events are due to the production and decay of the neutral intermediate vector boson Z^0 according to reaction (1). The transverse momentum distribution is shown in Fig. 38, compared with the observed distributions for the $W^\pm \to e\nu$ events and with QCD calculations. The muon events and one of the electron events (event B) have visible jet structure. Other events are instead apparently structureless.

From our observation, we deduce a mass value for the Z^0 particle:

$$m_{Z^0} = (95.2 \pm 2.5) \text{ GeV/c}^2.$$

The half width based on the four electron events is 3.1 GeV/c^2 (< 5.1 GeV/c^2 at 90% c.l.), consistent with expectation from the experimental resolution and the natural Z^0 width, $\Gamma_{Z^0} = 3.0$ GeV. At this point it is important to stress that the final calibration of the electromagnetic calorimeters is still in progress and that small scale shifts are still possible, most likely affecting

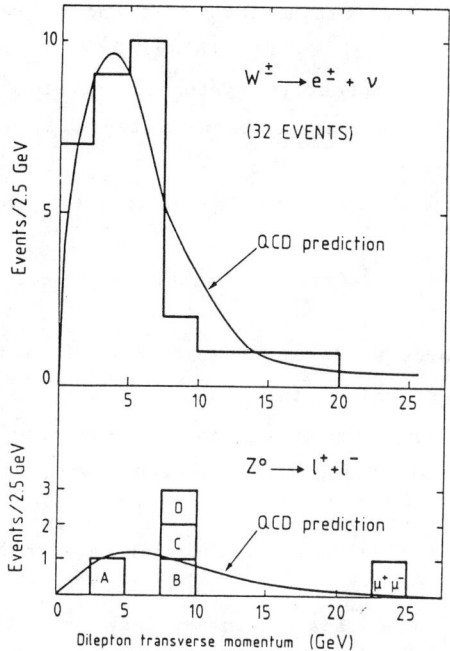

Fig. 38 Transverse momentum spectra: a) for W → eν events, and b) $Z^0 \to \ell^+\ell^-$ candidates. The lines represent QCD predictions.

<u>both</u> the $W^{\pm}$ and Z^0 mass values. No e.m. radiative corrections have been applied to the masses.

5. COMPARING THEORY WITH EXPERIMENT

The experiments discussed in the previous paragraph have shown that the W particle has most of the properties required in order to be the carrier of weak interactions. The presence of a narrow di-lepton peak has been observed around 95 GeV/c^2. Rates and features of the events are consistent with the hypothesis that one has indeed observed the neutral partner of the $W^{\pm}$. Statistics at present are not sufficient to test experimentally the form of the interaction, neither has parity violation been detected. The precise values of the masses of Z^0 and $W^{\pm}$ now available constitute however a crucial test of the idea of unification between weak and electromagnetic forces and in particular of the predictions of SU(2) x U(1) theory of Glashow, Weinberg and Salam. Careful account of systematic errors is needed in order to evaluate an average between the UA1 and UA2 mass determination.

The charged vector boson mass given in the present work is

$$m_{W^\pm} = (80.9 \pm 1.5) \text{ GeV/c}^2 \quad \text{(statistical errors only)}$$

to which a 3% energy scale uncertainty must be added. In the present report a value for the Z^0 mass, $m_{Z^0} = (95.1 \pm 2.5)$ GeV/c^2 has been given. Neglecting systematic errors, a mass value is found with somewhat smaller errors:

$$m_{Z^0} = (95.6 \pm 1.4) \text{ GeV/c}^2 \quad \text{(statistical errors only)}$$

to which the same scale uncertainty as for the $W^\pm$ applies. The quoted errors includes i) the neutral width of the Z^0 peak, which is found to be $\Gamma < 8.5$ GeV/c^2 (90% confidence level), ii) the experimental resolution of counters, and iii) the r.m.s. spread between calibration constants of individual elements. In Fig. 39a we have plotted m_Z against m_W. The elliptical shape of the errors reflects the uncertainty in the energy scale. One can see that there is excellent agreement with the expectations of the SU(2) x U(1) Standard Model. One can also determine the classic parameters:

$$\sin^2 \theta_W = \frac{38.5 \text{ GeV/c}^2}{m_W} = 0.226 \pm 0.008 \; (\pm 0.014),$$

$$\rho = \frac{m_W^2}{m_Z^2 \cos^2 \theta_W} = 0.925 \pm 0.05,$$

where the number in parenthesis is due to systematic errors. The result is shown in Fig. 39b, again in good agreement with expectations and published results [4,3].

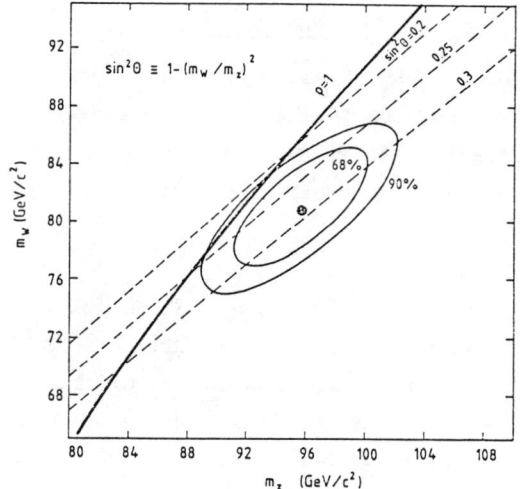

Fig. 39 a) m_Z plotted against m_W determined by the UA1 experiment. The elliptical error curves réflect the uncertainty in the energy scale at the 68% and 90% confidence levels. The heavy curve shows the Standard Model prediciton for $\rho = 1$ as a function of the Intermediate Vector Boson (IVB) masses.

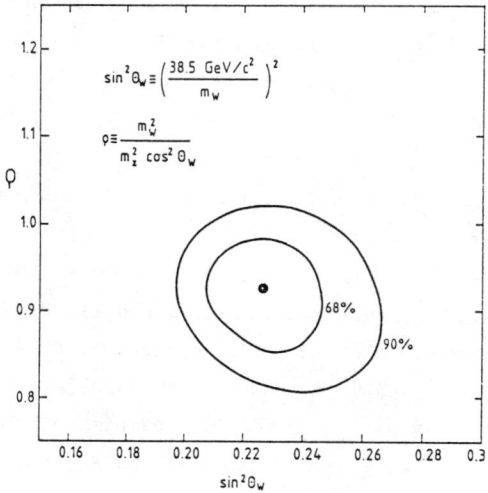

Fig. 39 b) ρ plotted against $\sin^2 \theta_W$ as determined from the measurement of the IVB masses. The 68% and 90% confidence level limits are shown.

ACKNOWLEDGEMENTS

These lectures are based on the work of the UA1 Collaboration team, to which I want to express all my appreciation for the remarkable work which has led to so many results. At present the following individuals are members of the collaboration :

G. Arnison, A. Astbury, B. Aubert, C. Bacci, A. Bezaguet,
R.K. Bock, T.J.V. Bowcock, M. Calvetti, P. Catz,
P. Cennini, S. Centro, F. Ceradini, S. Cittolin, D. Cline,
C. Cochet, J. Colas, M. Corden, D. Dallman, D. Dau,
M. DeBeer, M. Della Negra, M. Demoulin, D. Denegri,
A. Diciaccio, D. DiBitonto, L. Dobrzynski, J.D. Dowell,
K. Eggert, E. Eisenhandler, N. Ellis, P. Erhard, H. Faissner,
M. Fincke, G. Fontaine, R. Frey, R. Frühwirth, J. Garvey,
S. Geer, C. Ghesquiere, P. Ghez, K.L. Giboni, W.R. Gibson,
Y. Giraud-Heraud, A. Givernaud, A. Gonidec, G. Grayer,
T. Hansl-Kozanecka, W.J. Haynes, L.O. Hertzberger,
C. Hodges, D. Hoffmann, H. Hoffmann, D.J. Holthuizen,
R.J. Homer, A. Honma, W. Jank, G. Jorat, P.I.P. Kalmus,
V. Karimaki, R. Keeler, I. Kenyon, A. Kernan,
R. Kinnunen, W. Kozanecki, D. Kryn, F. Lacava, J.P. Laugier,
J.P. Lees, H. Lehmann, R. Leuchs, A. Leveque, D. Linglin,
E. Locci, J.J. Malosse, T. Markiewicz, G. Maurin,
T. McMahon, J.P. Mendiburu, M.N. Minard, M. Mohammadi,
M. Moricca, K. Morgan, H. Muirhead, F. Muller, A.K. Nandi,
L. Naumann, A. Norton, A. Orkin-Lecourtois, L. Paoluzi,
F. Pauss, G. Piano Mortari, E. Pietarinen, M. Pimiä,
J.P. Porte, E. Radermacher, J. Ransdell, H. Reithler,
J.P. Revol, J. Rich, M. Rijssenbeek, C. Roberts,
J. Rohlf, P. Rossi, C. Rubbia, B. Sadoulet, G. Sajot,
G. Salvi, G. Salvini, J. Sass, J. Saudraix, A. Savoy-Navarro,
D. Schinzel, W. Scott, T.P. Shah, D. Smith, M. Spiro,
J. Strauss, J. Streets, K. Sumorok, F. Szoncso, C. Tao,
G. Thompson, J. Timmer, E. Tscheslog, J. Tuominiemi,
B. Van Eijk, J.P. Vialle, J. Vrana, V. Vuillemin, H.D. Wahl,
P. Watkins, J. Wilson, R. Wilson, C.E. Wulz,
Y.G. Xie, M. Yvert, E. Zurfluh

REFERENCES

[1]   C. Rubbia, P. McIntyre and D. Cline, Study Group, Design study of a Proton-Antiproton Colliding Beam Facility, CERN/PS/AA 78-3 (1978), reprinted in Proc. Workshop on Producing High-Luminosity, High Energy Proton-Antiproton Collsions (Berkely, 1978), report LBL-7574, UC34, p 189.
Proc. Inter. Neutrino Conf. (Aachen, 1976) (Vieweg, Braunschweig, 1977) p. 683.

[2]   The Staff of the CERN Proton-Antiproton Project, Phys. Lett. 107B, (1981) 306.

[3]   UA1 Proposal: A 4π solid-angle detector for the SPS used as a proton-antiproton collider at a centre-of-mass energy of 540 GeV, CERN/SPSC 78-06 (1978).
M. Barranco Luque et al., Nucl. Instrum. Methods 176 (1980) 175.
M. Calvetti et al., Nucl. Instrum. Methods 176 (1980) 255.
K. Eggert et al., Nucl. Instrum. Methods 176 (1980) 217 and 233.
A. Astbury, Phys. Scr. 23 (1981) 397.
UA1 Collaboration, The UA1 detector (presented by J. Timmer) in Proc. 18th Rencontre de Moriond, Antiproton-Proton Physics, 1983, to be published.

[4]   UA1 Collaboration, G. Arnison et al., Phys. Lett. 123B (1983) 115.

[5]   Transverse Energy Distributions in the Central Calorimeters, CERN Internal Report EP/82-122 (1982).

[6]   M.J. Corden et al., Physica Scripta 25 (1982) 468.
C. Cochet et al., UA1 Tech. Note TN 82-40.

[7]   G. Arnison et al., Phys. Lett. 107B (1981) 320.
G. Arnison et al., Phys. Lett. 123B (1983) 108.

[8]   G. Wolf, DESY Report EP/82-122 (1982).

[9]   F.E. Paige and S.D. Protopopescu, ISAJET, BNL 31987.

[10]   M. Della Negra, Physica Scripta 25 (1982) 468.
        R.K. Bock et al., Nucl. Inst. Meth. 186 (1981) 533.

[11]   Z. Kunszt and E. Pietarinen, CERN preprint TH 3584 (1983).

[12]   Z. Kunszt and E. Pietarinen, Nucl. Phys. B164 (1980) 45.
        T. Gottschalk and D. Sivers, Phys. Rev. D21 (1980) 102.
        F. Berends et al., Phys. Lett. 103B (1981) 124.

[13]   R. Branelik et al., Phys. Lett. 114B (1982) 65.[16]

[14]   UA1 Collaboration, Jet fragmentation at the SPS $\bar{p}p$ collider - UA1 experiment (presented by V. Vuillemin), in Proc. 18th Rencontre de Moriond, Antiproton-Proton Physics, 1983.

[15]   G. Arnison et al., Phys. Lett. 118B (1982) 173.

[16]   B. Combridge et al., Phys. Lett. 70B (1977) 234.

[17]   B. Combridge and C. Maxwell, preprint RL-83-095 (1983).

[18]   UA1 Collaboration, G. Arnison et al., Phys. Lett. 132B (1982) 214.

[19]   F. Berends et al., Phys. Lett. 103B (1981) 124.

[20]   J.C. Collins and D.E. Soper, Phys. Rev. D16 (1977) 2219.

[21]   D. Drijard et al., Phys. Lett. 121B (1983) 433.

[22]   R.K. Ellis et al., Nucl. Phys. B173 (1980) 397.

[23]   N.G. Antoniou et al., Phys. Lett. 128B (1983) 257.

[24]   H. Habramowicz et al., Z. Phys. C12 (1982) 289.

[25]   G. Arnison et al., Phys. Lett. 122B, 103 (1983).

[26]   G. Arnison et al., Phys. Lett. 129B (1983) 273.

[27]   UA2 Collaboration in Proceeding of the International Europhysics Conference , Brighton, July 1983, page 472.

[28]   A. Nakamura, G. Pancheri and Y. Srivastava, Frascati preprint LFN-83/43 (R) (June 1983).

[29]   P. Aurenche and J. Lindfors, Nucl. Phys. B185 (1981) 274.

[30]   F.E. Paige and S.D. Protopopescu, ISAJET program, BNL 29777 (1981). All cross-sections are calculated in the leading log approximation assuming SU(2) x U(1).

[31]   F. Berends et al., Nucl. Phys. B202 (1982) 63, and private communications.

[32]   M. Jacob, Nuovo Cimento 9 (1958) 826. We thank Prof. M. Jacob for very helpful comments on the subject.

[33]   K. Eggert et al., Nucl. Instrum. Methods 176 (1980) 217 and 233.

[34]   For more detailed information, see for example : UA1 Collaboration, Search for Isolated Large Transverse Energy muons at $\sqrt{s}$=540 GeV, in Proc. 18th Recontre de Moriond on Antiproton-Proton Physics, La Plagne, 1983 (Editions Frontières, Gif-sur-Yvette, 1983),p.431.

[35]   G. Arnison et al., Phys. Lett. 118B (1982) 167.

[36]   Calculation based on : M. Banner et al., Phys. Lett. 122B (1983) 322.

[37]   In the maximum likelihood fit, the measured quantities of each event are compared with computed distribution functions, smeared with experimental errors. A Breit-Wigner form is assumed for the W mass [with a width (FWHM) of 3 GeV/c^2], and Gaussian distributions are used for the transverse

and longitudinal momenta of the W (with r.m.s. widths of 7.5 GeV/c and 67.5 GeV/c, respectively). In the W centre of mass, the angle θ^* of the emitted positive (negative) lepton with respect to the outgoing antiproton (proton) direction is generated according to a distribution in $\cos \theta^*$ of $(1+\cos \theta^*)^2$ as expected for V($\pm$ A) coupling.

[38] J.T. Carrol, S. Cittolin, M. Demoulin, A. Fucci, B. Martin, A. Norton, J.P. Porte, P. Ross and K.M. Storr, Data Acquisition using the 168E, Paper presented at the Three-Day In-Depth Review on the Impact of Specialized Processors in Elementary Particle Physics, Padua 1983, ed. Istituto Nazionale di Fisica Nucleare, Padova, (1983) p. 47.

[39] Electron-pion discrimination has been measured in a test beam in the full energy range and angles of interest. The muon tracks have the following probabilities : i) no interaction : 2×10^{-5} (4×10^{-5}) ; ii) interaction but undetected by the calorimeter and geometrical cuts : 10^{-4} (4×10^{-4}) ; iii) decay : 10^{-3} (0.7×10^{-3}).

[40] S.D. Drell and T.M. Yan, Phys. Rev. Lett. $\underline{25}$ (1970) 316.
F. Halzen and D.H. Scott, Phys. Rev. $\underline{D18}$ (1978) 3378. See also ref. 6.
S. Pakvasa, M. Dechantsreiter, F. Halzen and D.M. Scott, Phys. Rev. $\underline{D20}$ (1979) 2862.

[41] R. Kinnunen, Proc. Proton-Antiproton Collider Physics Workshop, Madison, 1981 (U. of Wisconsin, Madison, 1982):
R.W. Brown and K.O. Mikaelian, Phys. Rev. $\underline{D19}$ (1979) 922.

[42] T.G. Gaisser, F. Halzen and E.A. Paschos, Phys. Rev. $\underline{D15}$ (1977) 2572.
R. Baier and R. Rückl, Phys. Lett. $\underline{102B}$ (1981) 364.

F. Halzen, Proc. 21st Int. Conf. on High Energy Physics, Paris, 1982 [J. Phys. (France), No. 12 t $\underline{43}$ (1982)], p, C3-381.

F.D. Jackson, S. Olsen and S.H.H. Tye, Proc. AIP Dept. of Particles and Fields Summer Study on Elementary Particle Physics and Future Facilities, Snowmass, Colorado, 1982 (AIP, New York, 1983), p. 175.

[43] J.E. Kim et al., Rev. Mod. Phys. $\underline{53}$ (1981) 211.

HOW DOES THE STANDARD MODEL STAND UP TO THE REAL WORLD?

C.H. Llewellyn Smith
Oxford University, Oxford, England

TABLE OF CONTENTS

1 Introduction 299
2 QCD 299
3 SU(2) x U(1) 303
4 Beyond the Standard Model 312
5 Conclusion 318
6 Bibliography 318
7 Footnotes and References 318

HOW DOES THE STANDARD MODEL STAND UP TO THE REAL WORLD?

C. H. Llewellyn Smith
Oxford University, Oxford, England

1 INTRODUCTION*

I take the standard model to mean quantum chromodynamics, the theory of the strong force, and $SU(2) \times U(1)$, the partially unified theory of the electroweak force. The difficulty of discussing how well this model stands up to the real world is that everybody in the audience already knows the answer: there are no glaring failures or you would have heard about them. However, there are some minor difficulties, there are many parts of the model which are untested, and there are also parts of the model which we don't really understand theoretically. For example, we don't understand how QCD leads to the observed spectrum of hadrons, although nobody doubts that it does. Furthermore, the standard model has numerous theoretical shortcomings and it simply doesn't explain enough to be regarded as a complete final theory. In this talk I shall consider in turn QCD and $SU(2) \times U(1)$, recalling the original reasons for believing them, examining some of their successes and failures, and identifying further theoretical and experimental challenges which they pose. I shall then discuss some of the shortcomings of the standard model and give a brief overview of some of the cures that have been proposed. Finally I shall focus on the Higgs mechanism, which is the weak link of this model, and the related hierarchy problem, and I shall argue that its resolution requires new physics at energies below 1 TeV.

2 QCD

Quantum chromodynamics (QCD) is the theory of colored quarks interacting through the exchange of colored vector (spin one) bosons, known as gluons. The quark model is, of course, very well established, and there is also excellent evidence for the color degree of freedom, which is needed

(a) to allow us to make the Δ^{++} out of 3 up quarks in the same quantum state,

(b) to account for the cross-section for $e^+e^- \to$ hadrons, which is observed to be 3 times what it would be without color, and

(c) to account for the rate of $\pi^0 \to \gamma\gamma$.

Given color, it follows that the force between the quarks must be color dependent. From the three possible colors of quark and the three possible colors of antiquark, nine possible colors of pion can be constructed, whereas only one is seen. There must therefore be

---

*This is an edited version of the lecture given at the 3rd U.S. Summer School on High Energy Particle Accelerators. No attempt has been made to give a complete set of references; a very short bibliography is included at the end.

some dynamical mechanism which lifts 8 of those states to much higher energy, perhaps to infinite energy, relative to the color singlet pion that we observe. In other words, the forces must be color dependent. There is also very good evidence that these forces are mediated by spin one exchange. The evidence for this is the success of chiral symmetry. Chiral transformations are independent isospin rotations on the two different helicity states of the up and down quarks. Under a normal isospin rotation, the left-handed and the right-handed states are rotated identically. Chiral invariance corresponds to invariance under independent rotations of these states. If this symmetry were exact, it would lead to the list of predictions in Table I,[1] whose origin I am not going to explain.

Table I

	Prediction from exact chiral symmetry	Experimental value
$\dfrac{M_\pi^2}{M^2}$	0	0.03
$1 + \dfrac{2 M g_A}{f_\pi g_{np\pi}}$	0	0.08±0.01
$M_\pi^2 a_{\pi N}^{1/2, 3/2}$	0.16, −0.079	0.17±0.05, −0.088±0.0
λ_{Ke3}^o	0.021±0.003	0.019±0.004
$\Gamma(\pi^o \to \gamma\gamma)$	7.87 eV (for three colors)	7.95±0.55 eV

Comparing these predictions with experimental data, you see that they don't work exactly. However, they all work to something like 10% or better. I think you'll agree that this impressive agreement can't possibly be an accident. It tells us that the world is very close to the limit of exact chiral symmetry—it seems that if it were exact, the world would only be changed by less than 10%. Now, in the framework of the quark model the only way to get exact chiral symmetry is to have vector forces coupled to massless up and down quarks. Presumably in order to be within 10% of this world, quark binding must be due to vector forces and the up and down quark masses must be small.

We see then that the existence of quarks and color and also the facts that the forces are color dependent and are vector in nature are more or less dictated directly by the data. If you put together these ingredients you have QCD. This is the motivation for QCD; to me it's almost inconceivable that it could be wrong. What are the successes

of this theory? Probably the first success was the discovery, soon
after QCD was written down, that it's an asymptotically free theory.
This means that if we look at the force between say a quark and an
antiquark at a distance r and write it as $\alpha_s(r)/r$ then $\alpha_s(r)$, which
represents the square of the strong coupling constant or the strong
charge, decreases and goes to zero when the quarks approach each other
as $(\log r)^{-1}$. Conversely, $\alpha_s(r)$ grows at large distances. I won't
dwell on the origin of this effect, which was explained in the lecture
by Bill Marciano (see AIP Conf. Proc. No. 127). Any successful theory
of the strong force must have some property like asymptotic freedom.
We know that when you pull quarks apart, the force gets very strong,
perhaps infinitely strong, making it impossible to separate quarks.
However, the force must become weak when the quarks come close together, because we know that if we subject a quark to a strong im-
pulse, it responds at short distances essentially like a free non-
interacting particle. This is the basis of the highly successful
parton model. So the data show that the forces are strong at long
distances and weak at short distances, and QCD has this feature built
into it. Not only does QCD have this correct qualitative property,
but it enables us to say something quantitatively. Using the fact
that at short distances the coupling constant is weak, we can use perturbation theory to predict corrections to the parton model (for example the famous scaling violations seen in deep inelastic scattering)
and to make many other predictions which are in semi-quantitative
agreement with the data. It also allows us to predict new phenomena.
For example, the existence of three-jet events in e^+e^- anniliation--
in which the quark-antiquark pair radiates a hard gluon and the $q\bar{q}g$
system materializes as three jets--is a genuine prediction of QCD,
made before it was observed.

QCD has no real gold-plated successes--nothing like g - 2 for
QED--but it describes so much data satisfactorily that it's got to be
right. There are many semiquantitative successes and, more particularly, there are no real failures--although there are many places
where QCD could have gone wrong. That's not to say that we understand
everything: many things are not understood even qualitatively. For
example, in e^+e^- anniliation it's known that at PETRA energies baryon-
antibaryon production is about 0.7 of a $B\bar{B}$ pair per event, and nobody
has a model which explains this number.[2] That's not to say that it is
in contradiction with QCD. There's no reason to doubt that in principle it follows from QCD, but I don't believe anybody has the slightest
idea how to calculate this number at present.

So the evidence seems to be that QCD is correct but there are
many things we cannot yet explain. The challenges for the theory are
to explain these things and, indeed, to justify fully the parton
model. The outstanding challenges are to show that QCD quantitatively
reproduces the spectrum of known hadrons and that it leads to nuclear
physics (which I believe cannot be discussed accurately without reference to QCD), and to the spectrum of unknown hadrons, in particular to
the spectrum of the glueball states (whose quantum numbers are carried
purely by gluons and which are made only of gluons to first approximation) and also to the spectrum of hybrid states (which contain
gluons and quarks, the quantum numbers being carried by both). At the

moment we don't even have a qualitative picture of how these states are built. If you look at the literature you can find conflicting and incompatible suggestions for the spectrum of these states and also quite different physical pictures of what is going on. It is a major challenge to theory to understand these states, and also to experiment to identify them.

It has been claimed that progress has been made in deriving the spectrum of the known hadrons from QCD by using lattice gauge theories. In lattice gauge theory the fields are defined only at discrete points on a lattice in space-time, or on the links between them, instead of being defined everywhere as is done in classical electromagnetism or normal field theory. The hope is that if the distance between the points is small enough, much smaller than the system you are dealing with, then the discrete system will be a good approximation to the real world. Theoretically the point is that if fields are defined only at discrete points, the number of degrees of freedom is reduced to a finite number if you study a lattice whose overall size is finite, though much bigger than the hadrons you're interested in. You can therefore simulate the theory on a computer and get numerical results. This has been done, some of the most important work having been done here at Brookhaven. It's clear that we are going to learn a lot about field theory by studying it on a lattice. However, it seems to me that it is rather unlikely that we are really going to learn whether and how QCD leads to the known spectrum of hadrons in this way in the near future. The reason is simply a limitation of computer time. With a lattice in the three space and the one time dimensions, you soon get an enormous number of points to store in the computer. Typically people have used 8^4 lattice points, but the overall size of the lattice has been only about 1 fermi, the lattice spacing being about 0.12 fermi. You can argue that this gap between the lattice points is quite small compared with the size of a hadron, but the overall size of the lattice is actually less than twice the root-mean-square radius of the proton. So up to now the lattices that have been used are smaller than the system we're trying to study.
It's clear that it's going to be very difficult to remedy this difficulty since the amount of computing time goes up as the 4th power of the size of the lattice. Nevertheless in the long run this may be the route to understanding QCD.

So all seems well for QCD theoretically although we must still meet the main challenge, which is really to solve the theory and to find out how it works. What are the fundamental questions? The main question for QCD is: What sets the value of the parameters in the theory? The coupling constant is not a parameter. As already stated, it varies with distance or equivalently with energy. In fact

$$\alpha_s(E) = \frac{12\pi}{25 \ln E^2/\Lambda^2}$$

to leading order, so the value of the dimensionless strong coupling constant is determined by a quantity Λ which has the dimensions of energy. However, Λ is not really a parameter. In principle the ratio of Λ to the mass of a proton is calculable. In particular, ignoring heavy quarks, in the limit of exact chiral symmetry with zero up and down quark masses, this ratio is a pure number which can be calculated in principle. This means that what seemed to be a parameter of the theory (Λ) can be traded in for the mass of the proton, which is not a parameter as it can be used as the unit of energy. So the strong coupling constant is not in fact a parameter in quantum chromodynamics! The parameters are the quark masses, which have to be put in from outside QCD. The real question then in grander frameworks going beyond QCD is what sets the scale of the quark masses relative to Λ.

3 SU(2) x U(1)

Turning to the electroweak SU(2) x U(1) theory, I shall begin with the original motivation before I turn to its successes and some of the problems that it faces. The data on leptonic and semileptonic processes at low energy (muon decay, neutrino scattering, beta decay, etc.) can all be described by the following effective Lagrangian:

$$L_{eff} = \frac{4G}{\sqrt{2}} \left(J_\mu^+ J_\mu^- + (J_\mu^3 - \sin^2\theta_W J_\mu^{em})^2 \right) ,$$

$$\vec{J}_\mu = \sum_i \bar{\psi}_L^i \gamma_\mu \vec{T} \psi_L^i ,$$

$$\psi_L^i = \begin{pmatrix} \nu_e \\ e \end{pmatrix}_L, \begin{pmatrix} \mu \\ d' \end{pmatrix}_L \cdots ,$$

$$e_L = \left(\frac{1 - \gamma_5}{2} \right) e , \quad \text{etc.}$$

The word effective means that you just take the matrix elements of this operator and compare them with the measured experimental matrix elements without bothering about higher orders. So this operator is just constructed from the experimental data. G is the Fermi constant; $J_\mu^{+,-}$ are the isospin raising and lowering parts of the so-called weak isospin current $\vec{J}_\mu$ of which J_μ^3 is the third component; θ_W is a parameter; and finally J_μ^{em} is the ordinary electromagnetic current. The symbols, e, μ, etc., represent the fields of the electron, up quark, etc., with d' being the Cabibbo-rotated down quark, and L stands for the left-handed states projected out by $1 - \gamma_5$. Note that in the limit of $\sin^2\theta_W = 0$, the Lagrangian is proportional to $\vec{J}_\mu \cdot \vec{J}_\mu$. The most obvious feature of this Lagrangian is that it has some sort of SU(2) invariance.

I said that the data lead to this Lagrangian, but of course data have errors, and you have to ask up to what accuracy this is true. After all, if there were no data at all it would be compatible with L_{eff}! In fact it fits the data to 10% or better, i.e., allowing for the errors, you can add extra terms of strength something like 10% in amplitude depending on what the terms are. To show this I give 3 examples of modifications of that Lagrangian and ask: What strength could they have?

(1) Let $(J_\mu^3 - \sin^2\theta J_\mu^{em})^2$ be multiplied by a free parameter ρ, to be adjusted to fit the data. One finds $\rho = 0.992\pm.017$, so it's true that you could put $\rho = 1$ but you're allowed something like a 3% deviation.

(2) Add a term $4GC(J_\mu^{em})^2/\sqrt{2}$. Many alternative theories actually suggest that such a term might be there, so it's interesting to limit it experimentally. The data give $C < 0.02$ at 95% confidence level. That seems a very good limit (2%) but if you compare it with the strength of the $(J_\mu^{em})^2$ term that's there anyway, which has $\sin^4\theta_W \simeq 0.05$ in front, it doesn't look so good.

(3) You can ask how sure we are that the weak interactions are purely left-handed. Imagine adding an extra charged current term with exactly the same structure as the term written except that it's right-handed (i.e., wherever here you have V - A, you add an identical term but with V + A) and introduce a parameter η in front of it. The low energy data tell you that η has to be less than something like 0.1.

So L_{eff} describes all the semileptonic and leptonic data--deviations are allowed but they can't be more than something like 10%, depending on exactly what they are. So far I've mentioned only the leptonic and semileptonic data, but L_{eff} contains currents that change (say) an up quark into a down quark or an up quark into a strange quark. Therefore it contains operators which are capable of producing nonleptonic decays ($\Lambda \to N\pi$, etc.). One can ask therefore whether it accounts for the nonleptonic decays as well. That's a hard question to answer because in order to calculate the matrix elements between hadrons you have to understand everything about QCD and about the structure of hadrons. It does have all the right quantum numbers in it; it satisfies all the observed selection rules. However, the magnitude seems to be difficult to understand. There is a famous problem that naive estimates fail, especially for amplitudes which change isospin by a half unit (the failure is a factor of 20 in amplitude). However, in QCD with more care you can come some way towards understanding this factor. My own feeling is that this is not a problem for the weak interactions but rather one of the things that we have yet to understand in detail about the structure of hadrons in QCD.

The charged current part of L_{eff} has been well established for about 25 years. The neutral current part has been established for about 5 years. Theoretically speaking, why were we not just content with L_{eff} in the past? The reason is if you treat it seriously as a Lagrangian, you must consider not just the Born approximation, but also higher orders, and as soon as you do that you find nonsense. You get divergences--meaningless infinities which you can't sweep under the rug. Furthermore, even if you forget about the higher orders and just stick with the lowest order, you discover that L_{eff} predicts amplitudes which shoot up linearly with the center of mass energy and at rather small energy (around 300 GeV in the center of mass) come into conflict with the unitarity. So at most L_{eff} could be an effective theory which must have a very limited domain of applicability. It's been known for a long time that the cure for this is to imagine that the current-current interactions in L_{eff} are mediated by the exchange of massive vector bosons (the W and the Z) so that a process involving 4 fermions has an amplitude like $g^2/(q^2 + M_W^2)$. At low energies, the four-momentum transfer squared, q^2, is small, and the amplitude behaves as g^2/M_W^2, which has to be identified (up to a constant) with the Fermi constant. Thus if you're probing distances that are large compared with M_W^{-1}, the interaction will appear to be point-like. However, at high energies the propagators will come in and ameliorate the difficulties that we ran into before. Furthermore, to make the theory perfectly OK theoretically, you have to believe that there are WW self-interactions such as are built into gauge theories. At this point you find that the simplest theoretically satisfactory theory which reduces to L_{eff} and therefore fits the data at low energy is the SU(2) x U(1) gauge theory.

In SU(2) x U(1) the masses of the W and Z are predicted to be of the order of 80 or 90 GeV respectively. Up to six months ago all of this was just theoretical speculation but now of course the W and Z have been found with essentially the predicted masses. I'll come back later to precisely what the predictions are and how well they compare with the measured values. So it seems the theory is right, but we still have to ask the question: Are there are any bits of data which the model cannot describe? There do seem to be some minor difficulties. Next we have to ask: Is this the only model that fits these facts? Are there viable alternatives? We'll see that the answer is yes. Finally, we have to ask: What further tests are needed to eliminate these alternatives and really to confirm all the ingredients of this model?

I turn now to possible failures of the standard model. I know of only two as of today. The first is the observation in the 1979 beam dump experiments at CERN that the production of prompt ν_e's is less than the production of prompt ν_μ's:

Group:	ABCLOS	CDHS	CHARM
ν_e/ν_μ ratio (1979 expt)	$0.56^{+.35}_{-.21}$	$0.64^{+.22}_{-.15}$	$0.48^{+.12}_{-.10}$

The idea of these experiments is that you dump the proton beam into a thick target right in front of the shield so that the π's and K's that are produced cannot decay but are just absorbed. The only neutrinos produced are so-called prompt neutrinos, from the decays of heavy states which have very short lifetimes. In the standard model, there is μ-e universality, which means that heavy objects with short lifetimes should decay universally and that the ratio of prompt ν_e's to prompt ν_μ's should be equal to one. The results of the three different 1979 experiments from CERN (in the same beam) are not compatible with 1. As far as I know, this phenomenon has no explanation in the standard model and would mean that we have to add some new ingredient or in some way go beyond the standard model. People have proposed various things (for example, the existence of new heavy objects with peculiar decays), but I don't even know of any ad hoc phenomenological explanation of these data which really holds water and does not get into trouble somewhere else. On the other hand, although we seem to have three measurements in impressive agreement, it's important to remember that they used the same beam; one of the things you could fear is that the beam scrapes the beam pipe a little bit and produces π's and K's which decay predominantly into ν_μ's. However, at the moment I know of no reason to doubt these measurements. They have no explanation in the standard model, and this should be grounds for me now to take off and speculate on some alternatives. However, I'm not going to do that for the simple reason that there have been new experiments at CERN the results of which are expected to be announced at the Brighton Conference the week after next. So it doesn't seem like a very fruitful time to speculate when the rug may well be pulled out from under my feet. (The new results have confused the issue--see Ref. 3.)

The other fact that I know of which seems to be in conflict with the standard model is the rate of production of same-sign dimuon's by neutrinos:

$$\frac{\nu_\mu A \to \mu^- \mu^- \ldots}{\nu_\mu A \to \mu^- \ldots} \lesssim 10^{-3} .$$

It's been known for a long time that neutrinos produce $\mu^-\mu^+$ at the 1% level relative to single μ's; these events can be well explained by the production of charm particles which decay into μ^+. The charm model works very well and explains the rate and all the features of these events. On the other hand, although mechanisms do exist in the standard model to produce same-sign dimuons, it seems impossible to get the rate to be as big as 10^{-3}. Various mechanisms have been considered:

(1) The associated production of charm particles, for example from gluon radiation, as shown in Figure 1.

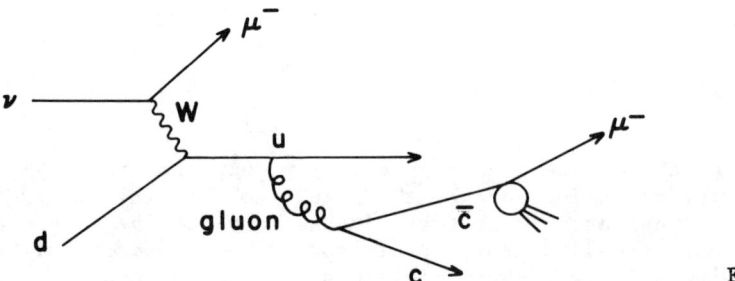

Fig. 1

However, this diagram gives a rate which is something like a factor of 500 lower than what's observed experimentally, and furthermore the characteristics of events due to this mechanism are quite different from those that are observed. You can try to patch this up by saying that you don't believe the gluon bremsstrahlung model and make some ad hoc phenomenological fragmentation model fit to the data.[4] Such a model is not ruled out by other neutrino data. However, it may possibly conflict with the cross-sections for muons and for hadrons to produce charm, and a rate of fragmentation to charm 500 times as big as the gluon bremsstrahlung model would be very hard to understand. Furthermore, since giving the talk, I have learned of limits on associated $c\bar{c}$ production from neutrino emulsion experiments[5] which seem to rule out this explanation.

(2) Another mechanism which must exist in the standard model and which gives rise to same-sign dilepton events is D^0-$\bar{D}^0$ mixing. The ordinary weak current produces charm particles and sometimes these might be D^0's which can mix, turn into $\bar{D}^0$'s, and then decay to a μ^-. The difficulty is that the amount of mixing expected in the standard model gives a rate which is too small by a factor of about 10^6. Taking a pragmatic attitude and treating the amount of mixing as a parameter doesn't work. The EMC experiment sees one $\mu^+ N \rightarrow \mu^+\mu^+\mu^+ \ldots$ event in a sample with 83 $\mu^+ N \rightarrow \mu^+\mu^+\mu^-$ events. If you interpret this event as being due to mixing (and there are other things that it could be due to), you get an empirical limit on mixing, and you predict that $\nu_\mu \rightarrow \mu^-\mu^-/\nu_\mu \rightarrow \mu^-\mu^+$ should be less than 0.03, whereas the experimental rate is about 0.1.

(3) A final possible mechanism is through the production of b quarks, from a diagram like Figure 2.

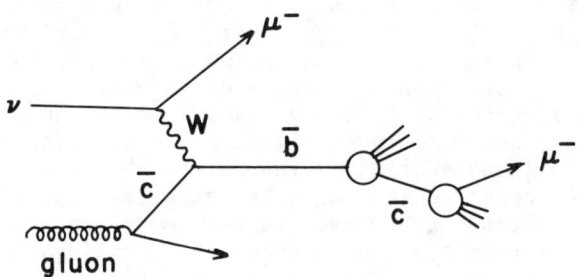

Fig. 2

This must happen at some level. However, the rate is too small by a factor of 100. (Since this talk the Mark II and MAC groups have reported[6] a b lifetime of order 10^{-12} sec. This implies a very small b-c coupling (either left- or right-handed) and completely rules out b production as a source of the same-sign dimuons.)

The upshot is that the standard model has mechanisms for producing same-sign dimuons but none of them can convincingly come within 1 or 2 orders of magnitude of explaining the data. However, given the difficulty of these experiments and the fact that this is just one piece of data out of the many hundred that the standard model does fit, I don't think it means that the standard model is necessarily wrong--although it is something we should worry about.

I shall now address the question of whether the standard model is a unique way of explaining the data or whether there are other viable models. One possibility which is quite popular is the idea that the vector bosons are not elementary gauge theory objects, as in the standard model, but are composites.[7] If leptons and quarks are composites of spinors (—) and scalars (---), then the weak interactions could be due to constituent exchange, as shown in Figure 3.

The constituents must have strong forces that bind them together, and the pair of particle-antiparticle constituents that are exchanged could be bound to some sort of vector meson that looks something like

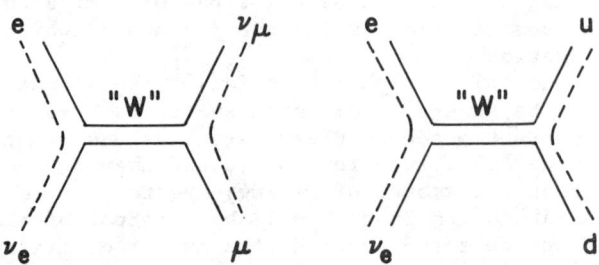

Fig. 3

the W. The weak interactions will be controlled at low energy by the spectrum of the states that can be exchanged. Suppose that the spectrum is dominated by a single isotriplet of objects W which couple to just the left-handed particles. The constituents of the W must be charged, so that the W^0 will couple directly to the photon and there will be some sort of mixing. This mixing automatically pushes up the mass of the $W^0(Z)$, and the model reproduces the results of the standard model at low energy. You may say this is not surprising--you put in an isotriplet of W's, isn't that just the standard model? Well, no, it's not, because in the standard model you put in the $\vec{W}$, the triplet of vector bosons from $SU(2)$, and you also put in a single vector boson B from the $U(1)$ theory, i.e., you put in two elementary neutral objects. However, in these composite theories the W^0 is not elementary, it's a bound state, although the photon is elementary. In terms of the new constituents the W is rather analogous to the ρ meson in the context of the quark model. In this model it can be argued, by using vector dominance type ideas about the couplings of the W's to the leptons and the quarks, that the charged W's and the neutral one, which I now call Z, would have approximately the standard mass.

So composite models can naturally more or less reproduce the successes of the standard model. However, they would not agree exactly, and this points to the importance of testing the standard model to really high accuracy, which I'll come back to later. First of all, the masses won't be exactly the standard ones. In the standard model, one can calculate the mass of the W and the Z in terms of other parameters perturbatively, just as you can calculate something like g - 2. Composite models will certainly not give exactly the same results. A second point concerns a possible difficulty for these models. The constituents are supposed to bind together to make the analog of the ρ, giving an isotriplet of W's which is supposed to dominate the spectrum. Why is there no analog of the ω? Well, advocates of these models wave their hands a lot and say that for some reason it's much heavier. However, presumably there is some sort of isoscalar, and I think you should expect the low energy isoscalar neutral current couplings to differ somewhat from those in the standard model. Furthermore, you would not expect that ρ parameter, which I discussed before, to be exactly 1, nor that coefficient C of the extra $(J_\mu^{em})^2$ term that I added to the Lagrangian to be exactly zero. So these models can approximately produce the standard model, but they won't produce it exactly, and they point to the importance of testing the standard model as accurately as possible.

Not only are there non-gauge theories that still fit the facts, there are other alternative gauge theories. For example, Barger et al.[8] have recently looked at an $SU(2)_L \times SU(2)_R \times U(1)$ model which fits the facts but is nevertheless basically quite different from the standard model. In such models there are two neutral W's, one coupled to left-handed and one to right-handed currents, which mix with the neutral U(1) boson, so that one ends up with the photon and two Z's. It turns out that it's possible to arrange, by artifically fiddling the parameters, that the lighest Z is identical to the one of the standard model, so that by doing experiments on resonance at LEP or SLC you will never tell this theory from the standard model. However, there is a second Z at 200 GeV. This is achieved by exploiting the possible 10% deviation from the atandard model at low energy. As you go up in energy and approach the first Z, these deviations will get less and become zero on resonance! Nevertheless the second Z can be as low in energy as 200 GeV, so that it might be produced at Fermilab in 2-TeV collisions of protons and antiprotons. So there still is scope to exploit the deviations at low energy in order to make models which, although they might look very standard at SLC, will look very nonstandard at maximum LEP energies. Again this shows the importance of making high precision tests, getting below the 10% level at low energy.

Having stressed that we need further tests of the theory, I'd like to examine what these tests might be. First of all, I shall consider various ingredients of the model which are not tested even qualitatively, and then I'll come to precision tests. The most obvious missing ingredient is of course the top quark, which has got to exist in the standard model. We can only hope that it's discovered soon at the CERN $p\bar{p}$ collider.

The next thing I want to discuss is the origin of CP violation. Let me remind you that so far we've seen CP violation only in the K^0-$\bar{K}^0$ mass matrix. It is therefore possible that CP violation is a purely $\Delta S = 2$ effect which is due to some new superweak interactions. However, this is certainly not right in the standard model, in which CP violation arises from a phase in the mixing matrix that describes the mixtures of quarks that appear in the weak interactions in terms of the mass eigenstates. This generates CP violation in $\Delta S = 1$ processes. Thus there is CP violation in the decay amplitude for $K \to \pi\pi$ which makes the standard parameter $\varepsilon' = \frac{1}{3}(\eta_{+-} - \eta_{00})$ non-zero, in contrast to the value of zero given by the superweak model. It's been shown recently by Gilman and Hagelin and others that ε'/ε is expected to be greater than 1/500, the present limit being about 1/50. So it will obviously be extremely interesting to go to higher precision CP violation measurements and really see whether there is CP violation in the amplitude as there must be in the standard model. (The long b lifetime, reported since this talk, makes it possible that ε'/ε is as large as 0.02 or more.[9])

Of course, we'd like to test not just that the W and the Z exist with the right masses, but that they have the right sort of properties. One of the most obvious things you could consider is their decay branching ratios. With just the known quarks and leptons plus the top quark, we can predict everything about the branching ratios so we can ask whether the data allow new channels. Suppose that there exist very heavy charged leptons--let's say with masses of 500 GeV, so we're not going to discover them in the next 10 years. The corresponding neutrinos might be very light, maybe even massless, in which case they would show up in Z decay and change the branching ratios. If the only ingredient you add to the model lighter than the W and the Z comprises new neutrinos, each neutrino contributes 6% to the branching ratio so you change the branching ratios from the standard ones by $(1 + 0.06N)^{-1}$, where N is the number of these neutrinos. The present data already show[10] that the total number of neutrinos is less than 18 at 90% confidence level. This may not seem a very good bound, but the fact is that from particle physics the best bound we had previously on the number of extra neutrinos was about 1000! Of course, there are limits on the number of neutrinos from cosmology but those assume that the neutrinos are very light. If they weigh 1 or 2 GeV then the cosmological bounds don't apply, and the limit from Z decay is a totally new result, much better than existed before. Also from the collider we might learn whether there are other new things into which the W and Z can decay. For example, if there is a new charged lepton lighter than the W, then the W will sometimes decay into the heavy lepton and its neutrino. The heavy lepton will decay into, say, an electron or a muon or a tau, and one can try and see these events. So W and Z decays provide a window onto new particles such as new quarks or leptons or supersymmetric particles.

Another thing which we'd really like to test is the gauge nature of the W and the Z. As I said before, in a gauge theory there is a characteristic three-boson coupling between the photon and the W and between the Z and the W, which has to have a very specific form in order to render the amplitudes well behaved at high energy to stop them crashing through the unitarity barrier or, if you like, to make the theory renormalizable. Undoubtedly the cleanest way to study these couplings is to measure $e^+e^- \to W^+W^-$. But that's tough to do and we haven't yet got the money out of the taxpayer. Another possible way is to look at $e^+e^- \to W^-\nu e^+$ in the first stage of LEP. In the LEP study we looked at the rate for this process to ask if the W could be discovered in this way and we found that it was really marginal. However, it was assumed that the W was the standard one. If the coupling is nonstandard, the cross-section may be bigger. I don't know how much bigger because I don't think the calculation has been done,[11] but it seems to me quite likely that, knowing that the W exists, one can use the nonobservation of this process to put a useful limit on the difference of the coupling from the gauge theory value. Another conceivable way of looking at the gauge coupling is to look at $p\bar{p} \to W\gamma$ ... . With the particular coupling predicted by gauge theories, there is a characteristic dip in the angular distribution. However, at the CERN machine you need[12] an integrated luminosity of something like 3×10^{38}, which requires 10^5 days running at the present luminosity!

The final ingredient which we'd like to check in the standard model is the existence of the Higgs boson. I'll come back to the theoretical status of the Higgs mechanism later. Now, I shall consider only the phenomenology. The mass of the Higgs boson is not predicted.

The only thing that is predicted is that its coupling to a particle X is always proportional to M_X. So its coupling to the top quark is proportional to M_t and to the τ lepton is proportional to $M_τ$. This has the consequence that if you can produce the Higgs boson, it will decay mainly to the heaviest particles available. If it's heavier than $2M_W$, it will decay into 2W's. If it's not so heavy, it may decay into $t\bar{t}$. If it's lighter, it will decay into $b\bar{b}$, and so on. So the signature is to see it decaying into heavy particles. Also for this reason it is produced in association with, or by coupling to, heavy particles. It might be found:[13]

(1) In $t\bar{t}$ ("toponium") → $H^o γ$, which would produce a monoenergetic photon in coincidence with the debris of the H^o decaying into heavy particles. It seems that with LEP you can get limits on M_H up to something like 80 or 90% of the toponium mass.

(2) In $Z → H^o \bar{e}e, H^o \bar{μ}μ$. In this way you can go up to something like 45 GeV at LEP.

(3) If the Higgs is much heavier, the best way is probably to go to superhigh energy e^+e^- collisions and look for $e^+e^- → ZH^o$, which contributes something like a unit to R so it's very easy to see, but of course the difficulty is to get high enough energy.

Finally I want to discuss high precision tests of $SU(2) \times U(1)$. This is something that I've worked on myself but I'm not going to spend much time on it because Bill Marciano has also worked on it a lot and will talk about it in his lectures (see AIP Conf. Proc. No. 127). The model should fit all the data ($νe → νe$, $νN → νX$, $ep → eX$, $\bar{e}e → \bar{μ}μ$, M_W, M_Z, etc.) with essentially three parameters--$α$, the fine structure constant, Fermi's constant, and $\sin^2 θ_W$. I say "essentially" because the calculations also depend, although not very sensitively, on the mass of the top quark and on the mass of the Higgs boson. How well do we do with present data? One way of asking this is to take different quantities (say $νN → νX$, the parity-violating asymmetry observed at SLAC in electron-deuteron scattering, M_W, and M_Z) and use them to construct a value of $\sin^2 θ_W$. The electroweak theory gives the values in Born approximation, shown in the first column of Table II. However, when you include the next-order corrections, you discover that they're capable of changing these values by amounts that are comparable with the experimental errors here. Until you've calculated them, it's not clear that the second-order corrections aren't going to be bigger than the errors; in fact, it's surprising that the uncorrected values agree so well, and you get worried that the corrections may drag them apart! When you do the higher-order corrections and extract the value of $\sin^2 θ_W$ to second order, the result depends on how $θ_W$ <u>is</u> defined. The values in the second column are for the so-called $\overline{MS}$ scheme. However, this doesn't matter: whatever definition you use, you've got to get agreement! You see that it's not the case that the numbers are dragged apart. You also see from Table II that we're getting to the point where the errors are such that we can test the theory up to second order.

Tests that work to second-order accuracy are the analogues for $SU(2) \times U(1)$ of the Lamb shift for QED. The whole point of these renormalizable theories is to enable us to calculate higher-order quantum corrections (loop diagrams), so we want to test that the the-

Table II

Experiment	$\sin^2\theta_W$ (Born approx)	$\sin^2\theta_W(M_W)$ (with 1st-order corrections)
$\nu N \to \nu X$	0.227±0.015	0.215±0.015
$e_L d - e_R d$	0.223±0.015	0.215±0.013
$M_W = 81 \pm 2$	0.212±0.011	0.226±0.012
$M_Z = 95.2 \pm 2.5$	0.189±0.12	0.207±0.013

ory does give them correctly. Just as in testing QED where we don't want only the Lamb shift but also g - 2, etc., as well, we want to test SU(2) x U(1) in as many different ways as possible because the possible deviations we could get, from nongauge structures or from possible additional contributions from new particles, are different in different processes. Possibly a more dramatic way of saying that we are now sensitive to second-order effects is to use neutrino data to predict M_W and M_Z. If we analyze the neutrino experiments in Born approximation and we then calculate using the lowest-order mass formulae, we get M_W = 78.2±2.6 GeV, M_Z = 89.0±2.1 GeV. On the other hand, if we include the higher-order weak electro-weak corrections when analyzing the experiments and use the second-order mass formulae, we get M_W = 83.1±3.0, M_W = 93.8±2.4. We see that the shifts are getting comparable with or bigger than the errors. We are therefore getting to the point where we can tell the difference between gauge theories and nongauge theories, and the data are becoming sensitive to new contributions, for example from heavy quarks or from new light particles, such as those that exist in supersymmetric theories.

4 BEYOND THE STANDARD MODEL

The standard model works and is probably true. Are we then in the position that Lord Kelvin thought we had reached nearly 100 years ago when he said (just before J.J. Thomson discovered the electron) "There is nothing new to be discovered in physics now: all that remains is more and more precise measurement."? I would answer emphatically--no. The standard model cannot be the final theory. It simply doesn't explain enough. It has too many parameters, especially those associated with the Higgs mechanism, and too much arbitrariness. In particular it's got three coupling constants for the three forces. It doesn't explain charge quantization. It's got several arbitrary parameters associated with the Higgs boson, which determine the mass of the W and the Z. It's got a parameter you must put in for every quark mass that you get out, and a parameter for every Cabibbo-like mixing angle. When you count them up you find that the theory has something like 25 parameters, which are just put in and adjusted in an ad hoc way to explain the data. Surely that is not good enough.

What are the problems and the things we'd like to improve? Presumably we'd like some sort of further unification between the three forces. We'd like to address the problem of families: that is, the problem of why the muon exists, or why Nature is not more parsimonious and is not just content with the electron and its neutrino and with the up and the down quarks, which seem to be all that it needs to make normal matter. We'd like to understand the origin of mass. Is it really the arbitrary, ugly, complicated Higgs mechanism, or is it something else? We'd like to incorporate gravity. If we find a non-unified model based on a larger group, we'd like to know what dictates the choice of that group.

I'm certainly not going to discuss all the ideas that have been proposed in thinking about these problems. I'll just concentrate on a couple of specific questions. First of all, for further unification the obvious thing is to go to a grand unified theory which unifies $SU(3)$, $SU(2)$, and $U(1)$. However, grand unification answers only the one problem which it was designed to answer and it introduces a new problem of its own, the so-called hierarchy problem. The hierarchy problem is that if you introduce new physics at some higher mass scale (Λ), there's a tendency for the W mass to get a little piece of this mass mixed in ($\delta M_W \sim g\Lambda$). So if you don't want to mix something bigger than the observed value into M_W, you are led to the conclusion that, if there is new physics, the associated scale has to be less than or of order 1 TeV. I'll return to the hierarchy problem later.

The idea in grand unification is that we should take the 1, 2, 3 gauge theories and put them into a single grand theory based on a group G which describes all forces as different manifestations of a single underlying force. In such a theory there is fundamentally no difference between strong and electroweak forces. In this case, it must also be true that there's no real difference between quarks and leptons or between hadrons and leptons. After all, what's the definition of a hadron? A hadron is something which feels the strong force and a lepton is something which doesn't feel the strong force. But if the strong force is no different from the other forces, then this definition doesn't make any sense. So in grand unified theories there's no real distinction between quarks and leptons, and it's very natural to put them into the same families of the gauge group. Once you put quarks and leptons into the same family of the gauge group then there will be symmetry transformations which turn quarks into leptons, and in a gauge model this means that some of the new gauge bosons (those not associated with the subgroups $SU(3)$, $SU(2)$, and $U(1)$, which are called generically X) couple quarks to leptons and may mediate proton decay.

How do these theories work? How can we combine the strong force, which is after all strong, with the electroweak force which is weak? The trick, of course, is asymptotic freedom. As we go up in energies, the strong coupling goes down very quickly, and it's possible that above the mass of the heavy vector boson X, where the symmetry will effectively become exact, it merges with the $SU(2)$ and $U(1)$ couplings. If you work in a crude approximation in which you say that above the mass of the X it's a good approximation to set $M_X = 0$ and below it's a good approximation to set $M_X = \infty$, then the coupling constants should meet at M_X. The evolution up to this point will be controlled purely by the 3, 2, 1 theory and you can see whether this idea works without knowing what the group G is. It turns out that the couplings do meet, at a point around 10^{15} GeV, or, putting it the other way around, if we demand that they meet we can predict a relationship between two of them at low energy. The prediction is that $\sin^2\theta_W$ is of order 0.21. This is of course quite right, which is strong support for some sort of grand unification.

To do better we have got to specify what the grand unified theory is, and the simplest choice is SU(5). We can then calculate $\sin^2\theta_W$ more or less exactly and here SU(5) scores its great triumph. It gives a number $\left(\sin^2\theta_W(M_W) = 0.215 \pm 0.006\right)$ which exactly agrees with the measured value. We can also calculate the mass of the X boson more or less exactly, apart from some uncertainties in some of the ingredients, in terms of the mass scale of QCD:

$$M_X = (1.3^{+.9}_{-.6})\Lambda_{\overline{MS}} \times 10^5 .$$

We then then calculate the lifetime of the proton:

$$\Gamma^{-1}(p \to \pi^0 e^+) = (6^{+90}_{-?})\left(\frac{\Lambda_{\overline{MS}}}{350 \text{ MeV}}\right)^4 \times 10^{30} \text{ years}.$$

I've pushed the error to be as big as possible,[14] and I've normalized $\Lambda_{\overline{MS}}$ to the biggest value I dared, which is around 350 MeV. The latest limit for this mode is 10^{32} years, which means that the simple SU(5) model seems to be ruled out unless Λ is even bigger than the value I've taken and there's a real conspiracy of all the various uncertainties that went into the error, pushing them all in the same direction up to the top end. However, it's easy to construct other grand unified theories which can accommodate the value of 10^{32} years or longer, although you then lose the very precise prediction of $\sin^2\theta_W$ and open up a Pandora's box of any number of new theories. My gut conclusion is that grand unification is a very nice idea and there is some evidence that the coupling constants come together at high energies. However, if we abandon the minimal theory, the predictions of the proton lifetime become very imprecise and it's going to be very hand to test grand unified theories.

Let me now return to the weak link of the electro-weak gauge theory--the Higgs system. In a locally gauge invariant theory, it seems that the mass of the W and the Z are zero, which is certainly wrong experimentally. Furthermore if the gauge theory is $SU(2)_L \times U(1)$, the fermion masses would have to be zero. From a theoretical point of view the role of the Higgs system is to allow us to escape from these seeming consequences of local gauge invariance by hiding the gauge symmetry and somehow allowing the generation of masses. The way this works is that a scalar Higgs field (ϕ) is introduced which has some sort of potential: $V(\phi) = -\mu^2\phi^2 + \lambda\phi^4$ such that at the minimum of V, which represents the ground state or vacuum of the theory, φ is non-zero. Consequently, the vacuum is not an eigenstate of the gauge transformation and this hides the symmetry. This has the result that you can generate a mass for the W via the mechanism shown in Figure 4.

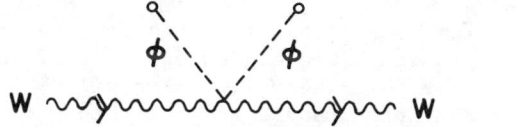

Fig. 4

The propagating W can interact with the φ (which is in fact a weak isodoublet) in an isospin, gauge-invariant way, but the φ can then just vanish into the vacuum because its vacuum expectation value is non-zero. This generates inertia for the W and hence generates a mass. At the same time φ can couple left-handed to right-handed fermions, which is what you want to get a mass because a mass is something which connects the left- and right-handed states. It can do this in a gauge-invariant way and then by vanishing into the vacuum it can generate a fermion mass, as shown in Figure 5.

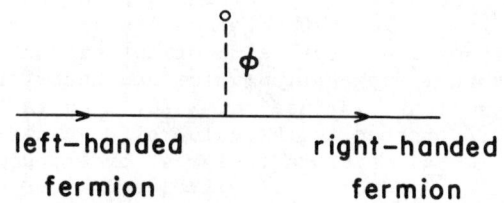

Fig. 5

If you believe in local gauge invariance, then you've got to have the Higgs mechanism in order to generate a mass for the W and the Z and for the fermions. On the other hand, you could take a more robust view and say to hell with local gauge invariance, there's nothing sacrosanct about that. Let's just put a mass in by hand and see what happens. By seeing what goes wrong, we can find the physical role of the Higgs boson or, if you prefer, the role of gauge invariance. First of all we find that higher-order quantum corrections get out of control and the theory becomes non-renormalizable. But that's a fancy theoretical concept--can we find some physics that goes wrong? Consider WW scattering for longitudinally polarized W's. If you look at the lowest-order amplitude for this process in the theory with the mass put in by hand, with no Higgs bosons, and take the center of mass energy $\sqrt{s}$ to infinity at fixed angles (s/t fixed), you find that the amplitude grows like s and that this growing amplitude runs into conflict with a rigorous bound from unitarity (from the conservation of probability) at something like 1 TeV in the center of mass. This doesn't necessarily mean that this theory is wrong, but what it does mean is that at 1 TeV perturbation theory fails: nonperturbative phenomena must set in and something new, which we're not at the moment able to calculate, must start. Now with the Higgs mechanism, there is a new diagram in which the Higgs boson is exchanged. This generates an amplitude which nicely cancels the piece which grows like s and makes the theory well behaved up to much higher energies.

So there are three alternatives:

(1) There is no Higgs boson and something else generates masses. In this case new phenomena must come in at or below 1 TeV in the center of mass.

(2) There is a Higgs boson, but its mass is greater than 1 TeV. In this case its contribution isn't big enough to perform the required cancellation except at very high energy. There will be a region around 1 TeV where we run into trouble and the theory must become a strong coupling theory. Some new phenomenon must therefore occur at around 1 TeV.

(3) The Higgs boson really is there below 1 TeV. The whole system is ugly, artificial, and complicated, but there's no fundamental problem provided SU(3) x SU(2) x U(1) is the end of the story. However, if the standard model is not the end of the road and there is new physics at some scale Λ, then there's the famous hierarchy problem (unless Λ is below 1 TeV). So in all cases you come to the conclusion that there's a reasonable chance that something new is going to happen below 1 TeV.

Let me finish by returning to the hierarchy problem. Recall that in the potential $V(\phi)$, which generates masses by allowing ϕ to develop a vacuum expectation value, the value of the parameter μ, which sets the scale of the vacuum expectation value and therefore sets the scale of the weak interactions, is just put in by hand in the classical theory. If you're interested in the value of μ which is relevant to physics at a momentum scale p, then to lowest order, $\mu(p^2) = \mu_0$, the bare (classical) value. However, it gets quantum corrections. Symbolically they are generated as shown by diagrams like Figure 6.

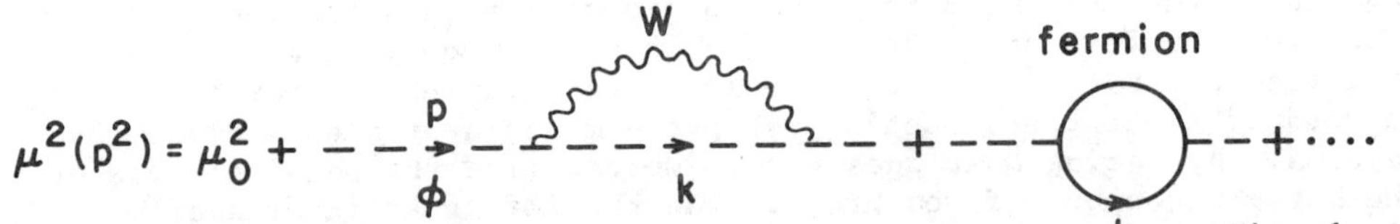

Fig. 6

The first thing you find is that when you calculate these quantum corrections, which involves integrating over the unobserved momenta (k) of virtual particles, you get a divergent integral. This is the familiar divergence problem of field theories which arises because we were too ambitious in trying to calculate the observable quantity μ in terms of a bare quantity. The relationship involves integrating up to infinite energy, but of course, when you go to infinity, presumably the integrand is all wrong so naturally there is trouble. The solution of this problem is the familiar one: you should be less ambitious.

Instead of trying to calculate in terms of some bare parameter, you can calculate the value at a scale p in terms of a given value at scale Λ by considering $\mu^2(p^2) - \mu^2(\Lambda^2)$. Taking the difference, the unknown bare parameter μ_0 cancels out. Very roughly

$$\mu^2(p^2) = \mu^2(\Lambda^2) + g^2 \sum c_i \int_{\max(p^2, m_i^2)}^{\Lambda^2} dk^2 + \ldots$$

where the omitted terms grow at most logarithmically for $\Lambda \to \infty$. I have assumed that $\Lambda \gg m_i$, the masses of virtual particles in the loops; effectively the integrals run from whichever is the larger of p^2 and m_i^2. Now you see the hierarchy problem. Suppose the scale Λ at which the parameter μ is given, by whoever wrote down the more fun-fundamental theory, is the grand unified scale of 10^{15} GeV. Then we've got an integral which runs from the pathetic energy that we're at now (a few 100 $(GeV)^2$ or less) up to 10^{15} $(GeV)^2$. So there's a contribution of order 10^{30} $(GeV)^2$. On the other hand we want $\mu(M_W)$ to be of order M_W. So we've got to cancel the contribution of order 10^{30} $(GeV)^2$ by putting in the parameter $\mu^2(\Lambda)$ with a value which has to be adjusted to something like 26 decimal places! This is the hierarchy problem. New physics at a scale Λ tends to drag μ up to the scale Λ through the integral. To get $\mu(M_W) \sim M_W$ naturally, we must demand that somehow the quantum corrections are not much bigger than the value that we want. Otherwise we'll have to invoke miraculous cancellations. There are basically four ways known in which this might happen:

(1) The new physics itself may be at a scale less than 1 TeV.

(2) The scalar particles (ϕ) could develop strong self interactions at around 1 TeV in the center of mass which might cut off the integrals. If there are strong $\phi\phi$ interactions at around 1 TeV, this would mean that the longitudinal part of the W (which in the standard model is just given by a component of the Higgs field) will become strongly interacting and funny things would happen in $e^+e^- \to W^+W^-$ at 1 TeV.

(3) The ϕ fields, and hence also the longitudinal W, might not be fundamental particles at all, in which case we shouldn't have been calculating the diagrams above. This is the so-called technicolor picture. It has a lot of technical difficulties but it's a very nice idea. Again in this picture the scale on which new things have to happen is around 1 TeV.

(4) The most fashionable possibility is that there may be some sort of conspiracy so that different contributions cancel out ($\sum_i c_i = 0$) and μ is not sensitive to Λ for Λ very large. That's what happens in supersymmetric theories. The trick is that, because of the fact that in a virtual fermion loop there's a minus sign relative to the contribution of boson loops, a cancellation is possible and will occur if there is a suitable symmetry which relates fermion couplings to boson couplings. The contributions from the upper ends of the integrals cancel and we're left with an integral which effectively runs from the lightest mass which occurs in the loops to the heaviest. Again we don't want quantum corrections to be much bigger than the value we're trying to get--we don't want to invoke miracles. Hence the mass splittings in the supersymmetric multiplets times the coupling constant cannot be much bigger than M_W; for example, the mass splittings between the electron and the scalar electron or between the photon and the spin-1/2 photon that must exist in such theories, must be small

compared with 1 TeV. So once again we find the magic number 1 TeV. For any solution of the hierarchy problem it seems that there has to be new physics below or around 1 TeV.

5 CONCLUSION

The standard model is in great shape, but there's much to be done. In the case of QCD the things that have to be done are mainly theoretical. In the case of $SU(2) \times U(1)$ they're mainly experimental. However, the standard model is not the final theory, and there are plausible arguments which suggest that whatever new physics comes in, it ought to show up below 1 TeV. Now this 1 TeV number is a bit vague. It's of order 1 TeV, so maybe theorists will back off and tell you it's actually 3 TeV. On the other hand maybe it's 100 GeV. We don't really know and we don't know what's going to happen there. All we can say is that we want to get to the biggest energy possible to find out and we are relying on you machine builders to get us to the next threshold.

6 BIBLIOGRAPHY

All the topics discussed in this talk were considered at the 1983 International Symposium on Lepton and Photon Interactions at High Energies (referred to in the references below as Cornell Conference, 1983); comprehensive references to the origin of the data and the ideas quoted here may be found in the proceedings. For more pedagogical discussions, from which the literature may be traced, see the following.

For perturbative QCD: G. Altarelli, Phys. Rep. 81 (1982) 1.
For Lattice QCD: J.B. Kogut, Rev. Mod. Phys. 55 (1983) 775.
For $SU(2) \times U(1)$ and beyond: J. Ellis, in Gauge Theories in High Energy Physics, Ed. M.K. Gaillard and R. Stora, North Holland, 1983.

7 FOOTNOTES AND REFERENCES

1. The relation between g_A, f_π, and the pion-nucleon coupling constant is the Goldberger-Treiman relation. $a_{\pi N}^{1/2, 3/2}$ are the πN scattering lengths for $I = 1/2$ and $3/2$; the predictions are due to Weinberg and Tomozawa. λ_{Ke3} is the slope parameter in the Dalitz plot for $K \to \pi e \nu$, for which the predictions is due to Callan and Treiman.
2. See, however, R.D. Field, Proc. Cornell Conf., 1983.
3. Now two CERN experiments (BEBC and CDHS) and one FNAL experiment (FMOW) find values compatible with one, but CHARM still finds a value less than one. See K. Winter, Proc. Cornell Conf., 1983.
4. R.M. Godbole and D.P. Roy, Phys. Rev. Lett. 48 (1982) 971; J. Smith and G. Valezuela, Phys. Rev. D28 (1983) 1071. Reviews of the 1982 status of the same-sign dimuon problem may be found in V. Barger, J. Phys. Paris 43 (1982) C3-32; F. Halzen, Ibid. C3-381.
5. For a review see T. Nash, Proc. Cornell Conf., 1983.

6. For a review see N.W. Reay, Proc. Cornell Conf., 1983.
7. For a review see R. Barbieri, Proc. Cornell Conf., 1983; D. Schildknecht, Max Planck Munich preprint MPI-PAE/Pth 23/83; H. Fritzsch, MPI-PAE/Pth 47/83.
8. V. Barger, E. Ma and K. Whisnant, Phys. Rev. D28 (1983) 1618.
9. F.J. Gilman and J.S. Hagelin, Phys. Lett. 133B (1983) 443. For a recent comprehensive review see A.J. Buras, MPI Munich preprint PAE/Pth 46/84 (to be published in Proc. Freiburg Workshop on the Future of Medium Energy Physics in Europe). Buras et al. find that for $M_t < 50$ GeV, usually accepted calculations of weak amplitudes plus the latest value of the b lifetime give a lower bound on ε'/ε which is only just compatible with the measured value of $(-4.6 \pm 5.3 \pm 2.4) \times 10^{-3}$ presented by B. Winstein at the 1984 Dortmund Neutrino Conference.
10. B. Sadoulet, Proc. Cornell Conf., 1983.
11. Since this talk, this calculation has been done by O. Cheyette (LBL preprint 17083, 1983), who finds that experiments at LEPI may be able to test the coupling.
12. J. Cortes, K. Hagiwara and F. Herzog, Madison preprints MAD/PH/102 and 108 (1983). Even if the luminosity is achieved, the dip which is characteristic of gauge theories may be filled in by higher-order QCD effects.
13. For a review and references see the paper by J. Ellis cited in the bibliography above.
14. For details see C.H. Llewellyn Smith, Phil. Trans. Roy. Soc. A310 (1983) 253.

THE FUTURE OF HIGH-ENERGY PHYSICS

J. D. Bjorken
Fermi National Accelerator Laboratory, Batavia, Il. 60510

TABLE OF CONTENTS

1. Introduction. 321
2. The Standard Model and Our Future 321
3. The Standard Model and Our Present. 322
4. The Intermediate Energies: 100 GeV. 326
5. High Energies 326
6. Nonaccelerator Physics. 328
7. Accelerators. 329
8. Summary 330

THE FUTURE OF HIGH-ENERGY PHYSICS

J. D. Bjorken
Fermi National Accelerator Laboratory, Batavia, Il. 60510

1 INTRODUCTION

Mel Month is responsible for the modest title of this talk. Lately I have spent time worrying about machines and experimental physics, and I have not had much time to keep up with theory. But I have tried to fulfill the charge. Given the impossibility of doing that, what you get here is a bland, broad-brushed overview. Sorry about that.

2 THE STANDARD MODEL AND OUR FUTURE

By now, almost everyone believes in the correctness of the standard model, at least in its broad outlines. The weak and electromagnetic forces are described by the SU(2) X U(1) electroweak theory, and QCD underlies the strong forces, and we essentially know the Lagrangian associated with those interactions. Nevertheless, the standard model has basic shortfalls and fundamental problems which we don't know how to handle. There are the three generations of building blocks and the many parameters of the model. The basic mechanism for the origin of W and Z masses of spontaneous symmetry breakdown is most likely right. But what in detail is going on in the Higgs sector really isn't understood — or at least trusted. Maybe the minimal Lagrangian picture of the Higgs sector with its single Higgs particle is correct, but it looks a little artificial and clumsy. The origin of the fermion masses and mixings is believed to be coupled to that problem. The strong CP problem is awkward and along with it the whole question of the origin of CP violation. These are believed to be all tied together but we cannot really prove that yet. These are the big problems that provide the background for the whole field, problems that all red-blooded theorists grapple with so unsuccessfully. We all believe that sooner or later the shortfalls of the standard model will yield. But it's still a judgment call on what mass scale is needed in order to crack the problem and begin to resolve these shortfalls. We don't have many clues. The most solid clue is the upper limit on the mass on the Higgs boson of around a TeV, which comes from unitarity considerations. The natural scale of the electroweak theory is really not the mass of the W and the Z, but rather the vacuum expectation value of the Higgs field (or the decay constant of the Goldstone modes) which is somewhat higher, of order 250 GeV.

There may even be not much to discover within the TeV mass range except the single Higgs boson, especially if it turns out to have low mass. This is a very minimalist view, which suffers from "hierarchy," "fine tuning," and other technical problems of the theorists. But, albeit unlikely, it is at least thinkable that there is nothing beyond a single Higgs boson until the fantastic grand unification theory

(GUT) scales of 10^{15} GeV. The evidence for that hypothesis really depends, first of all, upon making sure that the parameters at low energies are consistent with simple SU(5) grand unification, and whether or not the decay of the proton occurs on schedule. As of today it seems to be behind schedule.

3 THE STANDARD MODEL AND OUR PRESENT

The standard model has been so successful that theorists have become rather arrogant and experimentalists have become rather intimidated about possibly getting results which are in disagreement with it. The standard model needs more testing. This can be done at all energy scales. And, aside from the fundamental tests of the standard model, a lot of details and loose ends remain. They involve the kind of work that usually doesn't get on the front pages of newspapers but nevertheless forms the backbone of our subject. I have tried to make a short listing of loose ends: jobs to be done in the near future.

Everybody believes there is a tau neutrino but it would be nice to find one. Beam dump experiments, we hope, can do that. The lifetime of the tau ought to be measured well enough for us to see whether it behaves the way an ordinary heavy lepton should behave.

Spectroscopists might like to see the magnetic moment of the omega minus. Even the QCD lattice people might have a good time calculating it, and it may be easier to calculate than the proton moment.

Same-sign dimuons produced in neutrino interactions present a nagging problem. The yield is large compared with what theorists can estimate from charm production and known mechanisms. I believe that phenomenon is not understood. Confusion remains in the beam dump experiments looking for prompt neutrinos ν_e and ν_μ from charm production in beam dumps: some get a result of unity for ν_e/ν_μ and others don't. Such experiments must be done again so that it is absolutely clear what's going on.

All this is related to the problem of hadronic charm production, also a very confusing subject. A lot of leading charm is found at the ISR, and possibly in pion beams at the SPS, both with rather large cross sections; other experiments at Fermilab and SPS energies seem to be more consistent with central charm production, with leading charm being relatively small. The overall level of charm production doesn't seem to jibe very well with the simplest QCD calculations and again is not well understood.

A marginal measurement of the various asymmetry parameters in the beta decay of the Σ^- had been published in which the polarization asymmetry comes out with the wrong sign compared with Cabibbo theory. Were that experiment to hold up, there would be mass suicide among theorists; they will categorically deny the correctness of that experiment. This result obviously needs a follow-up experiment.

The phenomenology of semi-hard hadron collisions ($5 < p_T < 10$ GeV) is something of an embarrassment for theorists. It isn't very clear how to handle the median p_T range before the cleanliness of the hard QCD collisions that we see at the SPS collider emerges. There should be better understanding of that whole question.

Even in the relatively clean deep-inelastic phenomena of lepton-nucleon scattering, much is not understood. For example, can QCD describe the Regge behavior of the structure functions, especially the nonsinglet parts, where the Regge trajectory carries nontrivial quantum numbers? As one goes to the small values of x, theorists working within the modern QCD context are notably silent on predicting what goes on, although in the old days everybody knew how to talk about it.

Another major area comprises non-perturbative strong-interaction phenomena. It's not so many years ago that the measurement of the rise in the total pp cross section with energy was said to be one of the most important discoveries in 20th century physics. It is important, and I don't think we yet understand the reason for the rise. Does QCD predict the rise in the total cross section? I don't think that has been shown. In particular the whole relationship of Pomeron phenomena to QCD is something that very few people even work on. But I would guess in the future more people will. As easy problems in QCD like the mass of a proton get solved, people will go on to more difficult ones like nonsinglet Regge trajectories. Finally the ultimate challenge will be the Pomeron. It's a tough problem and sooner or later it will get attacked with more energy than now.

Another whole class of phenomena associated with high multiplicity or high transverse energy seems to get more important as energy goes up. KNO scaling is one manifestation of it. There are events at the SPS collider in which enormous numbers of particles (say 50 GeV of energy) emerge isotropically into a central calorimeter. I don't think these events are well understood. They may or may not be related to the question of whether or not high energy ion-ion and/or hadron-hadron collisions can make quark-gluon plasma. Ephemeral plasma production may occur even in a $\bar{p}p$ collision.

Of course production of quark-gluon plasma is itself an interesting field, no matter what the projectiles. Some people think it's a dirty business to slap a couple of ions together and watch thousands or tens of thousands of particles be produced. What can you learn from that? The skeptics may be right. It may be very hard to learn anything from it. On the other hand, it may in fact impact in very fundamental ways on QCD. For example, people work very hard to measure Λ, the scale parameter of QCD. They feel quite happy if they measure it to 50%. Now suppose you can convince yourself that there is a first-order phase transition between ordinary hadronic matter and quark-gluon plasma. Then the transition temperature is proportional to the QCD scale parameter, with perhaps small corrections due to the presence of fermions and finite quark masses. This transition temperature is one of the easier things to try to calculate from nonperturbative QCD theory, and the lattice calculators already give us the constant of proportionality.

If there's any truth to this idea of quark-gluon plasma and a phase transition, then the transition temperature between normal hadronic matter and the plasma is somewhere between 100 and 300 MeV, because at 100 Mev one clearly has dilute pion gas, and at 300 MeV the energy density is so high that it's hopeless to imagine that there are still hadrons swimming around rather than quarks and gluons. Also, common sense says that this is a good scale just from estimating the characteristic momenta of the constituents, whether they be hadrons or quarks and gluons. Therefore, just by waving of the hands, I can tell you that the transition temperature is 200±100 MeV. Given that lattice theorists can tell me what the proportionality coefficient is that connects it to Λ, I can tell you what Λ is to 50%. If I can do this without any calculation at all, then with a lot of hard work — including some measurements that convince us that this idea of plasma isn't just a figment of our imagination — it may be that measurement of the transition temperature could become the most accurate measurement of Λ that we can get.

Of course other features of heavy ion collisions may be even more fun than quark-gluon plasma. Some people think that fractional charge is more easily liberated if it is in a lot of boiling colored soup. There may be other exotic objects, such as high density metastable hadronic matter. Those are very speculative ideas. It is very hard to know how to weigh their importance. But I think they're not completely out of the question and therefore should be factored in when one is thinking about how productive relativistic heavy ion physics might be.

Another interesting area concerns phenomena having to do with large longitudinal distances, such as A-dependence of high energy collision processes. This tells us about the nature of the evolution of processes in space-time, i.e., how the initially formed hadronic matter evolves. Its interaction with nuclear matter during the evolution tells something about the early stages of collision processes.

Also, a lot of spectroscopy remains to be done. Heavy quark spectroscopy is relatively clean from the viewpoint of QCD. But it's good to have light quark spectroscopy to compare it with and to give some idea of continuity or discontinuity between the properties of light and heavy quarks. And of course glueballs need a lot of attention.

Many of the fundamental parameters of the standard model need to be learned from low energy experiments. It is very important to get an accurate measurement of the weak mixing angle at energy scales very small compared to the electroweak mass scale of 100 GeV. The angle at low energies should be compared with that obtained at the electroweak scale via direct measurements of properties of the W and the Z. The comparison of the two classes of measurements may provide information on the radiative corrections involving higher orders of the weak interactions. These may be sensitive to what is going on at a mass scale much higher than even the natural electroweak mass scale, just as accurate radiative-correction work in QED leads to information about very short distances. This kind of precise, careful work may in fact have a lot of leverage.

The mixing parameters of the quarks, i.e. the generalized Cabibbo angles which in the six-quark world become the elements of the K-M matrix, are obviously very fundamental parameters. It may be hard to measure them accurately, and will probably take a lot of work. For that, copious charm and bottom production is needed, perhaps not only at the e^+e^- machines but also in the fixed target machines. If one can learn how to handle hadronic production of charm and bottom in the presence of a very large background, Tevatron II is especially attractive.

CP violation parameters are vital: is the phenomenon milliweak or is it superweak? There is healthy activity now on the neutral kaon system. It is very deserving and has to go on.

Are there neutrino masses and mixings? Are the Russians correct in their tritium beta decay endpoint experiments? Of course, this is a very important topic right here at Brookhaven.

Proton decay will tell us whether or not "naive" SU(5) is alive or dead. Little more need be said on the significance of that kind of experiment, except that the big proton-decay detectors may provide some information about high energy cosmic rays too.

Data can show in many ways phenomena that deviate from the standard model. This can happen even with the lowest energy experiments. Rare decays such as $K \to \mu e$, $K \to eee$, etc., or rare charm and bottom decays may occur. The history of the kaon is germane; the more intense the kaon beams were made, the more we learned, and that continues to this day. I believe it could be the same with charm and bottom decays. We need the most intense pure charm and bottom beams we can get, whether they come from e^+e^- machines or hadron machines. Maybe we will be surprised by mixings of the neutral D's and neutral B's as we were with the neutral kaons.

There may be axions. There may be extra neutrinos or other neutral fermions which can be produced in beam dump experiments. There may be low mass supersymmetric particles, e.g. photinos, which likewise could be produced in beam dump experiments or otherwise.

The question of right-handed currents comes up in everything from low energy experiments to ep colliders. There may be a right-handed Fermi constant which is smaller than a left-handed one by the square of the mass ratios of right-handed to left-handed gauge bosons. This would give electroweak and milliweak interactions involving right-handed currents; again the scope of searches for such things is very broad.

To summarize, clearly much remains to be done at existing energies, and it must be done. It is important, and new directions of study are there. Ion-ion and maybe even electron-ion collisions should be considered as well. I think this is a very exciting direction in which to explore and would hope that it will become a very serious Brookhaven future option during the coming months and year. Charm and bottom physics requires an e^+e^- machine with as high a luminosity as possible at the psi and upsilon regions. Neutrino experiments need to be done at all energies, for example, for the long base-line physics of neutrino oscillations as well as for short base-line dump experiments. We must always remember to emphasize this part of our future. In many ways it's the lifeblood of the field, even though the bulk of the results don't end up on the front pages of newspapers.

4 THE INTERMEDIATE ENERGIES: 100 GeV

The intermediate energy range of center-of-mass energies, of the order of 100 GeV for the processes of interest, will be the natural habitat of TeV I, SLC, LEP, HERA, as well as the SPPS. This is familiar territory, already well covered at this school. It is the focus of such broad attention these days that I will not belabor it much here. The next step after W and Z seems to be to find the top quark. Maybe that one will come soon, but a fourth generation may have to wait awhile — if there is one. If a fourth generation is "typical" in the sense of having a light neutrino and its lepton, along with a pair of quarks which are not degenerate in mass, then it really belongs in the intermediate range category. A fairly convincing theoretical argument can be made for a ceiling on the masses that those fermions could have. Heavy fermions contribute vacuum-polarization radiative corrections to the mass of the Z and W, so that above a mass scale of ~300 GeV these fermions cannot exist unless they become degenerate in mass or decouple from the electroweak interaction.

5 HIGH ENERGIES

High energy is the name of the game. We can't just stop in the intermediate range, and it is undoubtedly essential to go to higher energy to really get at the fundamental problems of the standard model. A lot can be done without the high energies, but I doubt that it is sufficient. History offers an interesting analogy. Compare electroweak interactions now with strong interactions in the early sixties. We understood a lot about the symmetries. Pions were exchanged to give the basic long-range force. Looking retrospectively, we see that all the machinery needed to understand QCD and the nature of the strong interaction - the gauge principles, the ideas of chiral symmetry and chiral symmetry breakdown, spontaneous symmetry breakdown - was in place quite early in the sixties. What wasn't there, at least until late in the sixties, was the quarks. People used the gauge principles to try to gauge the flavor symmetry, and thus they went off on the wrong track because essentially one had to get down to the constituent picture first. We had to understand the need for color and the inevitability of the quarks before the next step could be made.

Now comes the question: could we have found QCD from observations only at low energy? For that matter would, say, the AGS, SLAC, the Bevatron, and SPEAR have been sufficient to give us <u>all</u> of the standard model? We got a lot of the standard model from them, but I really doubt that this was enough to put it all together in a convincing way. The high energy experiments were extremely supportive and essential in helping guide the way. It would have been much more confusing without them.

And what about the converse: would <u>only</u> the highest energy machines have been sufficient? The answer <u>is</u> the same: with only the high energy experiments available we would have had a very hard time.

The low energy ones were vital. I particularly have in mind the baryon spectroscopy done in the sixties. A series of individually uneventful and tedious experiments came together to provide a very powerful piece of evidence in favor of quarks.

When going to high energies, again and again one comes back to the question of the Higgs sector. I think the existence of a Higgs boson or something akin to it is almost inevitable. No matter how elaborate one imagines the final explanation of the Higgs phenomenon to be, the family of objects envisaged as necessary to give the real world usually includes scalar particles of one sort or another.

The argument for the limit of 1 TeV mass scale for this Higgs phenomenon is as follows. For W W scattering, consider the amplitudes obtained from gauge couplings alone. Add them all up and look at the amplitude in the $J = 1$ angular momentum channel. At very high energies the amplitude grows in strength to a point, about a TeV in the center of mass, where unitarity is violated. If one takes the full electroweak theory with its Higgs particle and adds the exchange from the Higgs in the various channels, everything is smoothed out and unitarity is not violated. This is true provided the mass of the Higgs is not too high. Otherwise it doesn't have enough clout and doesn't avert the catastrophe. Upon going to the supercollider energy scale of, say, 20 TeV on 20 TeV (at a decent luminosity), the cross-sections of quark-antiquark annihilation to W pairs in the $J = 1$ channel are big enough to be observed out to the TeV mass scale. If there are large W W scattering phase shifts or crazy things (such as opening of unexpected inelastic channels) going on, one may be able to see them.

In the search for the standard Higgs, which will clearly be a central problem in the years to come, various mass scales have to be considered. For a Higgs mass of less than 40 GeV we may hope to count on SLC or LEP I to find the standard Higgs through the Z decay into the Higgs and lepton pair. This decay has a good signature, with a decent branching ratio, if the mass is no larger than that. If the mass is somewhat larger but under 100 GeV the process is slightly different. The first step is e^+e^- annihilation into a real Higgs and a real Z. Again the Z can decay into a dilepton or maybe a dijet and there is a good signature. Then an embarrassing mass region is encountered between 100 GeV and the W pair threshold, where everybody has a hard time. For a Higgs mass above the W pair threshold, hopes are rising that a hadron collider with high enough energy and high enough luminosity will produce the Higgs in an observable way. For this, the hadron beams should be regarded not even as quark beams, but as gluon beams. [The gluon is already a commonplace, and in the future we will be into gluon-gluon collision processes in a big way.] The process is resonant annihilation of a gluon pair to a Higgs particle, which then decays with a large branching ratio into a pair of intermediate bosons. This is a fairly good signature. The basic Feynman diagram goes through a virtual process involving the famous triangle diagram with a top quark loop (or, if there's a heavier quark, the heavier one). The calculation is done in the Snowmass Summer-Study Proceedings, and the rates look good. But it is not clear how high a mass can be reached even with a very high energy collider. As the Higgs mass approaches one TeV, the width of the Higgs boson grows rapidly. (It must, because unitarity is about to be violated. Somewhere around a TeV the width of the Higgs is comparable with its mass.) If it is so broad, then the resonant peak in the mass distribution of W pairs starts sinking into the background, and a detection problem arises. Exactly where that occurs is not yet very clear.

So much for the standard model Higgs boson. The Higgs phenomenon may have indirect manifestations such as technicolor. In the technicolor scenario, the Higgs boson is a complicated object, a bound state of fermion pairs, not an "elementary" particle. The analogy is to the pion and 0^+ sigma meson, which are bound states of quarks. The Higgs would have many family members and new strong interactions. Again, arguably this whole family should be below the TeV mass scale.

Another possibility is supersymmetry. If supersymmetry is connected with the underlying problems of the standard model, then all the known particles could have many partners. But the masses are unknown. Many schemes have been suggested for what the spectrum should look like, and some of the particles are arguably below the 1 TeV mass scale.

I think it's fair to say, however, that these ideas and others that I haven't talked about are in trouble. You hear only generalized discussions of these notions because specific models don't work. Were there good specific models, then you would have certainly heard much more. Thus these ideas are beautiful "in principle" ideas. Maybe they're even right, and the right combination just hasn't yet been found. Maybe supersymmetry is right in the same sense that the gauge principle in 1960 was right, but that it's not being applied in the right way. Maybe new ideas are needed, preons or constituents or something else to get us out of the present impasse. Everyone will agree that, if one could get experiments on the TeV mass scale, it would help a great deal.

6 NONACCELERATOR PHYSICS

Before finishing, I should mention the nonaccelerator experiments. These are sensitive to all energies. I've always believed that, when the dust settles, the big new instruments underground built to look for proton decay may well come up with cosmic-ray phenomena comparable in excitement with their original goal. This hasn't happened yet, but I keep waiting for it. Also, the Fly's Eye in Utah which scans very high energy air showers is also a nice direction to go. The Utah cosmic-ray group was forced upwards, both in energy and altitude, by the rapid increase in collider energy scales. They are now looking at scintillation light from air showers at 10^{17-19} electron volts propagating through the atmosphere. Recently they have become interested in the possibility of detecting upward-going showers caused by neutrinos interacting in the surface of the earth. The neutrino comes through the earth, interacts near the surface, and sends a shower up through the atmosphere. There are even rumors of candidate events. But, regardless of candidates, interesting calculations have been done on rates. Calculations using the standard model and standard neutrino cross sections give results indicating that the mean free path of a neutrino at these energies is reasonably well matched to the diameter of the earth. Furthermore, for an electron neutrino the effective thickness of the surface of the earth which can be used as a target is much larger than one would naively think at low energies. The reason for this is what is called the Landau-Pomeranchuk-Migdal effect, which is an increase of the effective radiation length with increasing energy scale. A couple of orders of magnitude are gained from this mechanism, with the net result that detection of such neutrino events at these colossal energies may be within reach. I think that's a most happy development and I hope it bears fruit.

Carlo Rubbia taught me that searching for other stable heavy relic particles is the ultimate in high energy physics. If one ever found a monopole and could bottle it, and then found an antimonopole and bottled it, and then brought them together — and if their mass were at the grand unified scale or anything like it — one would really have high energy physics. A single event would be a radiation hazard. If a monopole-antimonopole pair annihilates, say, into an X or Y boson of the grand unified theory (plus other particles), with a mass of 10^{14} GeV, and the X or Y decays into an electron, and that electron is headed your way, watch out. It's a rather deadly particle and — again because of the Landau-Pomeranchuk-Migdal effect — it penetrates.

7 ACCELERATORS

Finally, a few words on the future of accelerators. First I simply want to congratulate Mel Month on organizing these schools as an excellent way of stimulating what is really needed. If the SSC takes off and we have 40 TeV in the center-of-mass in the foreseeable future, where do we go from there? That machine is hard to beat with conventional technology. To go beyond it we need high gradient linacs (or maybe monopoles). The urgency of doing something radically new becomes greater if we do increase the slope of the Livingston curve and get to high energy faster than we thought before. My own prejudice is that the specifications for e^+e^- colliding linacs should be center-of-mass energies of at least 2 TeV, to compete with the physics done in the SSC. For proton linacs my specifications are enormous: 10 GeV per meter gradients, i.e., one electron volt per Angstrom. That is enough to destroy the accelerator every time you pulse it. But why not? Bob Palmer has already proposed that with his laser grating accelerator. I don't think we should be <u>ab initio</u> afraid of destroying an accelerator every time it's pulsed. But if you do that, you are making plasma out of it and so you have to know about plasma physics. A laser will probably be used to do the destruction, and so you have to know about laser physics. It is therefore likely that a much closer liason between our community and plasma and laser physicists will be needed.

8 SUMMARY

In summary, I have emphasized the importance of low and medium energies. Diversity exists at present in the program. It is very much needed and must be protected, even as we push aggressively to the highest possible energy. But the name of the game is energy and the push to higher energies must continue. The mandate which this year's Woods Hole Subpanel has given us is a very exciting one, one very much worth uniting behind. We must do our very best to make our hopes for such a high energy machine a reality.

AN ULTRAHIGH ENERGY HADRON COLLIDER
Round Table Discussion

TABLE OF CONTENTS

Introduction, M. Month, BNL 332
Particle Physics of Multi-TeV Collisions, J.D. Bjorken,
    Fermilab 334
        Comments 335
Accelerator Technology for a Multi-TeV Collider, M. Tigner,
    Cornell 336
        Comments 344
Experiments at Multi-TeV Energies, C. Rubbia, CERN/Harvard .. 345
        Comments 348
Prospects of an Ultrahigh Energy Collider: A View from
    Washington, N.D. Pewitt, Office of Science and
    Technology Policy 352
        Comments 355
Planning for a Super Collider, P.J. Reardon, BNL 357
        Comments 361
Discussion of $p\bar{p}$ vs. pp and the Question of Aperture and
    Luminosity 362
Appendix: The SSC Two Years Later, P. Dahl, BNL 364

AN ULTRAHIGH ENERGY HADRON COLLIDER
Round Table Discussion

INTRODUCTION*
M. Month, BNL

The world high energy physics community has for years stimulated a great deal of discussion on ways to achieve great leaps in particle collision energies. Formally, discussions have centered in the International Committee for Future Accelerators. Today, this dream seems more probable than ever before -- a hadron collider some 50 to 100 miles in circumference. The optimism stems from the great advances in accelerator technology over the past decade, the conceptualization and realization of great new hadron detectors culminating in the discoveries of the Z and W particles, and the great feats in particle theory which have changed dramatically how we perceive and interpret the physical world. To illustrate the serious nature of considerations on constructing such a device, we quote from the 1985 United States President's Budget submission to Congress:

> The High Energy Physics Advisory Panel (HEPAP) and its 1983 Subpanel on New Facilities unanimously recommended as highest priority "the immediate initiation of a multi-TeV high luminosity proton-proton collider project with the goal of physics experiments at the earliest possible date" and stressed that project preparation must begin as soon as possible in order to meet the goal of completion in the first half of the 1990's. The new facility, designated as the Superconducting Super Collider (SSC), is envisioned as a proton-proton collider with an energy of 20 TeV per beam and high luminosity, up to 10^{33} cm^{-2} sec^{-1}. Recent rapid progress in physics understanding and technology, including the great success of the CERN laboratory in their successful construction and operation of a proton-antiproton colliding beam capability and the subsequent discovery of the W and Z particles, has emphasized the importance of physics experiments in the several trillion electron volt (TeV) mass range where new phenomena, including new forms of matter with unique properties, are firmly expected. The SSC would have immense physics potential for truly major advances in understanding the fundamental nature of matter in an energy domain where there are no existing or planned facilities in the world and would assure a world competitive U.S. high

---

*In the several years elapsed since the Summer School of 1983, the R&D and preparation steps for realizing the SSC have made much progress, rendering some of the discussion at the Round Table somewhat dated. Nevertheless, for historical reasons we have opted for editing the discussion as little as possible, but adding an appendix bringing the SSC story up to date.

energy physics program from the mid-1990's into the next century. Such a project lies within the reach of present technology and has been made possible by the recent outstanding successes resulting from the substantial U.S. investment in superconducting magnet technology. The physics importance of exploration of the TeV mass scale coupled with the technical feasibility, leads to the conclusion that U.S. resources in manpower and funds are best utilized by concentrating on this highest priority project with its crucial physics importance.

Progress in high energy physics research is based on a continuing interplay of advancements in experimental methods, particle theory, and accelerator technology. In addition, the increasing size of the new accelerator-based projects means a growing involvement of the funding agencies and a pressing need for increased understanding and use of management and behavioral techniques. Thus, there are five essential areas which are needed to define the nature of such a large project as is being conceived. This Round Table on an Ultrahigh Energy Hadron Collider was organized with these five central aspects in mind. The speakers and topics are listed above in the Contents, and the essence of their talks, and the comments that followed, are presented in the following pages.

G. A. Voss (DESY) acted as chairman of the Round Table and iterated in his opening statement the fundamental questions under discussion: Why do we need ultrahigh energy colliders (what is the physics motivation), and how do we realize such challenging projects (what are the technology, government, and organizational needs)?

PARTICLE PHYSICS OF MULTI-TeV COLLISIONS
J. D. Bjorken, Fermilab

I will address the physics discussion in terms of the SSC as defined at Cornell and the Woods Hole panel deliberations, namely a roughly 20 on 20-TeV hadron collider of high luminosity, able to probe for subprocesses, or reach a mass scale of the order of a TeV. The question is, what sets that scale and why is it important? In my mind, the best answer to that is the traditional one that we really don't know. The push to high energy leads to great surprises, and in this particular case, as I mentioned this afternoon, I think there are compelling reasons why we suspect those surprises are there. However, I would bet that the kind of language that we use today may well be rather obsolete by the time the experiments are done, what with supersymmetry, technicolor, and similar concepts being the buzz words of the day. When these experiments are done I think it likely that there will have surfaced other evolutions of the theory, some of which may be created by the experiments themselves.

I'd like to draw the analogy with the early days of e^+e^- storage rings -- early 60's, late 60's. The first one (conceived but not constructed), at Princeton, was followed by the ring at the end of the old linac on the Stanford campus; it was an electron-electron machine, and was devoted to probing quantum electrodynamics at short distances. The thrust and motivation for the follow-on machines, the first e^+e^- machines, was largely the same. The 50's was a period when Feynman diagrams and all of that were elucidated and developed, and the theory became very sophisticated and precise. The experiments also became precise, everything worked, and there was a desire to keep pushing onward to test the theory beyond the scale of the proton and into much shorter distances, and that provided a lot of the motivation for the e^+e^- rings. I remember a discussion at Stanford -- I think it was with Matt Sands -- who was asking whether there was any reason to go from something like 2 GeV per beam to 2.5 GeV per beam because, after all, the electrodynamics tests weren't very sensitive to changes in energy by that small amount.

The strong interaction studies, on the other hand, promised to be very very different, because it was no longer possible to measure the elastic form factor of the proton, and this aspect of hadron physics was clearly the main incentive for building those e^+e^- rings. The importance of the inelastic cross sections and so forth was not apparent at the time, and in fact for quite a while thereafter it was thought that this famous r parameter that is now in use would fall as $1/q^2$ or $(\log q^2)/q^2$, or faster, and learned theorists were proving theorems to the effect that this might well be the case. Consequently, the strong interactions were supposed to wither away, and the tests of QED were the primary motivation. There were several scales, that of the strong interactions itself and that of the weak interactions, which one could point to as landmarks without any definitive arguments as to how things would work out when one got there.

Now we deal with hadron colliders. It could be e^+e^- again at the TeV scale, but let me talk about hadron colliders -- that is, broad-band colliding gluon beams and colliding quark beams, on the TeV scale. Again, we can point with great definiteness to tests of the standard model as the reason for doing this. We wish to probe ever shorter and shorter distances, and there's good reason to believe that there might be sensitivity to breakdown at the TeV scale, the weak interaction scale, without being able to point out exactly how. Thus, once again the standard model is in a condition similar to that of QED in the early 60's. It's worked, it's working very well. It's a precise framework and there's no more reason to doubt it, I think, than there was to doubt QED in those days. Of course, that may be a debatable point; on the other hand, there's no more reason to accept it blindly than there was to accept QED in those days. The search for breakdown in QED, has, after all, finally borne fruit in the fact that it hasn't exactly broken down, but we have a new language with the electroweak theory and a synthesis of the electromagnetic and weak interactions at the highest energy, and that was the way it happened. Although totally unforeseen, it nevertheless came about in a way that's consistent with the early landmarks. I wouldn't want to push it any further than that, and I leave it open for further discussion later on.

Comments

It was mentioned that the SSC, because of its unprecedented scale, may have to be located in a relatively uninhabited area of the western U.S., and this has led to its being referred to as the "desertron." This name is doubly appropriate as the SSC will explore the beginning of the energy region suggested by some theorists to be a "physics desert," devoid of new phenomena below energies of 10^{15} GeV. Responding to a question on the likelihood of such a physics desert, Bjorken noted that, in his opinion, the likelihood for, say, a single isolated Higgs boson, with hard collisions prevailing but no new phenomena emerging, is very small.

Rubbia concurred, arguing that, in any case, the question of compositeness -- i.e., a further substructure of matter -- would itself justify a bold step in energy and years of scientific effort. How bold a step is arguable, however. Rubbia noted that roughly 8 on 8 TeV could ultimately be realized (with superconducting magnets) in the LEP tunnel, akin to a "junior desertron." Do we foresee something like a "threshold" in energy, and is the choice of 20 TeV really optimum? Bjorken could only point to historical examples, some cases where a factor of 2 or 3 in center-of-mass energy was crucial (e.g., from ADONE to SPEAR), others where it was not (from BNL to Serpukhov). Candidate processes, such as the standard Higgs boson in the top mass range, could be examined in terms of various potential facilities covering a range of energies, but the question has no simple answer.

Although it can be argued (Palmer, Smith) that one should "go for broke," i.e., 100 on 100 TeV instead of 20 on 20 TeV, the latter machine appears economically feasible with technology at hand -- the former probably not.

ACCELERATOR TECHNOLOGY FOR A MULTI-TeV COLLIDER
M. Tigner, Cornell

Before we address the technology for a multi-TeV collider in depth, we have to define what we're talking about. I assume it is an accelerator which will allow physics access to the 1 to 2-TeV mass scale in a shortest possible time in calendar years. It could, in principle, be an electron-positron machine of 1 to 2-TeV beam energy, or a hadron collider of 10 to 20-TeV beam energy. What are the possibilities for this kind of accelerator? Although you've been regaled with various ideas under the vague rubric of laser accelerators, we're certainly not ready to proceed with one of those at present, I'm very sorry to say. You've heard about the plasma beat-wave accelerator -- actually another kind of laser accelerator; we're not ready to build one of those today either. You've heard about the wake-field accelerator, and while that's coming along with tremendous rapidity, it's still not ready to produce a TeV beam today. You've heard about the 2-beam accelerator, and that's not ready to go today. The standard e^+e^- storage ring simply can't be expanded to 1 TeV or 2 TeV per beam. A nonstandard one that I'll tell you about next year will work like a charm, but we're not quite ready to build that yet. And, in fact, you can't do it with a standard storage ring. With regard to a superconducting microwave accelerator, we're certainly not ready to tackle that either. What about a single-pass collider with standard components -- something like a super version of the SLC? Perhaps something like that would work, and in fact Burt Richter has proposed certain scaling arguments suggesting that it might not be totally inconceivable to extend the SLC to 1 TeV -- but first we have to make this one work. In addition, a fantastic amount of work remains to be done to improve the cost effectiveness (the cost per GeV must be improved by at least a factor of 5) of the kind of components used on that kind of linac in order to compete with proton technology. Where does all this leave us?

It leaves us with proton storage rings based on superconducting magnets, and even there we have some work to do. We certainly have to develop the technology to reduce the costs. It thus seems fairly certain that if we want to get into the multi-TeV collider business, it will have to be with a hadron machine based on further developments of existing superconducting magnet technology. This, of course, is the origin of the present title of reference, the SSC or Superconducting Super Collider. As it seems highly desirable that we should have a luminosity in the region of 10^{33}, it will most likely have to be a pp machine rather than a p$\bar{\text{p}}$ machine. Having reached such a conclusion, one could immediately write down some parameter lists which display the kinds of technology choices that evidently have to be made at the outset. The essence of these choices can be summarized rather simply. We have to balance high-field, relatively expensive magnets, which would allow us to invest relatively less of the money for the facility in the civil construction, against a machine using rather low-field magnets of lower cost, but which would require relatively more civil construction. To put that slightly more quantitatively, I prepared a short parameter table for one example, a 20-TeV pp machine which you might imagine built at three different field levels, 2.5, 5, or 8 Tesla, and you can see that the radius decreases with increasing field, from 37 to 12 kilometers.

Possible Partial Parameter List
20 TeV + 20 TeV

p   p

$L = 10^{33}$ cm^{-2} sec^{-1}

B (T)	2.5	5	8
$\bar{R}$ (km)	37	18	12
Half cell (m)	200	180	100
N_p	10^{14}		
No. of IP's	6-8		
β (m)	2		
Aperture (cm)	2-3		
Tune	∼100		
Free Space at IR's	∼20 total		
Power	75 MW total		

I don't want to go into details, but merely to demonstrate that these parameters are not very farfetched. In fact, the most interesting thing is the conclusion that one could probably make such a machine work for 75 MW total power, which is comparable with present facility power levels.

To give some idea of the comparative sizes involved, the three rings are shown in Figure 1. For 20 TeV, the three rings corresponding to the three different field levels are all a good deal bigger than the Fermilab ring. Figure 2 shows that we're really talking about an extension of the things we do today: it is a picture of this SSC, and is actually a typical picture of a cascade accelerator. That little spur at the top is the linac. A 50-GeV accelerator almost disappears on this drawing, and a 1-TeV booster is pretty small. But if we are not concerned about the scale or the numbers, this is a kind of thing we are familiar with -- a cascade accelerator. It's interesting that within the range of overall parameters, i.e., a 2.5 to 8 or 10-Tesla base machine, rather conventional designs for the optics seem to be satisfactory. Also, the intensity requirements for 10^{33} luminosity do not place extreme demands on the technology. There's nothing obviously very difficult about stability, and the shielding requirements don't appear to be overwhelmingly severe. To show that what I said is true, I include an elementary lattice (Fig. 3) which uses very standard scalings of optical components at 20 TeV. There are even several designs for interaction regions which aren't too unreasonable (Fig. 4). There are 10 meters on each side of the

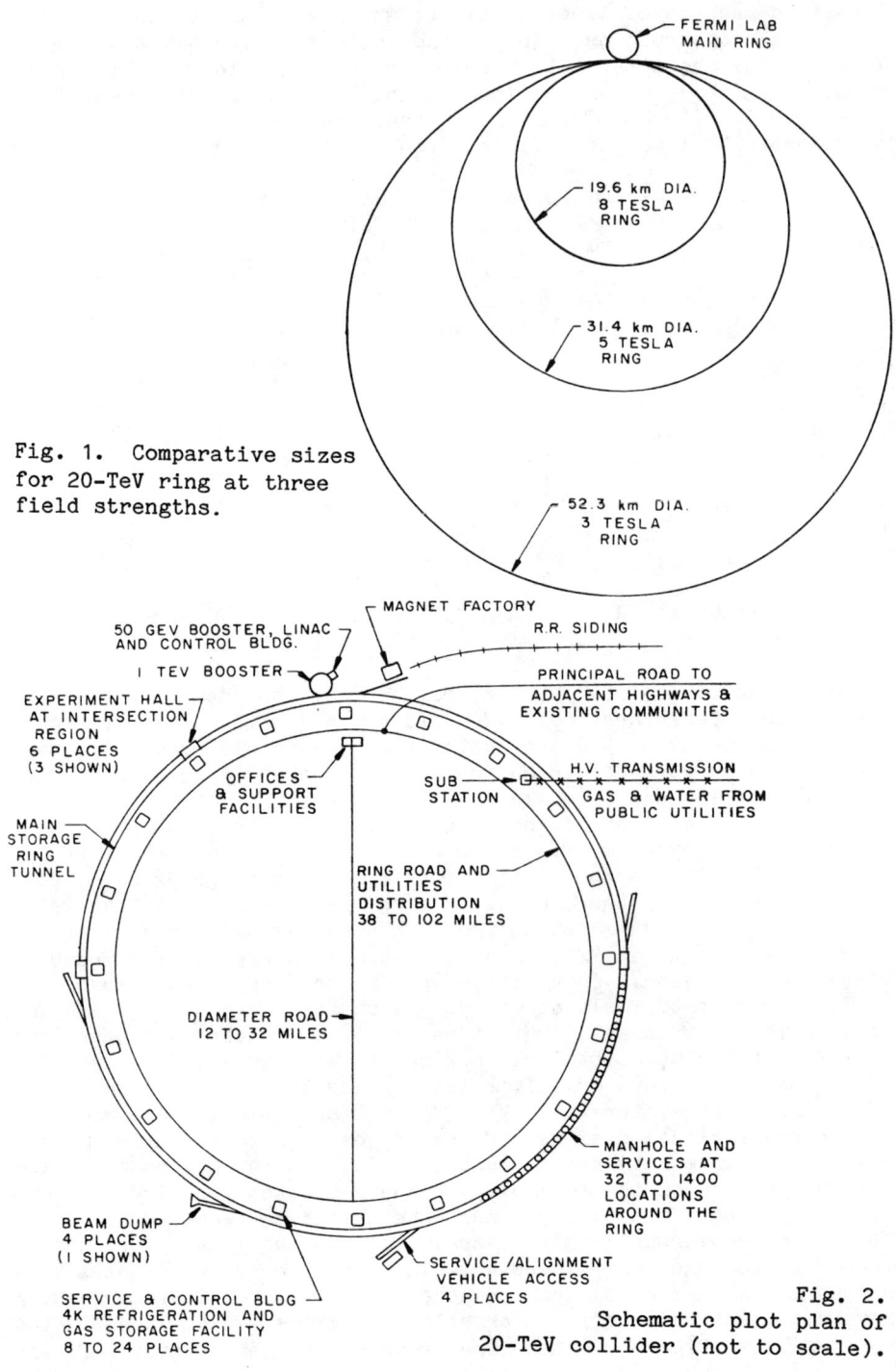

Fig. 1. Comparative sizes for 20-TeV ring at three field strengths.

Fig. 2. Schematic plot plan of 20-TeV collider (not to scale).

339

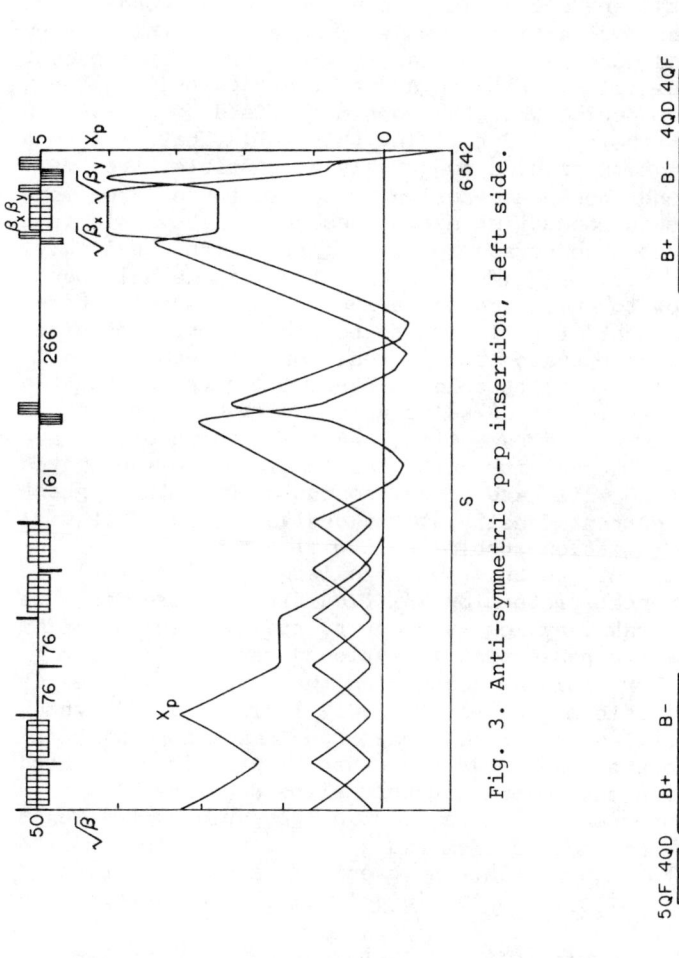

Fig. 3. Anti-symmetric p-p insertion, left side.

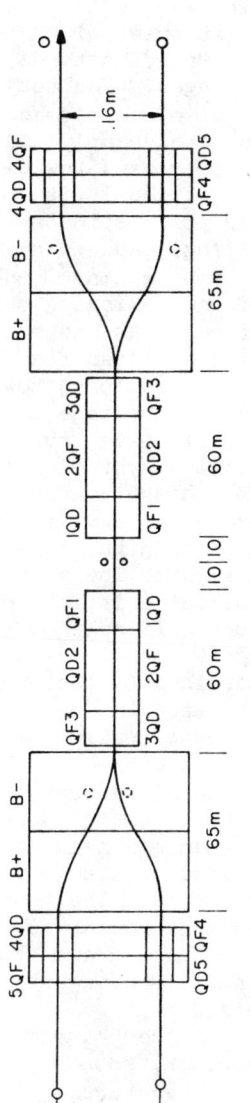

Fig. 4. p-p insertion region.

crossing point, flanked by some quadrupoles. You may think that a 60-m long set of quadrupoles is unusual, but it's not totally bizarre.

At this point you might say "Hey, this is duck soup!" How come we didn't do it long ago? Are there any real technical challenges? You bet there are. They can be summed under the rubric of cost and reliability. We will have to do things very much more cheaply and more reliably than we know how to at present. Let me give a few examples. With regard to the magnets, the first thing obviously is to decide what the optimum field will be in order to estimate the cost of building this SSC. That will be a challenge, and we have to proceed as quickly as possible. We need to combine into one coherent design good features from many different existing superconducting magnet designs. The principal feature that we must incorporate in the magnet design, obviously, is the smallest feasible bore. That is what the cost is all about. We have to learn how to build and use a machine with a small bore. To take advantage of that it will have to be a cold bore machine. It will have to incorporate cold iron, or at least the coil restraint system will have to be cold. There will have to be two beam channels in one cryostat if we're going to go with protons on protons and save money at the same time, and the magnets will have to have high manufacturability with automatic or semiautomatic equipment. The machine will have to use conductor with the highest feasible critical current density at the design field. That will be one difficult optimization problem. It will have to use the longest possible magnet units, and the length will probably be determined by the quench protection difficulties. Moreover, we must keep the heat leak very low in order to stay within the 75 MW for the overall facility power that I mentioned earlier.

There are a number of ideas concerning the magnets. One is the so-called superferric magnet for 2 or 2.5 Tesla (Fig. 5), which is 10 inches in overall diameter. It has iron laminations with the two beam channels located one above the other, and it is surrounded by liquid helium. [In the several years elapsed since Tigner's discussion, the superferric magnet design (prepared by the Texas Accelerator Center) has changed remarkably little, although its field has grown to 3.0 Tesla and overall diameter to about 17 inches. The magnet length, at this field level, is estimated to be 140 m.]

Another extreme example (Fig. 6) is based on niobium-titanium conductor, but is designed to go to very high field by operating at low temperature, 1.8 Kelvin. A type of medium-field magnet is shown in Figure 7; it is about a foot in diameter, and is supposed to have all those features mentioned in the preceding paragraph. [For several years preliminary designs of SSC magnets based on three possible field levels were pursued in the various laboratories: superferric magnets at the Texas Accelerator Center, 5-Tesla magnets based on available NbTi conductor technology (Fermilab), and 6.5-Tesla magnets based on improved NbTi conductor (a collaboration between LBL and BNL). In addition, a longer-range R&D study at BNL is directed at the possibility of utilizing Nb_3Sn

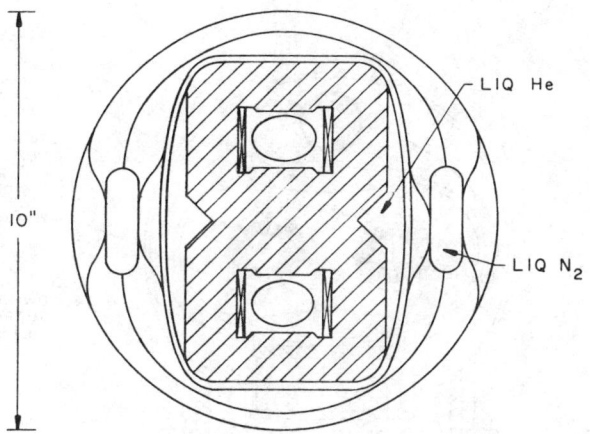

Fig. 5. Conceptual design of a 2.5 to 3-T dipole pair for a very high energy collider.

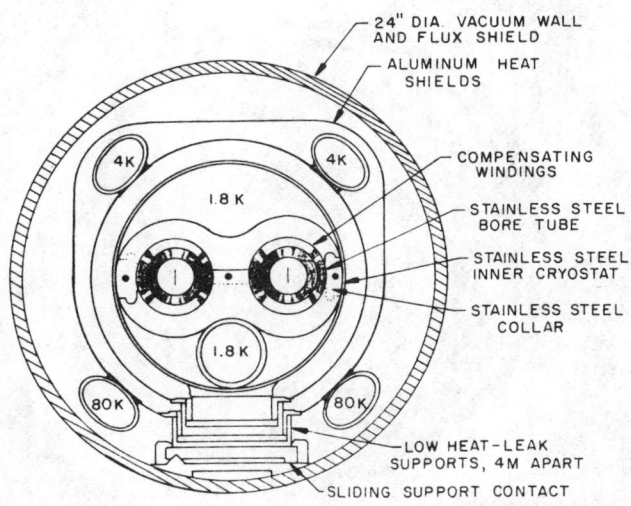

Fig. 6. Conceptual design of an 8-T two-in-one dipole pair for a very high energy collider.

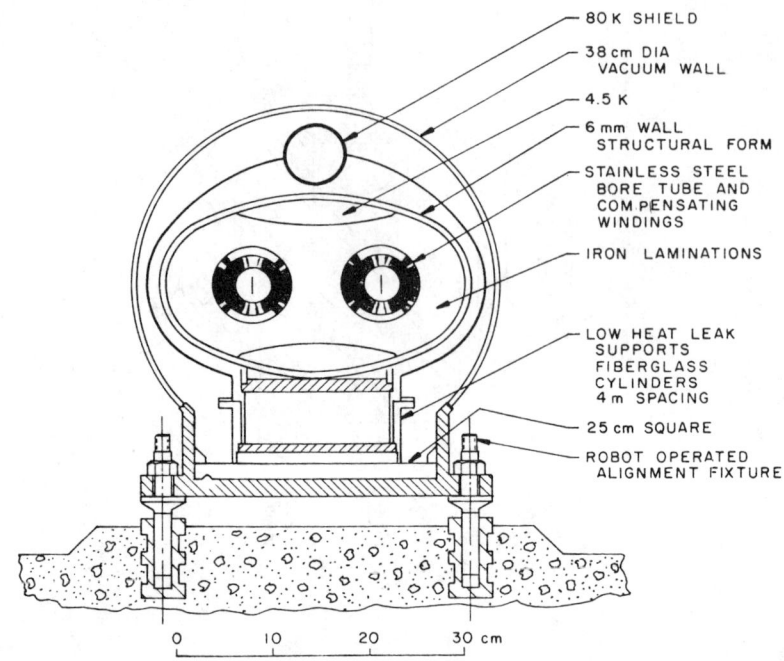

Fig. 7. Medium field magnet: 6.5 Tesla, 4.5 K, 4 cm coil i.d., 13.4 cm between beam centerlines.

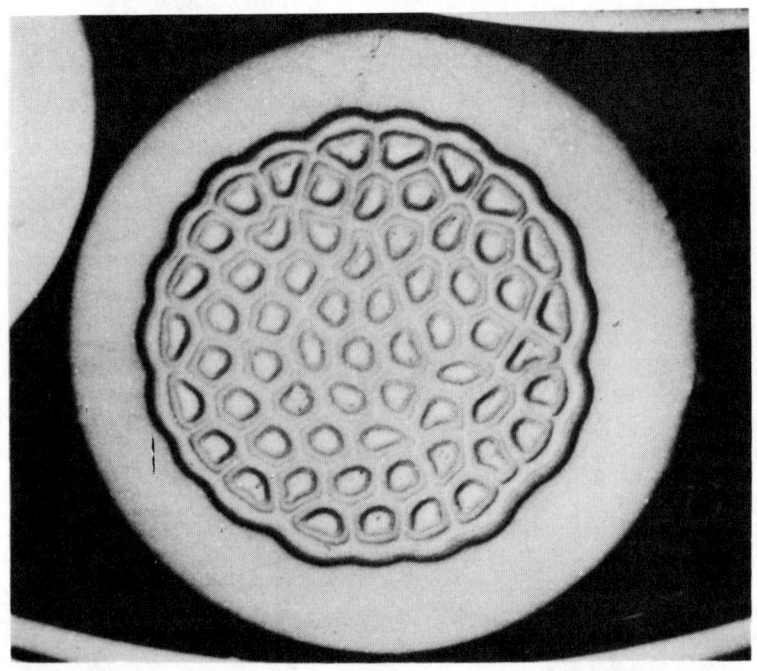

Fig. 8. Typical conductor.

conductor, a conductor with superconducting properties superior to those of NbTi but mechanically very brittle, in magnets of substantially higher fields -- perhaps 8 Tesla. In the spring of 1985 the three principal magnet candidates, based on NbTi, were reduced to two: the superferric magnet and a 6-Tesla magnet incorporating the various features of the LBL-BNL and Fermilab designs.]

I will not discuss the superconductor except to remind you that the secret of superconducting magnets is -- in bold and simple terms -- the extremely sophisticated conductor used to make them. The conductor shown in Fig. 8 is typical of those that industry now produces. It has an overall diameter of 1/16 inch. Each of the $\sim$61 little cells has 1100 filaments of niobium-tin in it. The conductor has tin and copper as well as a tantalum barrier, and the whole thing is made in a factory. This particular kind of conductor is only in pilot production, but there are factories that can make equally complicated conductors in quantities of hundreds of thousands of pounds. That's an enormous accomplishment, and it has happened in the last 10 years.

Another real advance needed in order to build the SSC is in the control instrumentation. We will have to make extensive use of absolutely the latest in electronic and computational techniques to provide the remote and flexible control of 100 kilometers of the machine and all its ancillaries, e.g., the refrigeration, the rf, the beam handling, and so forth. We will require complete automation of the beam position and orbit correction machinery. Continual surveying and adjustments on this enormous geological scale will be needed, and motions of the earth will have to be compensated almost continuously. We are really talking about something that our Russian colleagues talked about many years ago, the cybernetic accelerator; but today, with Texas Instruments and Motorola, we can actually do it.

Another thing that must be developed very hard is design technology. The size-related complexities and the need to be "right" about design choices, because of the large economic consequences of a mistake, mean that computer modeling and simulation will play a more crucial role in this accelerator design than they ever did before. Examples of what they will be needed for are dynamic aperture calculations (tracking) and optimization. We must learn to predict the operational effect of the beam-beam interaction. We will have to use simulations for stability calculations, and we will have to simulate all the systems so that we can understand the very complicated system behavior of these enormous cryogenic, electromechanical systems.

Similar remarks can be made about the civil construction technology. That will be difficult to find, as it will be difficult to understand how to specify sites that will allow use of the necessary economy-of-scale construction techniques. Hundreds of millions of dollars can be saved, according to the experts, if the site is properly chosen. Other hardware will require great emphasis on elegance and simplicity, for reliability and economy.

To sum up, the technological challenges of such a machine are great. We've met this kind of challenge in the past and we're going to do it again.

Comments

Reardon agreed that the design of a 20-TeV collider can proceed on the basis of extrapolations from present technology. However, certain key problems have not yet been solved. If its cybernetics problems are really fundamentally different, that may influence the size of the machine, its aperture, beam intensity, etc. Should the reliability issues be worse than now perceived, that could also interact with the magnet design, and a factor of 5 in magnet production cost could be fundamental. Thus, it should not be implied that the extrapolation is straightforward; it is an extrapolation of technology that must be significantly improved if the project is to be realized within a reasonable time frame.

Tigner noted that, although a fixed-target option had received little attention at the time of the Summer School, the experience gained in trying to extract a beam from the Fermilab Energy Saver will provide much valuable information (e.g., aperture requirements) relevant to this option. Beam aborts will be necessary in any case, so the SSC will certainly be designed for fast extraction. Slow extraction would be another matter. Excessive beam current would exacerbate the shielding problem. Concerning muon shielding generally, and its impact on site selection, Bjorken has performed rather careful calculations based on a "worst case" accident in which the entire beam is lost at a single point -- somewhere around the perimeter of the ring. A large hadronic cascade results, all of which is eventually absorbed except for muons; conservative estimates indicate a radiation level below roughly 100 mr 1.7 km downstream from the loss point. Since a muon beam doesn't broaden very much as it propagates through the earth, confining the beam at any point along the perimeter of the machine to below ground level by a few meters should be adequate. Locally produced neutrons from such an accident may require as much as 5 to 6 meters.

EXPERIMENTS AT MULTI-TeV ENERGIES
C. Rubbia, CERN/Harvard

I have a few sketches with ideas which I'd like to review. The real question is: Can you do physics with a machine which has this fantastic energy? Our own experience with, say, cosmic-ray experimental data of this type, is that the higher the energy the more complicated events become. For instance, if you think in terms of a machine of 40 TeV in the center of mass, the multiplicity and event complexity is absolutely enormous. The numbers of particles produced in these events will be similar to those we are accustomed to seeing in the most complicated cosmic-ray event. Thus, we are really facing a very tremendous multiplicity increase. The question, not entirely a trivial one, is: Can you do physics under all these particular conditions? The message I'd like to convey is that, in spite of all this, we have very good reason to think that physics will become simple again, and the reason is that, in spite of the tremendous multiplicity, the physics now is essentially reduced to a very few specific elements which the detectors are able to identify. First of all there are the leptons, which are electrons, muons, tau particles, and neutrinos. And then there are the hadrons. Of the hadrons, there are very many species. There will be hundreds of thousands of hadrons produced, but they will essentially reflect the presence of a very limited number of quarks and gluons manifesting themselves in the form of jets. Therefore, jets will become the real particles. You no longer have to define a detector which is conderned about what the pion does, what the kaon does, what the single particle does. You only have to worry about the energy flow. To illustrate that, I would like to show some of the things that start emerging even at the level of the CERN collider. For instance, the energy flow for a jet is a function of the phi angle, the radial angle as a function of the energy. The jet (Fig. 9) is a 10-GeV, 20-GeV, 30-GeV, 40-GeV jet. You can see that even for a 40-GeV jet (we're not going to talk about 400-GeV jets or 2-TeV or 4-TeV jets, although the phenomenon would be even more impressive), all the energy flow is concentrated in a very small cone of a few degrees. Therefore, although the particle density in there is enormous, our experiment has to be concerned only with energy flow, not the individual particles. This can be seen even more clearly in terms of a three-dimensional plot (Fig. 10), which shows the emergence of the energy flow around the jet axis. You can see that you're really dealing with very narrow fingers which carry all the energy, and all the complexity is buried in them. So I do believe that, in fact, this simply means that we will have to do simple physics again involving very few of these Manhattan towers, which will have to be understood in terms of physics, and all the apparent complexity is only superficial.

So much for a very optimistic point of view on how the physics can be done. Now, if we're dealing simply with jets, and if our interest is no longer in single particles but in energy flow, then

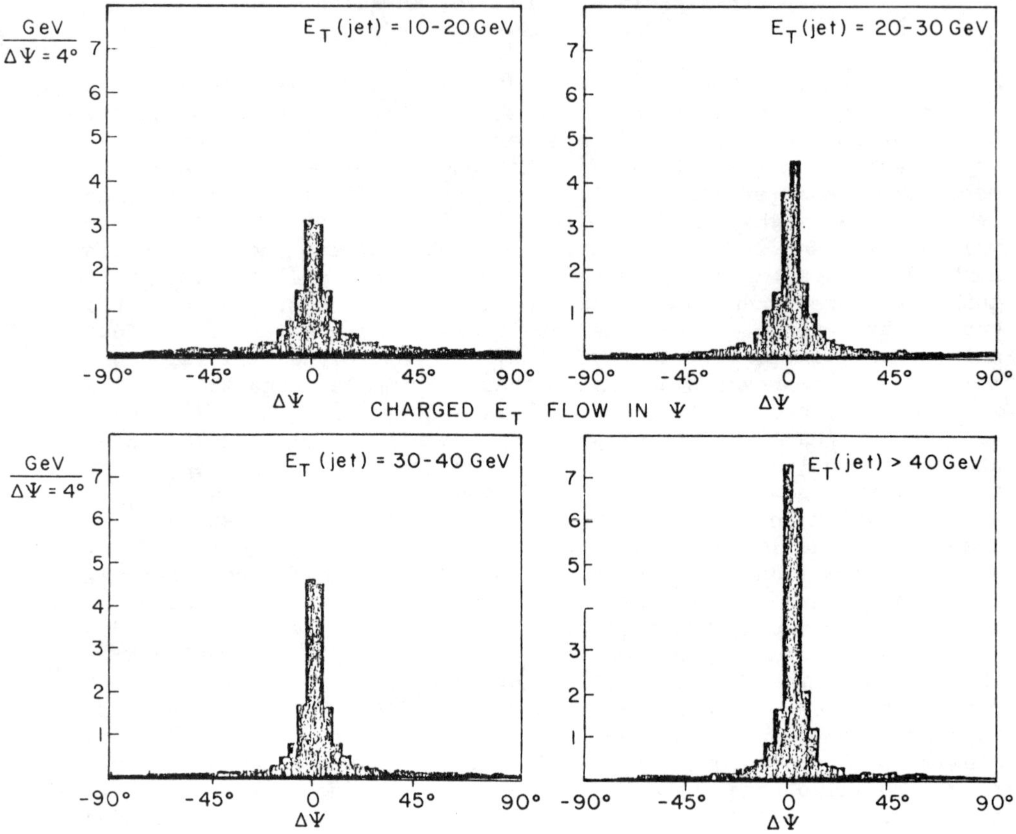

Fig. 9. Jet energy flow.

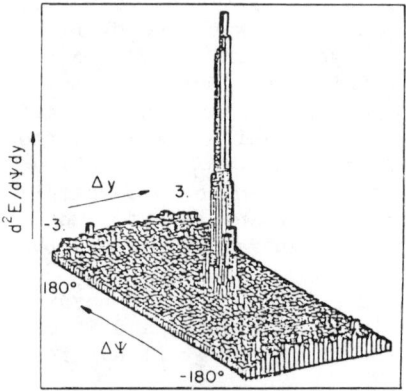

Fig. 10. E_T mean density around jet axis (calorimeters).

obviously (I think) for the first time in the history of our field, calorimeters will become the dominant instruments needed, and here some development will be necessary. Some important things have to be clarified because, as we heard this morning from Fabjan (AIP Conf. Proc. No. 127), the electromagnetic and hadronic showers in hadronic calorimeters are quite different and you get different responses; thus you really need the second-generation calorimetry. The measurements done by Willis and by Fabjan show that, if you use uranium as a calorimeter, you can use the fact that uranium is multiplying the neutrons from the hadron cascade in order to balance the response. Thus compensating calorimeters appear to be absolutely necessary if you rely exclusively on calorimeters to do physics.

Another important feature of those high energies is neutrino detection. You would like to detect emission of hard neutrinos, but hard neutrinos cannot be detected. That implies that your detector must be hermetic; that is, it has to be such that, if a neutrino is emitted, you'll be able to establish this fact from kinematic values. So you essentially end up with a very large uranium-compensated calorimeter which must cover the full angular region. A question that becomes extremely important in this energy domain is whether a magnetic field is still necessary. We have done most of our physics since the early days of, say, cosmic-ray experiments in the thirties by relying almost exclusively on magnetic fields. First cloud chambers, then bubble chambers, and now drift chambers. These require magnetic fields for particle identification, and the bubble chamber physics is the best example of how you do physics with a magnet as the essential experimental tool. Will a magnet retain its usefulness in this kind of energy domain? Obviously the amount of curvature expected in a reasonably sized detector with particles of several TeV, especially if the particles are leptons, essentially electrons and muons, is extremely small, and therefore perhaps the only thing you can hope to determine is the sign of the particle charges. It would be rather difficult to imagine that the magnetic field can compete in the electron determination of the calorimetry. That raises the question of how we cope with muons, because the muon energy is obviously not determined by calorimeter losses. Muons have to be analyzed with something like a magnetic field. Therefore you have two choices. Either you abandon muon detection and you somehow associate the muon with a neutrino through missing energy and momentum, or, if you really want to determine the momentum of the muon directly, you need immense $\int B\, dl$ in order to do it. An example of this is Ting's planned experiment for LEP -- a "low" energy machine. It involves enormous magnetic fields, extending over volumes as large as this room, and chambers with an accuracy of the order of a few microns, with laser alignments, and so forth. This sort of technology will have to be boosted much further if you wish to do physics. So I really perceive a problem as far as the muon is concerned. A third reason for the magnetic field, perhaps a very important reason, is the big difference between the soft stuff produced by spectator particles and the very hard jets produced by the interesting physics. And it would be convenient if this type of detector were able to eliminate all the soft tracks. Perhaps the magnet could remove all the soft emitted pions -- for instance, by generating strong solenoid fields of sufficient strength to prevent the 300-MeV/c particles from reaching the detector. Thus you may need to use a magnetic field for sweeping away low-momentum particles in order to conserve the hard simple events and eliminate all the debris produced by the spectator particles.

Another very important problem to be addressed is the question of high luminosity. Tigner just said that he would like to see a machine operating at 10^{33}; that is why he prefers a proton-proton collider. It's all very well, except that we have never operated a collider with such luminosity, and today many people still believe that 10^{33} is out of the question, with 10^{32} more likely. An added difficulty is that these machines lose a fair amount of energy by synchrotron radiation. The loss must be compensated for by some type of radiofrequency system, and the machines may end up being of the bunched beam type. That leads to a very complex interaction between the requirements of the physicist to get clean, smoothly flowing crossing events and to be able to handle high luminosity, with the need to resort to some radiofrequency to compensate the beam for hours for the losses from synchrotron radiation. These problems are not trivial. I think the high luminosity will not be trivially absorbed in the system, especially because of the complexity of events.

To summarize, it seems to me that the physics is essentially rather straightforward. It is tremendously exciting that the simplicity of the event analysis will really be there, and that reasonable extrapolation from present technology can allow you to do most of the desired experiments. The outstanding questions are how to deal with momentum analysis, which becomes really marginal, and how to maintain high luminosity, which has to be learned the hard way.

Comments

Responding to a question on his apparent caution with regard to a luminosity as high as 10^{33}, Rubbia emphasized again that it is a matter of learning how to handle it. A luminosity of 10^{33} with a cross section of 100 milibarns would correspond to a rate of 10^8 to 10^9 events per second. Open a window of a few nanoseconds, and one might pick an event in around 2 minutes. In this case, pile-up and detector performance become major problems. The question of handling the high rate was reiterated, assuming one is restricted to calorimeters, simulators and the like, plus triggers for high transverse energy. In Rubbia's opinion, what makes the high rate problematic is the necessity for compensation to ensure adequate resolution. Uranium has an intrinsic slowdown time of the order of 50 to 100 nanoseconds for thermalizing neutrons. It is not easy to design a second-generation calorimeter having all the characteristics of a beam compensated in terms of electromagnetic and hadronic events and at the same time ensuring nanosecond speed.

With regard to tracking chambers inside the calorimeters, Rubbia suggested that perhaps one should abandon the approach of examining individual tracks, except for particles like electrons and muons: the dominating feature is energy flow.

It is not resolution that is really at issue because of the higher energy scale. It is the presence of very strong systematic fluctuation. Consider, for instance, the UA-1 calorimeter (a major detector at the CERN p$\bar{\text{p}}$ collider, celebrated for its role in the discovery of the W and Z bosons). A hadronic cascade emanating from a jet which is completely electromagnetic will have a certain response. If the same event is completely hadronic, the response will be different. These two response states differ by a factor of 1.4 -- i.e., a π^0-dominated event produces 1.4 times higher light yield than does a charged π-dominated event. This means enormous changes in the energy assignment, and, since we are dealing with rapidly falling functions, chances of error in energy assignment are high.

Rubbia dwelt on an important question that was raised, namely, particle identification at very high energies. First of all, one must distinguish particles like leptons, which are single, very "stiff" particles, from hadrons, which are essentially bundles of highly collimated jets. Here the question is, what is the flavor carried by the quark, and what is the flavor carried by the gluon? The first foreign particle identification might distinguish, for example, the gluon jet from a quark jet. Figure 11 illustrates the problem. It shows the Z distribution, or a fragmentation function for the jets, obtained at Tasso (a large detector at the electron-positron storage ring Petra). The dots are quark jets and the crosses are points obtained at UA-1 from gluon scattering, i.e., gluon-dominated events. The nearly identical appearance of the gluon and quark jets clearly makes it very difficulty to distinguish a jet produced by a quark from one produced by a gluon

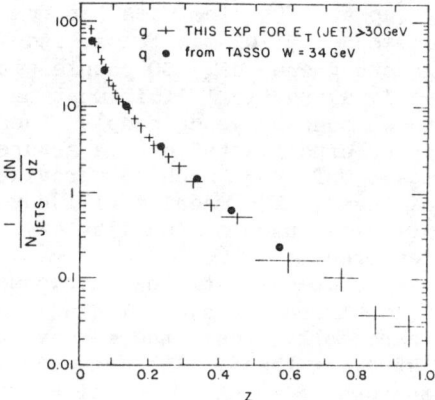

Fig. 11. Z distribution at Tasso.

(assuming the experiment is valid) or, that is, a quark from a gluon. The question then arises: in view of our inability to distinguish a quark from a gluon, can we distinguish a heavy quark from a light quark -- e.g., t, b? Here energy helps because the lifetime of the charmed quark is fairly long; the lifetime of the b quark also appears to be very long. The technology is developing very rapidly. The semiconductor detector is capable of micron resolution, and with a microvertex detector it should be possible to distinguish charmed and similar particles embedded in the event. Searching for single kaons or single pions is fruitless, because a jet of 100 particles will invariably be accompanied by 10,000 particles. Hence, a different approach is necessary; e.g., one must study the jets in terms of the higher hierarchy of flavors. The pattern of the event should allow identification of the V's in much the same way that strange particles were identified in the presence of a V in early cosmic-ray experiments. Very valuable physics was done with bubble chambers, for example, at BNL, in the 50's, simply by searching for V's (associated production). The basic technique still remains valid today, except for a change in time scale from 10^{-10} to 10^{-13} seconds. Although prospects for distinguishing a gluon from a quark are poor, prospects for recognizing heavy quarks and very heavy flavors are considerably brighter.

Regarding the important topic of user participation in the SSC, Rubbia agreed that the machine will have to accommodate a very large community of experimenters. Of necessity, one must face curtailment in operation of many existing facilities. This trend is already apparent at CERN in the case of LEP. The concentration of effort has meant phasing out the ISR; the big European bubble chamber (BEBC) is slated for closing, as is the muon facility. Clearly, a monumental project like the SSC -- ten times the size of LEP -- cannot be realistically contemplated without substantial reduction in other facilities. Therefore, the only way in which one can maintain a healthy program is by offering as many people as possible the opportunity of participating in experiments. The general tendency has been for experimenters to cluster in rapidly expanding teams. Thus, while a typical SPEAR team consisted of 30 people, the corresponding $p\bar{p}$ group at CERN includes approximately 100 participants. For LEP one expects 300-man teams, and further extrapolation of the trend suggests groups of 1000 people with the next generation machine. Clearly a practical limit must be found, Rubbia argued, if we are not to discourage young people; there must be a way in which a "democracy of experiments" can be assured. It is to be hoped that the very scale of the machine provides the solution, with widely dispersed teams. The machine will have to be a general purpose machine or genuine user's facility -- not a single experiment, as at the Tevatron.

Bjorken concurred, adding that, because of the enormous range of energies covered by the SSC, with broad-band colliding beams and very large cross sections for hard collisions, there may be all sorts of intensive ways of studying these. The argument is very similar to that put forth in support of an experimental program for the CBA: e.g., there should be room for spectrometers of small aperture, and for experiments aimed at investigating the high multiplicity events for which the cross sections are enormous and an understanding of the physics is almost totally lacking. Thus, there would seem to be room for a program as broad as the early ISR program -- in its time not at all uninteresting -- and room even for small experimental groups.

Pursuing this argument, Rubbia injected a reminder that, because of the large cross section, a 20-TeV beam incident on a fixed target will produce a copious yield of W particles; that is, a background of W events produced by beam-gas interactions. Means of exploiting this, as a sort of W factory, will surely be found. Clearly we are entering an entirely new domain in the application of multiple detectors, not simply 4π detectors, for measuring everything with a "steam engine" approach.

The last question to Rubbia concerned the cost of detectors. For most major general purpose detectors, it amounts roughly to a certain price per ton. (Interesting comparisons of large-scale SSC experiments can be made with proton decay experiments.) The basic question, the weight of a standard detector, was discussed at the Berkeley workshop in 1983 (sponsored by the Division of Particles and Fields of the American Physical Society), where some attention was also given to the configuration of such a detector. Essentially, a large magnetic detector (assuming such a detector is still applicable) should have a mass of $\sim$10,000 tons. For comparison, the CDF detector (the major detector under construction at the Tevatron) weighs 2000 to 3000 tons, about twice as much as the UA-1 detector at the CERN collider. Since a detector of the CDF type costs $\sim$\$50 million, the cost of a large detector for the SSC is probably somewhere in the range \$100 to \$150 million. Moreover, there is a linear relationship between the number of experimenters and the amount of money involved (i.e., each participant contributes a certain monetary fraction): at present \$100 to \$200 per physicist. The natural limitation of this trend (Rubbia suggested 10,000 tons as a limit) will tend to fragment experiments into more specialized devices. As a rule, a dedicated machine like LEP has an investment in experimental equipment distributed in major interaction areas (four in LEP) amounting to about half the total cost of the machine (\$250 million out of half a billion dollars in the case of LEP). Clearly this trend cannot continue for the next-generation collider, costing perhaps a total of \$5 billion.

PROSPECTS OF AN ULTRAHIGH ENERGY COLLIDER: A VIEW FROM WASHINGTON

N. D. Pewitt, Office of Science and Technology Policy*

We are in the midst of a critical period in high energy physics. What the community does over the next few years will have a great impact on what the field looks like a decade or more from now. We are faced with very, very difficult decisions about building the kind of immense accelerator needed to answer important and fundamental questions.

A machine on the order of 10 to 20 TeV challenges us on multiple fronts -- technically, politically, and financially. We're talking about a technological system that may be so complex that we have to go outside the physics community to find the skills to design and build it. Its size and impact on the area of the country where it might be built make it an instant political football. And the cost -- well, it makes all the previous problems of funding accelerators seem easy.

The High Energy Physics Advisory Panel has laid out the technical challenge for us. Its report earlier this week recommended "... the immediate initiation of a multi-TeV high-luminosity proton-proton collider project with the goal of physics experiments ... at the earliest possible date."

What are the chances that HEPAP's recommendation will be implemented? I don't know, and neither does Keyworth. This is an open question today. I can try to look at the future of such high-energy physics projects from several perspectives -- the Executive Branch of Government, the Congress, the physics community, and the science community in general.

I have to emphasize that any commitment to proceed with a next-generation accelerator will require that all those factions agree on its importance. It may seem pointless to suggest that anything should have to depend on that degree of unanimity, but it is a fact, and we can't pretend that this is simply a matter for high energy physicists, physicists in general, or even the scientific community.

Let me first give some indications of where I'm coming from personally on this issue. Unlike most of you, I was a late blooming - and I must admit an early-fading -- particle physicist. My membership dates back only to about a decade and a half, and I haven't had any real hands-on experience since the mid 70's. But for most of the past decade, as an official of the Executive Branch of Government, I've been viewing this community from a perspective different from that of most of my physics colleagues.

The march of physics facilities during this period has been instructive. In 1976 there were no construction projects authorized and under way. We then moved at SLAC through the building of PEP, and we've now secured the first funding for the Linear Collider. At Fermilab we've seen what was advertised as an

---

*Currently at Scientific Applications Inc., La Jolla, California.

$18 million accelerator upgrade converted into the nearly completed Energy Doubler project and then into the Tevatron I and II. Here at Brookhaven, we helped in the authorization and initial start of the Isabelle project.

Throughout this period we've watched the progress being made in Europe and tried to see how our own programs complemented and, frankly, competed with theirs. All of us have had our share of successes and disappointments over this past decade in high-energy physics. The Isabelle/CBA experience is of course the toughest situation we've faced. We too have lived with the anguish of having to declare CBA history -- made more difficult by the remarkable technical progress and sound management of the project over the last year or so. Now, however, it's time to put all of that behind us and proceed to develop our future.

What does it take to make a success, to convert an idea into experimental reality? How does one develop a commitment to a good idea and line up support?

The place to start is within the physics community. And let us recognize that approaches are different for different kinds of aspirations. In high energy physics we can't afford to fund a handful of mammoth new experimental facilities -- or even two. Only after an exhaustive analysis of needs and alternatives do we dare take our one shot at building a facility of this magnitude.

For that reason the Federal Government, which is the only conceivable funding source for large basic research facilities, relies on mechanisms like HEPAP to synthesize the judgement of the community. I wouldn't expect any federal initiative to be in strong disagreement with HEPAP's recommendation. But remember, even given the remarkable statesman-like history of HEPAP, especially evidenced in recent deliberations, its recommendations, although clearly necessary, are hardly sufficient to permit funding.

The unanimous recommendation of HEPAP is very gratifying. We must now gain the support of the general community of physicists and scientists. But, the more ambitious the recommendation -- and here I refer principally to its cost -- the more unambiguous the recommendation must be. Frankly, unless the country perceives the technical community as speaking with essentially one voice, there is little chance for public commitment to a major new accelerator.

Let's assume we do have unanimity among the physicists -- that is, that we have agreement on the project to push for and the approach to managing it, and that we agree to concentrate all efforts on that alone. The first challenge is to get the Executive Branch to agree to the proposal and make it part of the President's Budget. And while doing that is a major accomplishment, I've learned the hard way that there are still many possible detours after that.

The Executive Branch, perhaps to your surprise, has a quasi-rational process for making these kinds of decisions. In this particular instance, the agencies involved include the Department of Energy, the Office of Management and Budget, and the White House Science Office. I have worked in all these areas.

Believe me, each has to be convinced independently of the value of any new project. I think you all know that the Science Advisor's heart is with you on this, but he has other serious matters that demand his attention, and must himself depend on advice from others. Last night, for instance, he asked the White House Science Council to review the recommendation to proceed with the Superconducting Super Collider.

Besides considering the inherent value of the new accelerator, these senior officials have to weigh its impact on basic science. They have to ask whether a few billion dollars is best spent on a particle accelerator, even one that makes significant advances, or whether American science would be better served by using that political capital -- because that's what it is -- to increase funding in some other area. Maybe they are also being asked to consider expensive new space missions that would build on the success of the space shuttle and set the stage for future steps in the space program. Or maybe they are anxious to take advantage of rapid developments in molecular biology, a field that some think may rival the microchip revolution for impact in the coming years.

The problem is to convince these decision makers that their time and effort are most effectively spent on a high-energy particle physics machine -- and you can be sure they will consult with the broad scientific community in the process. This means, in all likelihood, that such an initiative is possible only with greater national commitment of resources to basic science.

However, if three conditions are met, I'm convinced it is possible to find additional funds for a truly important new project. First, the rest of the scientific community cannot feel too threatened; second, we must be perceived as being careful in using the large sums available to our field; and finally, most importantly, we must have a credible technical proposal and a sound plan for managing the project.

Clearly this last requirement will take some time. We can debate details, but a defensible project will likely be an issue for the next Presidential term. If we act effectively as a community, we can, however, lay out the elements of the necessary research and planning precursor activities in time for the Fiscal 1985 Executive Branch review.

The next step for an accelerator proposal would be the Congress. That branch of government is driven by different considerations, coming strongly under the influence of local constituency pressure and the political gain that can come from voting for popular causes. To get a multi-billion dollar project of any kind through the Congress there simply has to be a sense of national commitment. That means the strong support not only of the physics community but of a substantial fraction of the decision makers of this world as well. A Super Collider will cost at least as much political capital as anything American science has ever reached for before. It exceeds by far the commitment required to get Fermilab, and the Joint Committee on Atomic Energy that pushed the 200-BeV machine no longer exists.

The path through Congress can hardly be smooth. The Budget Committee has to find room in the general science account to fund the machine. And because of the makeup of that committee, siting of the facility will play an important part. Many scientists who get involved with presenting views to the Congress make the mistake of failing to recognize that real battles occur in the appropriations committees. Authorization committees are useful in providing support and a forum for discussion of issues, but they don't commit the money.

In any case, the combined committees of Congress feel they have done awfully well by the physical sciences for some time. And they have, especially for Fiscal 1984. But remember that, with the looming federal deficit, spending will keep Congress in a political hotseat for several years. It will be hard to convince the members that they should endorse a major facility like this. They'll need a lot of good reasons when the time comes for a major commitment to fund such a project in an area as esoteric as particle physics. It's much easier for them to say no to physicists -- and to the Administration -- and fund items more popular at home.

It will take a significant block of Congressional backing to generate enough support for such a project. Let me suggest this is one case where particle physicists might take a lesson from the nuclear physicists. Look at the kind of support the Southeastern Universities Research Association collected for their proposed $150 million electron accelerator. They had a well developed, winning proposal and managed to line up 13 states -- that's 26 Senators, and there are only 100 -- to back their proposal. They understand that a sound technical approach is needed for Executive Branch endorsement, that political support is necessary for the Congress, and that both are essential for the type of solid, enduring support necessary to get good science done. It's hard to imagine that the high energy physics community could succeed with a Super Collider with less. A broad base of in-depth support is needed.

I don't have a prescription for how all this can be achieved. It will be tough, but I am convinced it can be done. We can regain world leadership in particle physics facilities. Let's keep in mind that we have a unique technical strength upon which to base our approach. We can build on more than a decade of considerable effort and investment to bring superconducting technology to its present state. We are the only nation with this capability. It is ours. We should be proud of our accomplishments. The nation will be ill served if it does not benefit from this investment. We must exploit it; the Superconducting Super Collider offers promise of being the way.

Comments

The question of possible international collaboration was briefly discussed. Pewitt noted that this has been the subject of considerable discussions, including some in the White House Science Council. He also pointed out the existence of the International Committee for Future Accelerators (ICFA), established by the

International Union of Pure and Applied Physics in 1976. Reardon mentioned that international collaboration in this area is also under discussion as a result of the 1983 summit discussions at Versailles. Trivelpiece noted that "there is an enormous difference between inviting participation and setting it as a boundary condition."

Pewitt reminded the participants that the rough estimate for the cost of the SSC, $2 to $6 billion, should be compared with $47 billion expended overall per year in government funding for research and development in the U.S. The expenditures for the Apollo program amounted to approximately $20 billion at its peak (in 1983 dollars).

PLANNING FOR A SUPER COLLIDER
P. J. Reardon, BNL

We have heard about the standard model and its importance; about the machine we'd like to build; about some of the interesting physics, and how to sort out the detector issues; and, perhaps most importantly, about the stop-band resonances that can occur in the transition phase between project idea and project reality. I will follow up with the more nuts-and-bolts issues about how to implement such a project and what it means, with some general remarks.

I will not comment on any of the aspects of the design of the project, its cost, or its schedule, or the way to get it through the political process, or the HEPAP report, or the Subpanel Report. I want to put my remarks in the frame of reference of a project that has the support of a highly respected group of high energy physicists and accelerator physicists. And in that context I would share Tigner's optimism -- that it is not unthinkable to undertake such a project. This is my third decade in this business, and I remember that, when we were building 3-BeV machines (the Cosmotron), the AGS, a factor of 10 higher, was viewed as a tremendous undertaking. Certainly we couldn't do it the same old way; and then, when we set out to establish Fermilab in a cornfield with another factor of roughly 10 -- 400 GeV versus 30 GeV -- that seemed to be an impossible task, and clearly a fresh approach was again called for. Now we're talking about leaping from 1 TeV to 20 TeV, and again this is indeed a big extrapolation and an unprecedented endeavor, but it should by no means frighten us. However, we will have to change some of our ways of doing things, but we shouldn't be afraid to do that. We've done it for the last 50 years, ever since the invention of the cyclotron at Berkeley, and I think our field has always flourished by facing new challenges. My message, before I start on the more general planning issues, is that this is something not to be afraid of. We can do it, and if it's what everybody wants to do, we ought to put our shoulders to the wheel and make it go. We also shouldn't look for some easy system that is already established to do it. We have great laboratories here at Brookhaven and at Fermilab and other places, but they are not, as now set up, the kind of organizations that should do this. Also, we can't look to industry for guidance, because it has not taken on challenges of that nature in many years. The project will demand a much more innovative organization than we have seen before. Its very existence will bring to bear on the project the unique efforts that are required, assuming the right leadership is in place.

I should mention, in the light of what Pewitt said, that in 1979 I was involved in an ERAB long-range planning subpanel for fusion, and the panel came out with a recommendation that we must by all means proceed to build an engineering test facility at a cost of about a billion dollars. A year later this resulted in the Fusion Energy Act of 1980. About six months after that a new design group was established at Oak Ridge, and about six months later there was a round table, where I presented the design group as one of the criteria for a successful project. Subsequently the decision was made not to push fusion so fast, so that project never got off the ground. This demonstrates what Pewitt mentioned: that you've got to form a constituency in order for a project like this to get through the various systems, and the constituency has to start with each and every one of us. If a major project is to succeed, you first must have a sound proposal by a group whose credibility has been firmly established. In the case of SSC, that first criterion has already been met. There is a group, HEPAP, building on the good results of the Fermilab machine and our own results here at BNL on superconducting magnet development, and the machines existing throughout the world (the hadron colliders, including the ISR), which has established a sound technical base for saying this is an idea that can indeed succeed.

The next point is a little more complicated. I refer to what I call the Sears Roebuck mentality: if you can formulate the specifications as briefly as possible, certainly you can order it from Sears. I don't think Sears is going to build this one. You need to get an experienced group or institution, whose performance has been demonstrated, and whose credibility has been earned on similar large R&D enterprises. For this particular project, experienced people obviously are available, but they're not assembled in one group or institution at present. For the SSC we therefore need a new institution that will pull together the experienced people from various existing places. A key issue is the appointment of a strong leader who will remain decisive in spite of problems -- and there will certainly be problems with the SSC. They will vary widely, from administration to technical and scientific matters and to personnel, and more. You also require the ability to attract the best people to form a multi-disciplinary team able to communicate effectively with all the many disciplines and institutions involved. You need a first-rate team, brought together by a leader able to make the decisions and put up with the guff.

The organization, site, and group dynamics must be sufficiently attractive to obtain, retain, and satisfy highly competent and well qualified individuals, which means bright people who are difficult to manage but who will work under trying conditions, in spite of large bureaucracies and high visibility. They must hang in there when the going is rough, and that's the kind of group that builds great things, and that's what's needed. I believe, although many disagree, that you cannot get by with a small design group, with three or four people writing specs, with the work being done in ten other places. You need a very closely knit, highly motivated design group, working together in one place. It can certainly parcel out activities, even design activities, but the control of the design and the interface management should be done by a centralized group. An important issue that we sometimes forget is that we think an experimental physicist will write out a specification, but he goes away and works on an existing machine for 5 years and then comes back and asks for the sheet. That won't work. You must have constant dynamic interaction between those who are going to use the machine and those who are designing and building it. I repeat that the concept of a group off in the woods, even in Texas, wouldn't work. You must find a way for the people working on the machine to recognize the importance of rubbing elbows with the physicists who want to use it.

Another issue is responsibility and credit. We tend to point at each other when trouble develops and the question is: who's to blame or who took the risk? However, when the glory is passed out, it invariably goes to the people at the top. That's wrong. This project, like many others of similar scope, requires close coupling between the elements of government and the elements of science and technology, and those in government play at least as crucial a role as the people in the field. This project will require much close constructive interaction between the people in the federal system, who must not only get it approved but nurture it and review it, and those building it, so that the people involved on both sides not only share the risk but participate in the glory.

Since this is a school, I can offer a few general project principles. My view is that we now have a good idea that's been blessed by competent people, and that they employed the correct evaluation process. (I also thought they were pretty good people when I worked with them on other issues.) That step is behind us. Now we have to decide on our actual approach to the project. I cannot describe it precisely, but it should start with some ad hoc group of experts defining the various boundary conditions and the steps required to translate this concept, or preconceptual idea, into a working system. Such a group would meet for, say, 3 to 6 months and come up with the game plan to be proposed to the DOE.

The DOE, meantime, would be considering what steps to take upon receipt of this game plan. I think what they need to do is to tackle the problem of how to select the prime contractor that will see the project through from project definition through commission and operations. I'm a very strong spokesman on this subject, as most big activities don't start that way. Rather, a team is formed quickly for the conceptual design phase. When, finally, the conceptual design is in hand, a new group is formed. It immediately discards the conceptual design and starts over, and that causes cost overruns and further delays. I think, if it's important to us to have the machine running quickly, we ought to pick the prime contractor as soon as possible and give him the responsibility for the entire project, assuming of course that he does it well. If he doesn't do it well, fire him. That prime contractor, by the way, is most likely not an existing institution like Associated Universities, Inc. (AUI), or Universities Research Associates (URA). The size of this project is such that you might want a joint venture with an AUI or AUI/URA type contractor, a major architecture-engineer manager (such as Bechtel), a systems engineering organization (such as an aerospace company), and perhaps a manufacturing organization for hardware (such as General Electric). The contractor would not build each and every element of the project, but would agree to mold a group that would bring in the disciplines required to set up and run the project. That group, once it's selected, would establish a project concept definition team, and a director, and would present their plan to DOE. The project definition team would do the conceptual design and set up systems management and engineering.

Those are nice words, but what do they mean? A project like this will require an in-depth approach to reliability, availability, quality assurance standards to be used by everybody working on the project, and an interface and configuration management system so that all its elements are pulled together as one coherent system, and that's what I call systems engineering and management. This question of one prime contractor I think is important. That contractor could decide that he has one group of people for the design definition phase, another group for the design and construction phase, and a third group for the commissioning and operations phase. However, he should look at all three phases, and develop an integrated plan for the 10 to 12 years of project execution.

Site selection is not up to the prime contractor. The agreement between the DOE and the prime contractor should be that the design team will move to the site selected, although it will, with its architectural and engineering management capabilities in the site selection process, assist the DOE in determining the requirements of the machine, the geological conditions of concern, and so forth. Selecting the site and getting it approved is not the job of the scientists or even the managers; it is the job of the Federal establishment. Lobbying with a few Congressmen won't hurt.

I think the key question from the technical side (this is not the issue that Pewitt addressed) is how to get this project from a conceptual stage, which has taken weeks rather than months to accomplish, into a viable proposal with an acceptable cost estimate and schedule. You must get a prime contractor established as soon as possible. If he does the job right over the next three years, he sees it through; if he doesn't, you get rid of him.

Another kludge that has developed in recent years, sometimes not effectively, is to parcel out sub-elements of a project and require the various contractors to report to the government. I don't believe in that. I think one contractor should be responsible, with buildings, hardware, installation, and operation all assigned to him. He may, and should, parcel out sub-elements to other institutions -- laboratories, universities, and industry -- but there should be one responsible contractor, and the buildings and the hardware should not be done separately. I mentioned the need for a strong engineering and management group from the outset, and that group should have the continuing responsibility from conceptual design to operation, to ensure that the systems engineering aspects are well done. This ensures a high probability -- it's a very big macnine -- of success at initial turn-on. In my opinion, this systems engineering and management group must be some combination of existing laboratory people grouped under the new contractor, augmented by people with training in industry in some things such as quality assurance, configuration management, reliability, stress analysis, and so forth. And, right from that beginning, if the prime contractor has industrial participation in the board of directors, he will recognize this need for some form of cooperation between the scientists and engineers trained in the laboratory environment and those trained in the industrial environment.

R&D, of course, should not be done exclusively by this group. Existing facilities here at BNL and at Fermilab, Cornell, Berkeley, etc., should be used, but the management of the R&D, and the definition of its goals and requirements, should at some point be taken over by the systems management group of the new prime contractor. I am not suggesting a long wait; as soon as this group is activated, it should indeed assume the coordinating function for all of the R&D.

For a project this big, we may have to, and perhaps should, change our approach. As the central management gets organized, rather than designing in detail each and every part, it may let out certain major subsystems to industry on a competitive basis, for production of turn-key packages, for instance, the cryogenic system or the power supplies. This approach has been used very successfully in Europe and Japan, and I think is one which we could utilize more effectively as well. For a project of this size, it's folly to presume we can develop in-depth strengths in each and every professional approach to each and every problem. We need to parcel out more subsystems than we have in the past, without losing control of the integrated design.

Comments

Pewitt drew attention to the role of Bellcom in aiding NASA (Bellcom was formed by Bell Telephone Laboratories to aid NASA as an overall systems integrator in planning the Apollo project) as conceptually somewhat equivalent to the prime contractor discussed by Reardon. It is perhaps not generally known, "but the telephone companies switched this whole nation from rotary dialing to a digital phone system in a matter of a fraction of a second ... without us being aware of it." Moreover, it was not a technical person who came up with that approach to the Apollo project. The administrator of NASA had previously been Director of the Office of Management and Budget. Thus, it should not be assumed that the scientific community alone possesses all the talents necessary to see a project of the magnitude of the SSC through to completion.

DISCUSSION OF P$\bar{\text{P}}$ VS. PP AND THE QUESTION OF APERTURE AND LUMINOSITY

Voss opened a discussion on the relative merits of a proton-antiproton versus a proton-proton collider, noting that the choice of pp appears to be dictated mainly by its higher luminosity, yet Rubbia seems to express doubts about the wisdom of excessive luminosity (see the comments following Rubbia's presentation). Bjorken pointed out that at the SPS we are accustomed to subenergies from 40 GeV up to $\sim$200 GeV for the hard collisions of quarks and gluons, whereas for the SSC we are talking about a range from a few TeV up to perhaps 8 TeV, with a luminosity of 10^{33} -- an impressive change of scale, even by Tevatron I standards. He suggested that perhaps one should contemplate running the machine for even a few years at 10^{31} over a range of, say, 1 to 6 TeV. One should not underestimate physics at 10^{31}. Both technical and cost considerations will very likely be determining factors.

Rubbia recalled a paper by Huson et al. at the Snowmass Workshop (sponsored by the Division of Particles and Fields of the American Physical Society in 1982) which explains very clearly that 10^{31} is a viable luminosity, 10^{33} is not. As the energy increases, all kinds of problems must be faced: more particles emitted, rising cross sections and backgrounds. High luminosity experiments are very tempting, but in practice very difficult to perform. One of the great advantages of ep colliders is that they provide high rates for rare processes without the high background of low energy particles from shallow collisions characteristic of hadron colliders. Rubbia's suggestion to incorporate a magnetic field was an attempt at circumventing this very problem. In any case, Rubbia stresses the importance of not underestimating the long lead time and technology required for detector development -- comparable, in fact, with that required for the collider itself.

Returning to the question of p$\bar{\text{p}}$ vs. pp, from a technical point of view, Tigner recalled that at the Cornell Workshop (Technical Workshop on a 20-TeV Hadron Collider, March-April, 1983) he was initially inclined to regard a p$\bar{\text{p}}$ collider as obviously cheaper than a pp collider, and therefore thought it should be seriously considered. The conclusion at Cornell, however, was that up to a luminosity of 10^{33} a pp machine would cost about the same as a p$\bar{\text{p}}$ machine. The argument goes as follows. A p$\bar{\text{p}}$ collider requires rather large-aperture magnets to ensure that the beams are well separated and that the collisions are confined to a few locations around the ring where the beams suffer from tune shift. Since the cost of a magnet is a sensitive function of aperture size, one might consider instead two separate rings of very small-aperture magnets. Nevertheless, the p$\bar{\text{p}}$ machine costs about the same as a pp machine, when one takes into account the antiproton source ($\sim$\$100 million for a luminosity of 10^{31}), whereas in a proton collider luminosity doesn't cost very much up to $\sim 10^{33}$. Above 10^{33} the cost of shielding becomes a dominating factor.

Lindelbaum argued that, if the difference in cost is not excessive, it would be more logical to proceed with a high luminosity machine; a machine of luminosity 10^{33} can always be operated at 10^{29}. Moreover, the history of accelerators shows that the thirst for luminosity has always exceeded the original anticipation of the machine builders and experimental planners.

Rubbia agreed with Tigner's observation of the necessity for adequate aperture in a $p\bar{p}$ collider, but expressed considerable uncertainty as to what is adequate. Recently a machine study was undertaken at CERN to assess experimentally the effect of beam-beam interactions due to the coexistence of two beams in the same vacuum chamber. The results have been very encouraging, indicating that the estimate of $\sqrt{10}$ standard deviations assumed at Cornell is perhaps too conservative. Nevertheless, he agreed that, since this will be the ultimate machine, starting with a marginal luminosity would be a grave mistake. The matter requires much further study.

Danby recalled that, at Fermilab and CERN, the alternative of installing a second proton ring in the tunnel was originally considered in view of the projected cost of a $p\bar{p}$ source ($100 million at the SPS). It may be an add-on option for a future machine; in any case he thought it likely that the end product will be a pp machine. (Rubbia disputed the cost of the CERN antiproton accumulator, ICE, saying it was actually closer to $15 million. Its improvement will probably require an additional $20 million.)

Voss asked whether, at very high energies, a $p\bar{p}$ machine may not in fact be the cheaper solution, since here beam-beam interactions at a distance become negligible, while the aperture requirement remains the same because of the many distortions, etc. Tigner replied that this would be so only if a lower luminosity is acceptable; a luminosity increase of a factor of a hundred is simply not feasible technically. Reardon noted that it is not merely a question of cost trade-off between, say, two-in-one magnets (i.e., two coils in a common iron yoke) and the $\bar{p}$ production system and associated accumulator ring; i.e., not only the cost of the main ring must be considered but also that of the various injection systems. To proceed initially with a $p\bar{p}$ collider and later implement a pp system for high luminosity, Tigner observed, is technically feasible but expensive. Voss agreed; starting with one ring would only superficially appear to be much less expensive than starting with two rings. One participant even went further, noting earlier arguments that two (pp) beams in very small aperture two-in-one magnets actually appears cheaper than two $p\bar{p}$ beams in single magnets of larger aperture, to which Tigner agreed.

APPENDIX: THE SSC TWO YEARS LATER
P. Dahl, BNL

Since the Summer School in 1983, significant progress has been made in laying the foundation for the SSC. Following the HEPAP Subpanel's Woods Hole report issued in July of 1983, recommending the SSC as the highest priority for the U.S. high-energy physics long-term program, and the concurrence of HEPAP itself soon thereafter, a preliminary ad hoc inter-laboratory R&D panel was formed in August under M. Tigner to explore the technical possibilities in greater depth, even before DOE's formal acceptance of the recommendation. In September, at the request of DOE, a HEPAP Subpanel on Advanced Accelerator Research and Development (AARD), chaired by W. Panofsky of SLAC, was established to advise DOE on the planning of an R&D program and its management, particularly in FY 1984. The Subpanel heard presentations from various laboratories and universities interested in contributing to, or already engaged in, the R&D, and suggestions for unified management of the program to ensure it could push forward as fast as possible. As a result of these discussions, the Subpanel urged a series of further workshops to consider physics objectives of the SSC and accelerator-related issues. Moreover, it recommended an "interim line managerial structure" for coordinating the various exploratory designs and R&D on multiple technical approaches carried out in parallel by the different working groups at the various laboratories. Consequently, in January of 1984 a Reference Design Study Organization (RDSO) was drawn up and activated, under overall supervision of the four high energy physics laboratory Directors (Fermilab, Cornell, BNL, SLAC) and the leadership of M. Tigner. It was coordinated by representatives from Fermilab, BNL, and LBL (the Reference Design Study Office being located at LBL), and advised by a group of accelerator experts under P. Reardon.

The explicit responsibility of the RDSO was to establish a minimum number of accelerator reference designs (based on various assumed magnetic field levels, which impact significantly on the SSC ring size), and a corresponding number of reference facility designs, for the purpose of establishing technical feasibility, obtaining first-order cost estimates for SSC construction, and assessing further R&D requirements. The goal was to have the Reference Design Report in DOE hands by June, in time for the DOE Secretary's formal decision in August on further R&D funding for FY 1985.

From the many discussions leading up to RDSO phase ("Phase zero"), the primary parameters for the design of a pp collider were generally agreed upon among the various parties:

E_{beam} = 20 TeV
L up to 10^{33} cm^{-2} sec^{-1}
4 interaction regions developed initially, 2 undeveloped
E_{inj} = 1 TeV
$\leq$10 interactions/crossing

The reference design, as noted, would be based on three field levels, ranging from high (6.5 Tesla) to low (3 Tesla). One field value was for expediency, to be emphasized (6.5 Tesla, corresponding to a ring circumference of 90 km), with an estimate made of the range in machine cost implied by the other field levels. The reference accelerator systems design would encompass a corresponding self-consistent machine design (lattice), preliminary engineering of the major accelerator systems (e.g., magnet system, accelerator housing, injector system), and conceptual designs of sub-systems (intersection regions, etc.). The reference facility design would include support systems (buildings and utilities), and the experimental areas. Design of the various conventional facilities and site utility systems was to be carried out by an architect/engineering firm selected at the end of January 1984. Excluded from initial consideration and costing were site acquisition costs (it being assumed that the eventually selected site host would bear this cost), pre-construction R&D, and cost of experimental equipment. The end result was an estimated cost for an SSC facility based on the three magnet technologies ranging from $2.70 to $3.05 billion (FY 1984 dollars), including a contingency fund, suggested by DOE, of about 20%.

In August 1984 DOE gave approval for one year of R&D effort, formally launching the project. It was divided by DOE into two main phases. Phase I refers to the R&D and design activities prior to any construction, and is further divided into two parts: (Phase I-A (FY 1985 activities), and Phase I-B (activities into FY 1986 and beyond, up to start of construction). Phase II refers to actual construction. A Board of Overseers was appointed to hold primary fiduciary responsibility for the project, with the Universities Research Association, a consortium of 56 leading research universities, acting as parent organization to this board. The board established a Central Design Group, housed at LBL under M. Tigner as Director. It is formally charged with coordinating and directing the pre-construction R&D, providing technical support for the DOE SSC site selection process, and preparing a Conceptual Design Proposal as well as a realistic construction plan.

Among the ongoing tecnical activities in the spring of 1985 were those leading to the issuing by DOE of a Site Information Request and Preliminary Site Invitation, and the establishing of a number of technical review panels on various key technical questions concerning the SSC: magnet design type selection (to be completed by October 1985), plans for a large-scale R&D magnet test facility (the "magnet string test"), magnet aperture, vacuum (mainly the effects of synchrotron radiation), and machine commissioning and operation.

The goal of all these activities is to commence construction in October of 1987.

WHERE IS THE SSC TODAY?*

Maury Tigner
Supercollider, URA Design Center, LBL, Berkeley, CA 94720[†]

The SSC is a high luminosity pp collider designed to achieve 40 TeV in the center of mass. Depending on the final magnetic field chosen, the main ring will be between 90 and 165 km in circumference. Construction of the SSC has been recommended to the DOE by HEPAP for completion in the early 1990s. The Universities Research Association (URA) has been designated by the DOE to do R&D and prepare a design proposal, construction plan, and cost estimate. Model magnets are being tested, and a field level will be chosen before October 1985. A design proposal will be submitted in April 1986.

PHYSICS REQUIREMENTS

Penetration of the 1-TeV mass domain for elementary interactions is a major priority for experimental particle physics in the next decade and beyond. Extension of current understanding of elementary physics as well as surprises can be expected in this energy regime. If the Higgs mechanism is responsible for dynamic symmetry breaking and if its mass is 1 TeV or greater, the present electroweak dynamics will surely become a strong interaction at that energy and qualitatively new phenomena will occur at energies of 1 TeV and above. The other postulated means of dynamic symmetry breaking have similar consequences. Perhaps the quarks and leptons are not the ultimate constituents of matter. They may be composites of more basic entities (techniquarks, preons). Such ideas inevitably lead to families of yet undiscovered particles with masses in the range 0.1 to 2 TeV or to evidences of compositeness to be seen in collisional processes with subenergies in the same range. Perhaps these manifestations will be detected at the Tevatron Collider, but probably higher energies are needed. Already the CERN collider is providing hints of phenomena not explicable in the standard picture or a simple extension of it. Some physicists believe the peculiar events (so-called monojets)are evidence of supersymmetric particles, necessary consequences of a fundamental (broken) symmetry between fermions and bosons. Others have other explanations, conventional and unconventional. Regardless of the outcome, sub-energies well above 100 GeV = 0.1 TeV are clearly of vital interest. The $Sp\bar{p}S$ collider, with its total collisional energy of 0.6 TeV, and the Tevatron Collider, with 1.8 TeV available soon, can explore

---

* SSC-33
[†] Operated by Universities Research Association for the Department of Energy.

thoroughly the mass range up to about 0.3 TeV. The SSC, with its 20-TeV proton beams, will be able to extend this to roughly 3 TeV, more for some processes and less for others.

Technicolor, supersymmetry, and other theories can sometimes make rather specific predictions of the magnitudes of cross sections for new phenomena. Using the constituent-constituent energy distributions with the theoretical cross section, one can estimate the probability of creation of a Higgs or other particle per collision between protons (or protons and antiprotons) of a given total energy. [Some examples from a recent compilation[1] are shown in Fig. 1.] For a conventional heavy Higgs particle decaying into W pairs, the SSC can reach out to a Higgs mass of 1 TeV. For comparison, a proton-antiproton collider with a plausible luminosity of 10^{31} cm^{-2} s^{-1} and total energy of 10 TeV could not search successfully for such decays. For supersymmetric particles, the SSC could find a gluino (the spin-1/2 partner of the gluon) if its mass were less than about 1.6 TeV, whereas the 10-TeV, lower luminosity machine could probe up to 0.4 TeV; the limits for squark (spin-0 partner of quark) masses would be 1.5 TeV and 0.3 TeV, respectively.

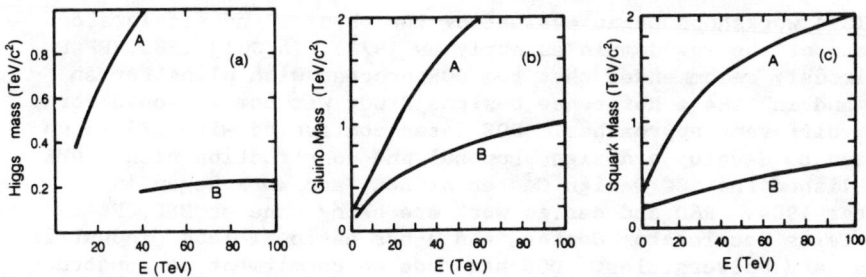

Fig. 1. (a) $H \to W^+W^-$. The SSC (curve A at 40 TeV cm) can plausibly discover any Higgs particle with a mass between 0.2 and 1.0 TeV; the lower-luminosity collider has a very narrow window for discovery of a heavy Higgs.

(b) Supersymmetric gluinos. The upper discovery limit is 1.6 TeV for the SSC, and 0.3 TeV (0.5 TeV) for a lower luminosity collider with 5-TeV (10-TeV) beams.

(c) Supersymmetric squarks. The upper limit in mass is 1.5 TeV for the SSC; the lower luminosity collider could reach to 0.4 TeV at the same beam energy.

The details of the theoretical estimates and machine assumptions can be challenged, but it is clear that effectively probing into the TeV mass range requires a very high-energy, high-luminosity collider. Quantitatively, a goal of 20 TeV per beam and a luminosity of 10^{33} cm^{-2} s^{-1} seems prudent; less energy and/or luminosity begins to compromise the discovery potential. The argument for the highest possible energies is reinforced by the possibility of discrete energy thresholds. If the maximum available energy lies just below the threshold for some new physics, that physics will not be discovered, regardless of the luminosity.

Figure 1 shows discovery limits for some new particles in hadron-hadron colliders. A discovery is defined as the creation of from 10 to 100 real and uniquely identified events in one year of data taking, after allowance for backgrounds and other spurious signals. The center-of-mass energy in the collider is plotted against the mass of the new particle. The curves represent approximate upper limits on the mass of particles of each type, discoverable with a collider of a given beam energy and luminosity. For curve A the collider luminosity is 10^{33} cm^{-2} s^{-1}, projected for the SSC; for curve B it is 10^{31} cm^{-2} s^{-1}, appropriate for a proton-antiproton collider.

ICFA workshops began evaluating the physics and accelerator aspects of the TeV domain as early as 1978. In July 1983 HEPAP unanimously recommended that the DOE proceed with plans for an SSC, and in 1984 a Reference Designs Study was commissioned for three different approaches. DOE later contracted with URA to do R&D and to develop a design proposal and construction plan. URA established the SSC Design Center at LBL, and work began in October 1984. R&D and design work are being done at BNL, FNAL, LBL, Texas Accelerator Center, and other national labs, industrial firms, and universities. DOE has made no commitment to construct the SSC.

REFERENCE DESIGNS STUDY (RDS)

The RDS was done in February - April 1984 by about 150 scientists and engineers from U.S. accelerator labs, universities, and industrial firms.[2] The RDS developed three approaches for the following primary parameters:

Beam energy	20 TeV
Luminosity	10^{33} cm^{-2} sec^{-1}
Particle	pp
Number of interaction regions	6

Designs of technical systems were based on three possible magnet designs: a 2 in 1 cold iron, cosine theta magnet at 6.5 T with horizontal beam separation; a 1 in 1 cosine theta magnet without iron immediately around the coil intended for vertical beam separation; and a 2 in 1 cold iron "superferric" magnet with vertical beam separation. (See Table I.) Detailed parameters are different for the three designs; those for the low field design are listed in Table II.

Table I. Major Design Features

	Design A	Design B	Design C
Central dipole field [T]	6.5	5.0	3.0
2-in-1 cryostat	yes	no	yes
Cold iron yoke	yes	no	yes
Conductor dominated	yes	yes	no
Field shaped by iron	no	no	yes
Magnetically-coupled apertures	yes	no	no
Saturated iron	yes	no	yes
Length of dipole magnet [m]	17.5	14	140
Inside coil diameter [cm]	4.0	5.0	2.5x2.4
[a]Main ring circumference [km]	90	113	164

[a]Pole face dimensions.

Typical lattice functions in the high field design for normal arcs and high luminosity experimental insertions are shown in Figs. 2 and 3 respectively.

For this large machine, the parameters are close to common experience. One possible exception is the synchrotron radiation from the proton beams, amounting to some 8 KW per beam in the high field design, which would not be remarkable if the beam tube were not to be cryopumped. The synchrotron radiation falling on the cryosorbed gas molecules might release them at a rate that could spoil the vacuum, although calculations indicate that this will not be the case. The physical mechanisms involved are complex enough, however, to require experimental demonstration to verify the calculation. Excessive gas desorption will necessitate an inner liner, permeable to gas but having a low transmission for scattered photons.

These studies established the technical feasibility of the SSC. The cost is a central issue, and considerable RDS effort was devoted to its estimation, guided by a detailed work breakdown structure. Industrial firms, engineering consultants, and national laboratory staffs experienced in building components and subsystems contributed to the estimates, which are detailed in the RDS[2] and summarized in Table III and Fig. 4.

Table II. Abridged Parameter List for Low Field Design

General Parameters

Maximum energy per ring [TeV]	20
Luminosity of each interaction region [cm^{-2} sec^{-1}]	10^{33}
No. of interaction regions	6
Injection energy [TeV]	1
Magnet type	superferric
Standard beam separation (vertical) [cm]	12.4
Peak magnet field [T]	3.0
Peak current [A]	10,000
Bunch spacing of overall design	variable
Bunch spacing used in this list [m]	10
No. of events per crossing (at a cross section of 100 mb)	3.3
Circumference [km]	164.4
Orbit frequency [kHz]	1.82
Orbit period [μsec]	548
No. of particles per bunch	1.45x10^{10}
No. of bunches per ring	16440
No. of particles per ring	2.38x10^{14}
Average beam current [mA]	69.7
Invariant transverse emittance [μm]	1.0
Beam-beam tune shift per crossing	0.0017

Magnets and Lattice

Amplitude function at interaction point [m]	1
Free space at interaction point [m]	±20
Crossing angle [μrad]	30-100
Length of interaction region insertion [m]	750
Phase advance for interaction region insertions [deg]	360
Length of standard half-cell [m]	150
Phase advance per cell [deg]	80
Magnetic (physical) length of dipole [m]	139.5 (140.2)
No. of dipoles per half-cell	1
No. of standard half-cells per ring	954
No. of half-cells per dispersion suppressor	4
No. of dipoles per dispersion suppressor*	2
No. of dispersion suppressors per ring	18
No. of dipoles per ring*	990
No. of standard quadrupoles per ring	1020
No. of utility insertions (abort, injection, RF)	3
Length of utility insertion [km]	2.0
Phase advance for utility insertion [deg]	200
Nominal tune (both planes)	121.76

*In units of standard length (139.5 m).
(Table 2 continued on next page)

RF System Related Parameters

Frequency [MHz]	360
Peak voltage per turn [MV]	35
Total cavity length per ring [m]	43.7
Acceleration period [sec]	1000
Momentum spread at injection, σ_E/E	2.5×10^{-4}
Momentum spread at 20 TeV	5.0×10^{-5}
Bunch length at 20 TeV-rms [cm]	7.0
Longitudinal emittance at injection (95%) [eV·sec]	1.1
Longitudinal emittance at 20 TeV (95%) [eV·sec]	4.4

Injection System

Linac energy [GeV]	1
Length [m]	200
Beam current [mA]	65
Ion species in linac	H⁻
Invariant transverse emittance [μm]	0.22
Low energy booster energy [GeV]	70
Circumference [km]	1.2
RF frequency [MHz]	60
Cycle time [sec]	3
Bunch coalescing frequencies (typical) [MHz]	60/k, k=1,2,...5
High energy booster energy [TeV]	1
Circumference [km]	6
RF frequency [MHz]	60
Cycled time [sec]	45

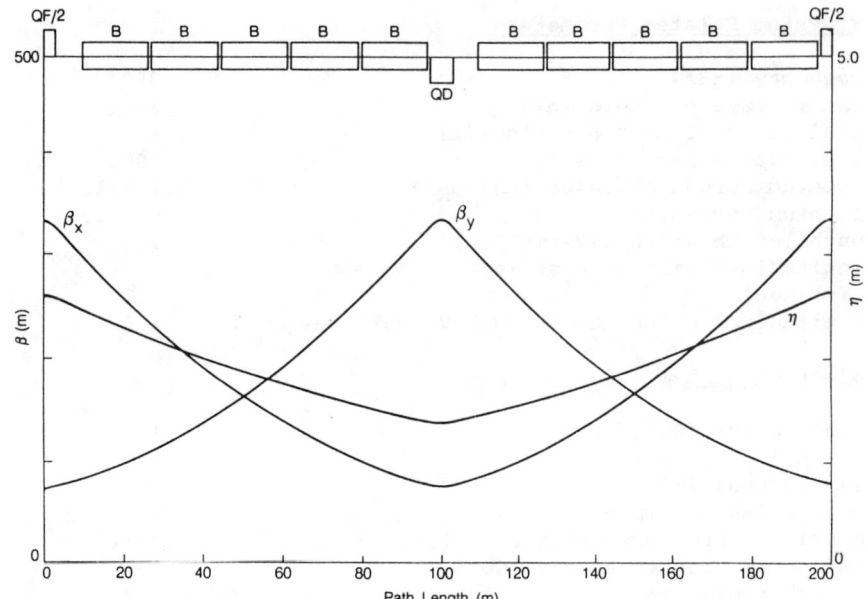

Fig. 2. The betatron amplitude functions (β_x, β_y) and the dispersion function (η) in the regular cells for the high field lattice.

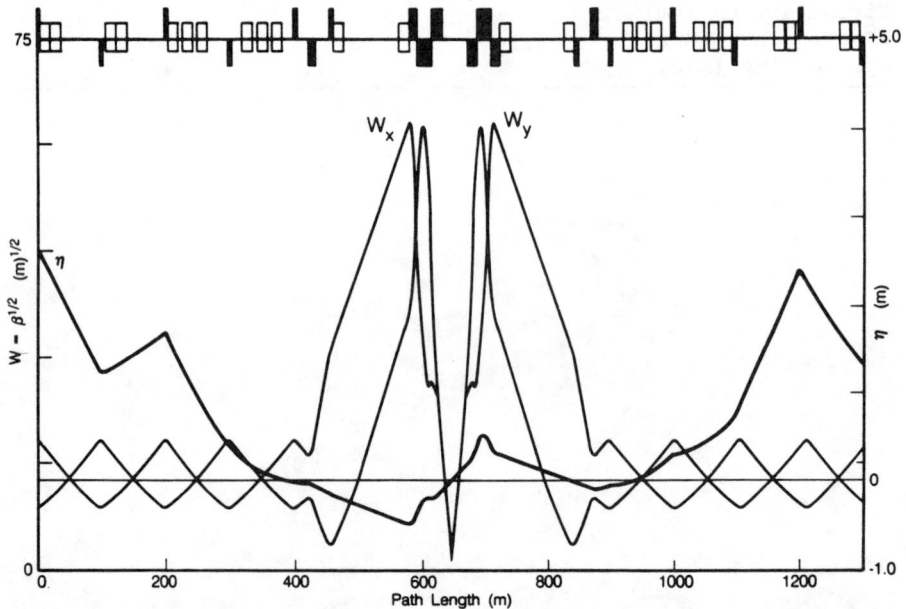

Fig. 3. The lattice functions in both planes (W_x, W_y) and the dispersion function (η) in the experimental insertions and the adjacent dispersion suppressors.

Table III. Total Project Cost Summary for High Field Design
(FY 1984 M$)

1.	SSC Laboratory		2724.9
	1.1 Project Management and Administration		113.5
	1.1.1 Construction Project Management	59.0	
	1.1.2 Laboratory Support Services	54.5	
	1.2 Central Laboratory Facilities		127.0
	1.2.1 Conventional Construction	86.0	
	1.2.2 Equipment	41.0	
	1.3 Injector Facilities		186.8
	1.3.1 Conventional Construction	39.6	
	1.3.2 Injector Systems	147.2	
	1.4 Collider Facilities		1402.4
	1.4.1 Conventional Construction	398.7	
	1.4.2 Collider Accelerator Systems	1003.7	
	1.5 Experimental Facilities		87.4
	1.5.1 Conventional Construction	87.4	
	1.6 Systems Engineering and Design		255.5
	1.6.1 Conventional Construction	97.9	
	1.6.2 Technical Components	157.6	
	1.7 Contingency		552.3
	1.7.1 Conventional Facilities	164.6	
	1.7.2 Technical Components	387.7	

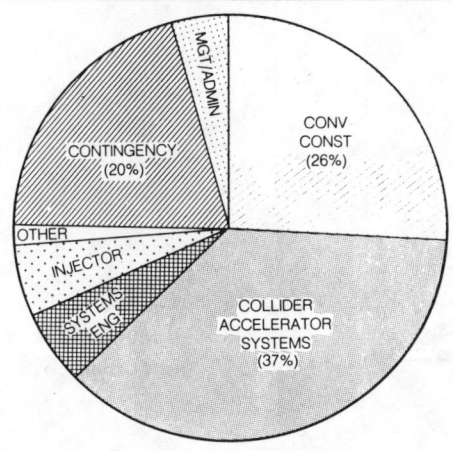

Fig. 4. Pie chart showing cost distribution.

OBJECTIVES AND ACCOMPLISHMENTS IN 1985 AND BEYOND

The single most costly technical sub-system is the superconducting magnet system. Conventional construction, largely of the tunnel, is also a principal cost element. These items received particular attention. Magnet costs were estimated with the important assumption that the critical current of the superconductor at 4.5 K, 5 T would be at least 2400 A/mm^2 by the time SSC magnets went into production. Low heat leak to the low temperature parts is also important for economical magnet design and was emphasized. Substantial progress has been made in these areas.

In the area of superconductors, significant advances have been made in current-carrying capacity of commercially available NbTi cable. Improved understanding of the fundamentals and application to improved processes has resulted from close university-laboratory-industry collaboration with DOE support. (See Fig. 5[3].) Already 2700-A/mm^2 material is commercially available and even higher densities are expected to be available soon.

In the area of low heat leak, significant demonstrations of efficient supports and insulation blanket schemes have been made. For example, FNAL,[4] in a realistic, 12-m-length model using multilayer insulation blankets and improved supports for the cold mass, measured a combined 4.5 K and 10 K static heat load of <1/4 W/m. Not only is this level of heat leak satisfactory but it agrees with the calculations, giving confidence in cost estimates for the cryogenic systems.

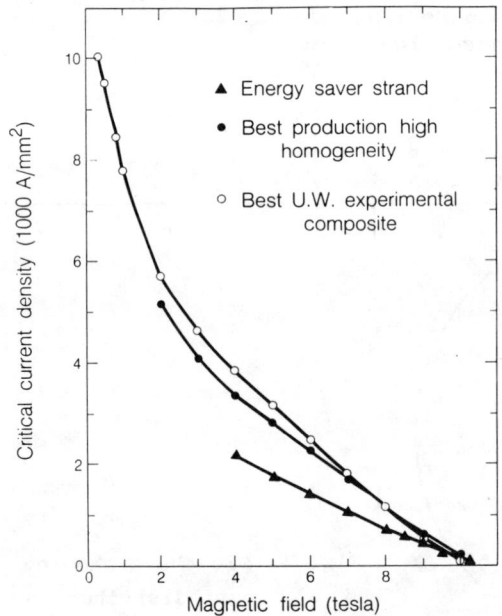

Fig. 5. History of superconductor performance.[3]

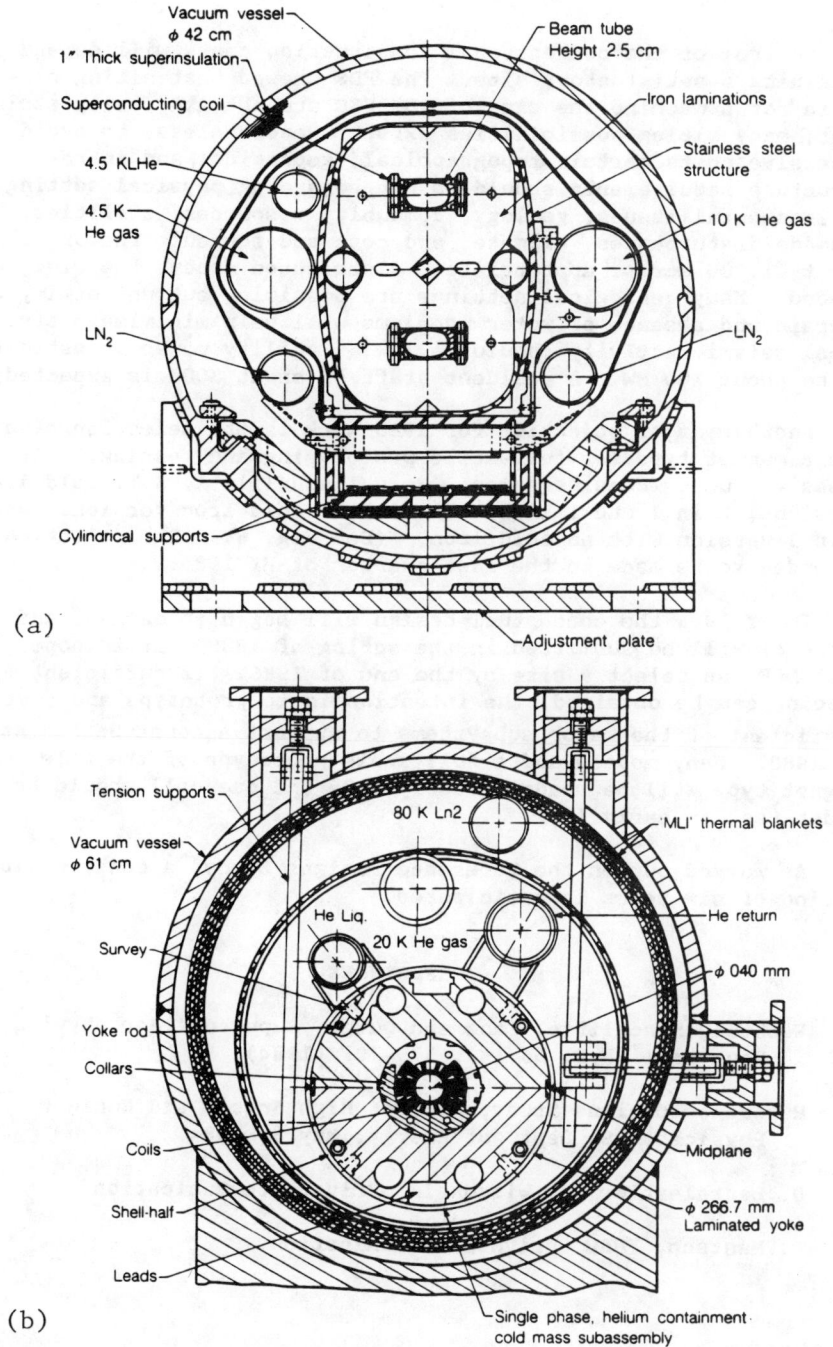

Fig. 6. (a) A low field superferric type of magnet; (b) high field cosine theta type.

Control of the conventional construction costs will depend on obtaining a satisfactory site. The RDS showed that siting criteria for a machine the size of the SSC are not abnormally stringent, many siting possibilities exist. Nevertheless, to avoid excessive costs, certain topographical, geological, and infrastructure requirements should be met regarding physical setting, environmental issues, geology, community resources, utilities, manmade disturbances, climate, and cost and schedule factors. About 11,000 acres, appropriately distributed around the ring, are needed. Many geological settings are possible, but uniformity of terrain and absence of water problems will help minimize costs. Local seismic activity should be low. Facility power is estimated to be about 100 MW. A resident staff of about 3000 is expected.

Another major objective for 1985 work is the selection of a basic magnet type for full-scale prototyping and testing. Five types are under consideration: 2 in 1 and 1 in 1, 3 T, cold iron versions; 2 in 1 and 1 in 1, 6 to 6.5 T, cold iron versions; and a 1 in 1 version with no cold iron. (See Fig. 6.) The selection is intended to be made in the last quarter of FY 1985.

In FY 1986 the conceptual design will begin in earnest and a proposal will be submitted in the spring of 1986. It is hoped that DOE can select a site by the end of 1986. If sufficient funding can be obtained, the intention is to prototype and test sufficient of the major subsystems to support a construction start in 1988. Many models and a full-scale prototype of the selected magnet type will be ready in early 1986 and one cell should be under test in early 1987.

As worked out in the Reference Designs Study, a construction period of six years is anticipated.

REFERENCES

1. Eichten, Hinchliffe, Lane and Quigg, Supercollider physics, Rev. Mod. Phys. 56 (4), 579-707 (1984).

2. Reference Designs Study, Div. of High Energy and Nuclear Physics, U.S. Dept. of Energy, May 8, 1984.

3. D. Larbalestier, U. Wisconsin, Private communication.

4. P. Mantsch, FNAL, Private communication.

AIP Conference Proceedings

		L.C. Number	ISBN
No. 1	Feedback and Dynamic Control of Plasmas – 1970	70-141596	0-88318-100-2
No. 2	Particles and Fields – 1971 (Rochester)	71-184662	0-88318-101-0
No. 3	Thermal Expansion – 1971 (Corning)	72-76970	0-88318-102-9
No. 4	Superconductivity in d- and f-Band Metals (Rochester, 1971)	74-18879	0-88318-103-7
No. 5	Magnetism and Magnetic Materials – 1971 (2 parts) (Chicago)	59-2468	0-88318-104-5
No. 6	Particle Physics (Irvine, 1971)	72-81239	0-88318-105-3
No. 7	Exploring the History of Nuclear Physics – 1972	72-81883	0-88318-106-1
No. 8	Experimental Meson Spectroscopy –1972	72-88226	0-88318-107-X
No. 9	Cyclotrons – 1972 (Vancouver)	72-92798	0-88318-108-8
No. 10	Magnetism and Magnetic Materials – 1972	72-623469	0-88318-109-6
No. 11	Transport Phenomena – 1973 (Brown University Conference)	73-80682	0-88318-110-X
No. 12	Experiments on High Energy Particle Collisions – 1973 (Vanderbilt Conference)	73-81705	0-88318-111-8
No. 13	$\pi\text{-}\pi$ Scattering – 1973 (Tallahassee Conference)	73-81704	0-88318-112-6
No. 14	Particles and Fields – 1973 (APS/DPF Berkeley)	73-91923	0-88318-113-4
No. 15	High Energy Collisions – 1973 (Stony Brook)	73-92324	0-88318-114-2
No. 16	Causality and Physical Theories (Wayne State University, 1973)	73-93420	0-88318-115-0
No. 17	Thermal Expansion – 1973 (Lake of the Ozarks)	73-94415	0-88318-116-9
No. 18	Magnetism and Magnetic Materials – 1973 (2 parts) (Boston)	59-2468	0-88318-117-7
No. 19	Physics and the Energy Problem – 1974 (APS Chicago)	73-94416	0-88318-118-5
No. 20	Tetrahedrally Bonded Amorphous Semiconductors (Yorktown Heights, 1974)	74-80145	0-88318-119-3
No. 21	Experimental Meson Spectroscopy – 1974 (Boston)	74-82628	0-88318-120-7
No. 22	Neutrinos – 1974 (Philadelphia)	74-82413	0-88318-121-5
No. 23	Particles and Fields – 1974 (APS/DPF Williamsburg)	74-27575	0-88318-122-3
No. 24	Magnetism and Magnetic Materials – 1974 (20th Annual Conference, San Francisco)	75-2647	0-88318-123-1

No. 25	Efficient Use of Energy (The APS Studies on the Technical Aspects of the More Efficient Use of Energy)	75-18227	0-88318-124-X
No. 26	High-Energy Physics and Nuclear Structure – 1975 (Santa Fe and Los Alamos)	75-26411	0-88318-125-8
No. 27	Topics in Statistical Mechanics and Biophysics: A Memorial to Julius L. Jackson (Wayne State University, 1975)	75-36309	0-88318-126-6
No. 28	Physics and Our World: A Symposium in Honor of Victor F. Weisskopf (M.I.T., 1974)	76-7207	0-88318-127-4
No. 29	Magnetism and Magnetic Materials – 1975 (21st Annual Conference, Philadelphia)	76-10931	0-88318-128-2
No. 30	Particle Searches and Discoveries – 1976 (Vanderbilt Conference)	76-19949	0-88318-129-0
No. 31	Structure and Excitations of Amorphous Solids (Williamsburg, VA, 1976)	76-22279	0-88318-130-4
No. 32	Materials Technology – 1976 (APS New York Meeting)	76-27967	0-88318-131-2
No. 33	Meson-Nuclear Physics – 1976 (Carnegie-Mellon Conference)	76-26811	0-88318-132-0
No. 34	Magnetism and Magnetic Materials – 1976 (Joint MMM-Intermag Conference, Pittsburgh)	76-47106	0-88318-133-9
No. 35	High Energy Physics with Polarized Beams and Targets (Argonne, 1976)	76-50181	0-88318-134-7
No. 36	Momentum Wave Functions – 1976 (Indiana University)	77-82145	0-88318-135-5
No. 37	Weak Interaction Physics – 1977 (Indiana University)	77-83344	0-88318-136-3
No. 38	Workshop on New Directions in Mossbauer Spectroscopy (Argonne, 1977)	77-90635	0-88318-137-1
No. 39	Physics Careers, Employment and Education (Penn State, 1977)	77-94053	0-88318-138-X
No. 40	Electrical Transport and Optical Properties of Inhomogeneous Media (Ohio State University, 1977)	78-54319	0-88318-139-8
No. 41	Nucleon-Nucleon Interactions – 1977 (Vancouver)	78-54249	0-88318-140-1
No. 42	Higher Energy Polarized Proton Beams (Ann Arbor, 1977)	78-55682	0-88318-141-X
No. 43	Particles and Fields – 1977 (APS/DPF, Argonne)	78-55683	0-88318-142-8
No. 44	Future Trends in Superconductive Electronics (Charlottesville, 1978)	77-9240	0-88318-143-6
No. 45	New Results in High Energy Physics – 1978 (Vanderbilt Conference)	78-67196	0-88318-144-4
No. 46	Topics in Nonlinear Dynamics (La Jolla Institute)	78-57870	0-88318-145-2

No. 47	Clustering Aspects of Nuclear Structure and Nuclear Reactions (Winnepeg, 1978)	78-64942	0-88318-146-0
No. 48	Current Trends in the Theory of Fields (Tallahassee, 1978)	78-72948	0-88318-147-9
No. 49	Cosmic Rays and Particle Physics – 1978 (Bartol Conference)	79-50489	0-88318-148-7
No. 50	Laser-Solid Interactions and Laser Processing – 1978 (Boston)	79-51564	0-88318-149-5
No. 51	High Energy Physics with Polarized Beams and Polarized Targets (Argonne, 1978)	79-64565	0-88318-150-9
No. 52	Long-Distance Neutrino Detection – 1978 (C.L. Cowan Memorial Symposium)	79-52078	0-88318-151-7
No. 53	Modulated Structures – 1979 (Kailua Kona, Hawaii)	79-53846	0-88318-152-5
No. 54	Meson-Nuclear Physics – 1979 (Houston)	79-53978	0-88318-153-3
No. 55	Quantum Chromodynamics (La Jolla, 1978)	79-54969	0-88318-154-1
No. 56	Particle Acceleration Mechanisms in Astrophysics (La Jolla, 1979)	79-55844	0-88318-155-X
No. 57	Nonlinear Dynamics and the Beam-Beam Interaction (Brookhaven, 1979)	79-57341	0-88318-156-8
No. 58	Inhomogeneous Superconductors – 1979 (Berkeley Springs, W.V.)	79-57620	0-88318-157-6
No. 59	Particles and Fields – 1979 (APS/DPF Montreal)	80-66631	0-88318-158-4
No. 60	History of the ZGS (Argonne, 1979)	80-67694	0-88318-159-2
No. 61	Aspects of the Kinetics and Dynamics of Surface Reactions (La Jolla Institute, 1979)	80-68004	0-88318-160-6
No. 62	High Energy e^+e^- Interactions (Vanderbilt, 1980)	80-53377	0-88318-161-4
No. 63	Supernovae Spectra (La Jolla, 1980)	80-70019	0-88318-162-2
No. 64	Laboratory EXAFS Facilities – 1980 (Univ. of Washington)	80-70579	0-88318-163-0
No. 65	Optics in Four Dimensions – 1980 (ICO, Ensenada)	80-70771	0-88318-164-9
No. 66	Physics in the Automotive Industry – 1980 (APS/AAPT Topical Conference)	80-70987	0-88318-165-7
No. 67	Experimental Meson Spectroscopy – 1980 (Sixth International Conference, Brookhaven)	80-71123	0-88318-166-5
No. 68	High Energy Physics – 1980 (XX International Conference, Madison)	81-65032	0-88318-167-3
No. 69	Polarization Phenomena in Nuclear Physics – 1980 (Fifth International Symposium, Santa Fe)	81-65107	0-88318-168-1
No. 70	Chemistry and Physics of Coal Utilization – 1980 (APS, Morgantown)	81-65106	0-88318-169-X

No. 71	Group Theory and its Applications in Physics – 1980 (Latin American School of Physics, Mexico City)	81-66132	0-88318-170-3
No. 72	Weak Interactions as a Probe of Unification (Virginia Polytechnic Institute – 1980)	81-67184	0-88318-171-1
No. 73	Tetrahedrally Bonded Amorphous Semiconductors (Carefree, Arizona, 1981)	81-67419	0-88318-172-X
No. 74	Perturbative Quantum Chromodynamics (Tallahassee, 1981)	81-70372	0-88318-173-8
No. 75	Low Energy X-Ray Diagnostics – 1981 (Monterey)	81-69841	0-88318-174-6
No. 76	Nonlinear Properties of Internal Waves (La Jolla Institute, 1981)	81-71062	0-88318-175-4
No. 77	Gamma Ray Transients and Related Astrophysical Phenomena (La Jolla Institute, 1981)	81-71543	0-88318-176-2
No. 78	Shock Waves in Condensed Mater – 1981 (Menlo Park)	82-70014	0-88318-177-0
No. 79	Pion Production and Absorption in Nuclei – 1981 (Indiana University Cyclotron Facility)	82-70678	0-88318-178-9
No. 80	Polarized Proton Ion Sources (Ann Arbor, 1981)	82-71025	0-88318-179-7
No. 81	Particles and Fields –1981: Testing the Standard Model (APS/DPF, Santa Cruz)	82-71156	0-88318-180-0
No. 82	Interpretation of Climate and Photochemical Models, Ozone and Temperature Measurements (La Jolla Institute, 1981)	82-71345	0-88318-181-9
No. 83	The Galactic Center (Cal. Inst. of Tech., 1982)	82-71635	0-88318-182-7
No. 84	Physics in the Steel Industry (APS/AISI, Lehigh University, 1981)	82-72033	0-88318-183-5
No. 85	Proton-Antiproton Collider Physics –1981 (Madison, Wisconsin)	82-72141	0-88318-184-3
No. 86	Momentum Wave Functions – 1982 (Adelaide, Australia)	82-72375	0-88318-185-1
No. 87	Physics of High Energy Particle Accelerators (Fermilab Summer School, 1981)	82-72421	0-88318-186-X
No. 88	Mathematical Methods in Hydrodynamics and Integrability in Dynamical Systems (La Jolla Institute, 1981)	82-72462	0-88318-187-8
No. 89	Neutron Scattering – 1981 (Argonne National Laboratory)	82-73094	0-88318-188-6
No. 90	Laser Techniques for Extreme Ultraviolt Spectroscopy (Boulder, 1982)	82-73205	0-88318-189-4
No. 91	Laser Acceleration of Particles (Los Alamos, 1982)	82-73361	0-88318-190-8
No. 92	The State of Particle Accelerators and High Energy Physics (Fermilab, 1981)	82-73861	0-88318-191-6

No. 93	Novel Results in Particle Physics (Vanderbilt, 1982)	82-73954	0-88318-192-4
No. 94	X-Ray and Atomic Inner-Shell Physics – 1982 (International Conference, U. of Oregon)	82-74075	0-88318-193-2
No. 95	High Energy Spin Physics – 1982 (Brookhaven National Laboratory)	83-70154	0-88318-194-0
No. 96	Science Underground (Los Alamos, 1982)	83-70377	0-88318-195-9
No. 97	The Interaction Between Medium Energy Nucleons in Nuclei – 1982 (Indiana University)	83-70649	0-88318-196-7
No. 98	Particles and Fields – 1982 (APS/DPF University of Maryland)	83-70807	0-88318-197-5
No. 99	Neutrino Mass and Gauge Structure of Weak Interactions (Telemark, 1982)	83-71072	0-88318-198-3
No. 100	Excimer Lasers – 1983 (OSA, Lake Tahoe, Nevada)	83-71437	0-88318-199-1
No. 101	Positron-Electron Pairs in Astrophysics (Goddard Space Flight Center, 1983)	83-71926	0-88318-200-9
No. 102	Intense Medium Energy Sources of Strangeness (UC-Sant Cruz, 1983)	83-72261	0-88318-201-7
No. 103	Quantum Fluids and Solids – 1983 (Sanibel Island, Florida)	83-72440	0-88318-202-5
No. 104	Physics, Technology and the Nuclear Arms Race (APS Baltimore –1983)	83-72533	0-88318-203-3
No. 105	Physics of High Energy Particle Accelerators (SLAC Summer School, 1982)	83-72986	0-88318-304-8
No. 106	Predictability of Fluid Motions (La Jolla Institute, 1983)	83-73641	0-88318-305-6
No. 107	Physics and Chemistry of Porous Media (Schlumberger-Doll Research, 1983)	83-73640	0-88318-306-4
No. 108	The Time Projection Chamber (TRIUMF, Vancouver, 1983)	83-83445	0-88318-307-2
No. 109	Random Walks and Their Applications in the Physical and Biological Sciences (NBS/La Jolla Institute, 1982)	84-70208	0-88318-308-0
No. 110	Hadron Substructure in Nuclear Physics (Indiana University, 1983)	84-70165	0-88318-309-9
No. 111	Production and Neutralization of Negative Ions and Beams (3rd Int'l Symposium, Brookhaven, 1983)	84-70379	0-88318-310-2
No. 112	Particles and Fields – 1983 (APS/DPF, Blacksburg, VA)	84-70378	0-88318-311-0
No. 113	Experimental Meson Spectroscopy – 1983 (Seventh International Conference, Brookhaven)	84-70910	0-88318-312-9

No. 114	Low Energy Tests of Conservation Laws in Particle Physics (Blacksburg, VA, 1983)	84-71157	0-88318-313-7
No. 115	High Energy Transients in Astrophysics (Santa Cruz, CA, 1983)	84-71205	0-88318-314-5
No. 116	Problems in Unification and Supergravity (La Jolla Institute, 1983)	84-71246	0-88318-315-3
No. 117	Polarized Proton Ion Sources (TRIUMF, Vancouver, 1983)	84-71235	0-88318-316-1
No. 118	Free Electron Generation of Extreme Ultraviolet Coherent Radiation (Brookhaven/OSA, 1983)	84-71539	0-88318-317-X
No. 119	Laser Techniques in the Extreme Ultraviolet (OSA, Boulder, Colorado, 1984)	84-72128	0-88318-318-8
No. 120	Optical Effects in Amorphous Semiconductors (Snowbird, Utah, 1984)	84-72419	0-88318-319-6
No. 121	High Energy e^+e^- Interactions (Vanderbilt, 1984)	84-72632	0-88318-320-X
No. 122	The Physics of VLSI (Xerox, Palo Alto, 1984)	84-72729	0-88318-321-8
No. 123	Intersections Between Particle and Nuclear Physics (Steamboat Springs, 1984)	84-72790	0-88318-322-6
No. 124	Neutron-Nucleus Collisions – A Probe of Nuclear Structure (Burr Oak State Park - 1984)	84-73216	0-88318-323-4
No. 125	Capture Gamma-Ray Spectroscopy and Related Topics – 1984 (Internat. Symposium, Knoxville)	84-73303	0-88318-324-2
No. 126	Solar Neutrinos and Neutrino Astronomy (Homestake, 1984)	84-63143	0-88318-325-0
No. 127	Physics of High Energy Particle Accelerators (BNL/SUNY Summer School, 1983)	85-70057	0-88318-326-9
No. 128	Nuclear Physics with Stored, Cooled Beams (McCormick's Creek State Park, Indiana, 1984)	85-71167	0-88318-327-7
No. 129	Radiofrequency Plasma Heating (Sixth Topical Conference, Callaway Gardens, GA, 1985)	85-48027	0-88318-328-5
No. 130	Laser Acceleration of Particles (Malibu, California, 1985)	85-48028	0-88318-329-3
No. 131	Workshop on Polarized ^{3}He Beams and Targets (Princeton, New Jersey, 1984)	85-48026	0-88318-330-7
No. 132	Hadron Spectroscopy–1985 (International Conference, Univ. of Maryland)	85-72537	0-88318-331-5
No. 133	Hadronic Probes and Nuclear Interactions (Arizona State University, 1985)	85-72638	0-88318-332-3